Bioterrorism and Infectious Agents: A New Dilemma for the 21st Century

Emerging Infectious Diseases of the 21st Century

Series Editor: I. W. Fong
Professor of Medicine, University of Toronto
Head of Infectious Diseases, St. Michael's Hospital

Recent volumes in this series:

INFECTIONS AND THE CARDIOVASCULAR SYSTEM: New Perspectives
Edited by I. W. Fong

REEMERGENCE OF ESTABLISHED PATHOGENS IN THE 21ST CENTURY
Edited by I. W. Fong and Karl Drlica

BIOTERRORISM AND INFECTIOUS AGENTS: A New Dilemma for the 21st Century
Edited by I. W. Fong and Kenneth Alibek

A Continuation Order Plan is available for this series. A continuation order will bring delivery of each new volume immediately upon publication. Volumes are billed only upon actual shipment. For further information, please contact the publisher.

Bioterrorism and Infectious Agents: A New Dilemma for the 21st Century

Edited by

I. W. Fong

University of Toronto, St. Michael's Hospital
Toronto, Ontario, Canada

and

Kenneth Alibek

The National Center for Biodefense
George Mason University
Manassas, Virginia

Editors
I.W. Fong
University of Toronto
30 Bond Street
Room 4179W
Toronto, ON M5B 1W8
Canada

Kenneth Alibek
Department of Medical Microbiology
and Immunology
George Mason Universuty
National Center for Biodefense
10900 niveĿsity lvd. B
Manassas, VA 20110
MS 1A8
USA

ISBN 978-1-4419-1265-7 ISBN 978-1-4419-1266-4 (eBook)
DOI 10.1007/978-1-4419-1266-4
Springer Dordrecht Heidelberg London New York

Library of Congress Control Number: 2009932897

Printed on acid-free paper

Springer is part of Springer Science+Business Media (www.springer.com)

Contributors

Kenneth Alibek The National Center for Biodefense, George Mason University, Manassas, VA

Harvey Artsob National Laboratory for Zoonotic Diseases and Special Pathogens, National Microbiology Laboratory, Health Canada, Canadian Science Centre for Human and Animal Health, Department of Medical Microbiology, University of Manitoba, Winnipeg, Canada

Thomas P. Bleck Neurological Surgery and Internal Medicine, and Neuroscience Intensive Care Unit, Charlottesville, VA

George W. Christopher Wilford Hall Air Force Medical Center, Lackland Air Force Base, TX

Theodore J. Cieslak San Antonio Military Pediatric Center, San Antonio, TX

David Allan Brett Dance Health Protection Agency, Department of Microbiology, Derriford Hospital, Plymouth, Devon, UK

David T. Dennis Division Vector-Borne Infectious Diseases, National Center for Infectious Diseases, Centers for Disease Control and Prevention, Atlanta, GA

Edward M. Eitzen, Jr. Department of Health and Human Services, Washington, DC

Heinz Feldmann National Laboratory for Zoonotic Diseases and Special Pathogens, National Microbiology Laboratory, Health Canada, Canadian Science Centre for Human and Animal Health, Department of Medical Microbiology, University of Manitoba, Winnipeg, Canada

Ignatius W. Fong University of Toronto, St. Michael's Hospital, Toronto, Ontario, Canada

Allison Groseth National Laboratory for Zoonotic Diseases and Special Pathogens, National Microbiology Laboratory, Health Canada, Canadian Science Centre for Human and Animal Health, Department of Medical Microbiology, University of Manitoba, Winnipeg, Canada

Lisa Hodges Infectious Diseases Section, Department of Medicine, Louisiana State University Health Sciences Center, Shreveport, LA

Steven Jones National Laboratory for Zoonotic Diseases and Special Pathogens, National Microbiology Laboratory, Health Canada, Canadian Science Centre for Human and Animal Health, Department of Immunology, University of Manitoba, Winnipeg, Canada

J. Michael Lane Smallpox Eradication Program, Centers for Disease Control and Prevention, Atlanta, GA

Catherine Lobanova The National Center for Biodefense, George Mason University, Manassas, VA

Maor Maman Israel Defense Forces Medical Corps Headquarters, Israel

Martin I. Metzer Office of Surveillance, Office of the Director, National Center for Infectious Diseases, Centers for Disease Control and Prevention, Atlanta, GA

Robert L. Penn Department of Medicine, Louisiana State University Health Sciences Center, Shreveport, LA

Serguei Popov National Center for Biodefense, George Mason University, Manassas, VA

Lila Summer A Human Writes, Atlanta, GA

Yoav Yehezkelli Israel Defense Forces Medical Corps Headquarters, Israel

Preface

Since the terrorist attack on the United States on September 11, 2001 and subsequent cases of anthrax in Florida and New York City, attention has been focused on the threat of biological warfare and bioterrorism. Biological warfare agents are defined as "living organisms, whatever their nature, or infected material derived from them, which are used for hostile purposes and intended to cause disease or death in man, animals and plants, and depend for their efforts on the ability to multiply in person, animal or plant attacked." Biological warfare agents may be well suited for bioterrorism to create havoc and terror in a civilian population, because they are cheap and easy to obtain and dispense.

Infectious or contagious diseases have played a major part in the history of warfare – deliberately or inadvertently – in restricting or assisting invading armies over the centuries. In 1346, the Tartars catapulted plaque-infected bodies into Kaffa in the Crimea to end a 3-year siege. Blankets contaminated with smallpox to infect North American Indians were used by British forces in the 18th century. More recently, the Japanese released fleas infected with plaque in Chinese cities in the 1930s and 1940s.

Biological research programs for both offensive and defensive strategies have been developed by the United States, Britain, the former Soviet Union, and Canada; several other nations are thought to have such programs.

Thus, it is important that physicians and health care personnel on the front line (Emergency physicians, Public Health personnel, Internists, Infectious Disease specialists, Microbiologists, Critical care specialists, and even General practitioners) be aware of the clinical manifestations, diagnosis, and management of these potentially deadly diseases. Awareness is the key to recognition of a bioterrorism attack. Thus, this volume will provide health care workers with up-to-date important reviews by world-renowned experts on infectious and biological agents that could be used for bioterrorism.

Contents

Chapter 1

Anthrax: A Disease and a Weapon
Kenneth Alibek, Catherine Lobanova, and Serguei Popov

Chapter 2

Plague as a Biological Weapon
David T. Dennis

Chapter 3

Tularemia and Bioterrorism
Lisa Hodges and Robert L. Penn

Chapter 4

Melioidosis and Glanders as Possible Biological Weapons
David Allan Brett Dance

Chapter 5

Smallpox as a Weapon for Bioterrorism
J. Michael Lane

Chapter 6

Hemorrhagic Fever Viruses as Biological Weapons
Allison Groseth, Steven Jones, Harvey Artsob, and Heinz Feldmann

1

Anthrax: A Disease and a Weapon

Kenneth Alibek, Catherine Lobanova, and Serguei Popov

1. HISTORY OF ANTHRAX

Anthrax infection was described in ancient literature and religious writings. Some biblical scholars have interpreted descriptions of two of the plagues that struck Egypt around 1500 B.C. as anthrax epidemics. The 5th plague, as described in the book of *Exodus*, is a plague of grievous murrain (an infectious and fast-spreading disease) that affected livestock (Witkowski, 2002). The description of the 6th plague, or the Plague of Boils, is very similar to the symptoms of cutaneous anthrax (raised inflammatory swelling and blisters also suggest anthrax) (CNN.com, 2001).

Descriptions of anthrax infection exist in ancient literature as well. For example, scholars have construed the "burning plague" described in Homer's *Iliad* (estimated date 950–700 B.C.) as an anthrax epidemic (Witkowski, 2002). In addition, Virgil detailed an outbreak in *Georgics* (29 B.C.). *Georgics* is set in the world of animal husbandry and describes a murrain that devastated farms and meadows, polluted water supplies, and contaminated food sources. Flocks of sheep were affected, and human victims died after suffering burning fever and shriveled flesh. Virgil made the connection between the murrain that affected the sheep and the disease killing humans:

> If anyone wore a garment made from tainted wool, his limbs were soon attacked by inflamed papules and foul exudates, and if he delayed too long to remove the material, a violent inflammation consumed the parts it had touched. (Witkowski, 2002)

Anthrax is the most likely infection to cause this.

Periodically, over the following millennia, there were outbreaks of anthrax worldwide. For example, there was a substantial outbreak in Germany in the 14th century. During the 17th century, there were large outbreaks in Russia and one in Europe that killed more than 60,000 head of cattle (CNN.com, 2001). During the mid-18th century, a panzootic consumed half of the sheep in Europe (Witkowski, 2002).

Bioterrorism and Infectious Agents
Edited by Fong and Alibek, Springer Science+Business Media, Inc., New York, 2005

Beginning in the late 18th century, researchers began examining anthrax more closely. In 1769, Jean Fournier classified the disease as anthrax or carbon malin (obviously because of the black lesions characteristic of cutaneous anthrax). Fournier also noted the link between people who worked with animal hair or wool and an increased incidence of anthrax (Office of Public Health Service Historian, 2001).

Although the disease anthrax dates back thousands of years, it was not recognized until the 1800s. Between 1827 and 1847, mohair (a type of wool) from Asia and alpaco from Peru were introduced into textile fibers. Deaths soon started occurring in those working with the products. In Bradford, England, an area particularly affected by anthrax, Dr. F.W. Eurich, a bacteriologist, and Elmhurst Duckering, a factory inspector, developed methods for decontaminating fibers (Witkowski, 2002).

Around the middle of the 19th century, Pierre Francoise Rayer, a French dermatologist, and Casimir Joseph Devaine, a parasitologist, reported the presence of small filiform bodies in the blood of sheep infected with anthrax. Rayer briefly mentioned these findings in a report delivered to the Société de Biologie in Paris. These findings were later confirmed by Franz Aloys Antoine Pollender (Office of Public Health Service Historian, 2001).

Pollender has been credited with discovering the etiologic agent of anthrax. His first experience with the disease was in 1841 when he unsuccessfully treated a flayer (a person who strips off skin) with a carbuncle. Eight years later, he purchased a microscope to examine the corpses of cows (with carbuncles) that had recently died in the area. He discovered what he called "chyllus corpuscles" in the blood, spleen, and fluid drained from carbuncles on the cows. In his findings published in 1855, he questioned whether corpuscles could cause anthrax (Witkowski, 2002).

Soon after, in 1858, Freiderich August Brauell noted that the bodies described by Pollender never appeared in healthy animals or in animals that had other diseases. One year later, German veterinarian Christian Joseph Fuchs published findings he originally made in 1842 describing granulated threads he discovered in blood from animals that had died of anthrax (Witkowski, 2002). Delay in publishing his findings seems to have cost him credit for the discovery. Researchers at the time generally agreed that anthrax was an infectious disease, but did not agree on the cause.

As researchers continued to debate the cause of anthrax, the disease continued to kill large numbers of animals and people. For example, in 1864 alone, more than 72,000 horses died of anthrax in Russia. Between 1867 and 1870, 528 men, as well as 56,000 horses, head of cattle, and sheep, died in Novgorod, Russia (Witkowski, 2002).

Finally, in 1876, Robert Koch, a German physician, took blood from the spleens of animals that had died from anthrax and injected it into healthy mice. All of the mice died. Mice that had been inoculated with the blood of animals without the disease survived. This confirmed that the blood of infected animals was infectious. Later, he isolated *Bacillus anthracis* in pure culture. Koch was also the first to inoculate animals with pure cultures of *B. anthracis*. He wanted to determine if bacilli that had never been in contact with an animal could cause disease. He grew the bacilli in pure culture using the aqueous humor of an ox's eye. Mice inoculated with the organisms died. Koch also discovered that the bacteria develop protective spores that enabled them to withstand unfavorable conditions to reemerge later when conditions improved (Nobel e-museum, 1967).

Shortly after Koch's discoveries, John Bell, a Bradford physician, linked anthrax with woolsorters' disease. Woolstorters' disease (or "ragpicker's" if in Germany) is inhalational anthrax related to textile industry work. Bell developed disinfection techniques that were expanded into overall preventive measures known as the Bradford Rules and were made into law in 1897. Creation of the Anthrax Investigation Board soon followed (Witkowski, 2002).

In 1880, William Greenfield successfully immunized livestock against anthrax though credit is usually given to Louis Pasteur who developed a similar vaccine a year later. Greenfield was the Professor Superintendent of Brown Animal Sanatory Institution in England and performed experiments before Pasteur, but political and financial influences prevented him from reporting his work. In 1881, Pasteur developed a vaccine based on a live, attenuated strain (Mikesell *et al.*, 1983).

Though a vaccine had been developed, anthrax continued to be a problem in the 20th century. In 1919, government officials in Bradford, England, passed the Anthrax Prevention Act, which stipulated that imports from certain countries must be decontaminated (Witkowski, 2002). During the last part of the century, the number of cases of industry-related infections in humans decreased dramatically as a result of improved industrial hygiene practices and restrictions on the importation of animal products. Vaccination of animals also contributed to this decline though outbreaks continue in depressed or unstable areas. For example, between 1978 and 1980 during the civil war in Zimbabwe, an outbreak of cutaneous and gastrointestinal anthrax occurred. The cause was likely the breakdown of veterinary care that resulted in the disease infecting animals (primarily cattle). Humans were subsequently infected by eating contaminated animal meat (Knights, 2002).

China has also been affected by persistent anthrax outbreaks. In 1989, 507 people were infected during an outbreak in Tibet; 162 of them died. The Chinese government has made significant attempts to reduce the incidence of outbreaks; however, outbreaks of one or more forms of the disease have occurred in at least 30 Chinese remote and pastoral provinces. To combat the problem, the Chinese government has focused on vaccine development, education, and regulation. Each spring, animals with a high incidence of anthrax are vaccinated. The government has also instituted national health regulations, such as the Diagnosis Criteria and Principles of Management for Anthrax and the Management of Anthrax Control, which is updated every 4 years. Additionally, there is a surveillance network in place that requires every new case of human or animal disease to be reported to local health authorities within 48 hours (ABS, 2004).

1.1. Anthrax in the United States

There have been numerous outbreaks of anthrax in the United States. The first documented infections in animals occurred in the early 18th century in the Louisiana Territory. The disease spread through the South and to the Northeast. By the 1820s, the first human cases in the United States were reported in Kentucky (Mikesell *et al.*, 1983).

In 1868, a Massachusetts physician, Silas Stone, reported treating 8 patients with an unusual disease involving malignant pustules over a 14-month period. All of the men had worked at a local animal hair factory. Stone treated the men with a mixture of iodine, iron, and quinine. The patients showed a variety of symptoms; some patients had general symptoms – such as pain, nausea, and vomiting – whereas others had mediastinal involvement, including chest pain, dyspnea, and tachypnea. Two patients died of the disease, apparently from meningitis. Stone noted that each patient had been exposed to hair or dirt from the factory (Macher, 2002).

During the 1950s, the Communicable Disease Center (which was later named the Centers for Disease Control and Prevention – CDC) investigated outbreaks of anthrax in Pennsylvania, Colorado, New Hampshire, and Louisiana. An outbreak involving five people also occurred over a 3-month period at a mill outside of Monroe, North Carolina. The source of the anthrax was traced to a shipment of wool from Iraq and Iran. The goals of the investigations were to discover the cause of the anthrax in the wool and animal hair industries; determine the particle size of anthrax in contaminated aerosols; study the epidemiology and epizootiology of anthrax; collect and study various strains of anthrax; and assess the effectiveness of an anthrax vaccine for humans.

At approximately the same time the CDC was investigating these outbreaks, another group within the center was testing a human anthrax vaccine at several mills in Pennsylvania and at the Arms Textile Mill in New Hampshire. In 1957 (during the trial), over a 10-week period, nine men at the Arms mill were infected with anthrax; five with the inhalational form and four with the cutaneous form. It was eventually discovered that all of the men, though they worked in various areas and departments at the mill (which turned goat hair into wool for lining clothes), had been in contact with a single shipment of black goat hair from Pakistan. Anthrax infection was not unusual at the mill, as there were more than 100 reported cases of cutaneous anthrax between 1941 and 1957.

As part of the vaccine trial, approximately half of the 600 mill employees elected to be vaccinated. Half of those who participated in the trial were given the vaccine; the other half were given a placebo. Four of the nine men who were infected with anthrax had participated in the trial. All were given the placebo, and all four died. The mill closed in 1968 and was later purposefully destroyed in an attempt to decontaminate the area (Belluck, 2001).

2. ANTHRAX AS A BIOLOGICAL WEAPON

The North Atlantic Treaty Organization lists 31 agents as potential biological weapons (Zimmerman and Zimmerman, 2003). Anthrax is one of the top four agents listed. Even though anthrax is not a contagious disease, there are certain characteristics of the pathogen that make it ideally suited for development into a biological weapon. The first characteristic is that anthrax is relatively easy to produce. Anthrax also has a long shelf-life and is stable in the environment. Anthrax spores have a very high survival rate and can be used in an explosive device. Anthrax has a high mortality rate – approaching 100% in the case of untreated inhalational anthrax. Even with treatment, the mortality rate is still high because there is no effective treatment for advanced forms of inhalational anthrax.

The potential threat of anthrax being used as a weapon is and has been well recognized. In fact, anthrax has periodically been used as a weapon for more than 8 decades. During World War I, for example, the Germans used anthrax (and glanders) to infect shipments of horses, cattle, and sheep destined for the battlefront from at least five countries, including the United States. Longshoremen pricked the animals with infected needles while they waited to be boarded onto ships (Barnaby, 2000).

In January 1917, a Norwegian aristocrat, Baron Otto Karl von Rosen, was arrested for possessing anthrax. Though Norway was officially neutral during the war, the government allowed the allies to transport horses across the country so they could be of use on the Russian front. Von Rosen planned to feed anthrax-laced sugar cubes to the horses he came in contact with (Barnaby, 2000).

Shortly before World War II, the Japanese government built its offensive biological weapons research, development, and production compound known as Unit 731. Among the numerous infectious agents examined by Unit 731 was anthrax. By the time Unit 731 was overrun by allied forces at the end of the war, the facility was capable of producing 1,400 pounds of anthrax annually. In addition, researchers and engineers designed a fragmentation bomb to disperse anthrax spores. Anthrax-infected chocolate was also developed by the unit (Mangold and Goldberg, 1999).

Other countries developed offensive biological weapons programs as well. Near the end of 1940, Great Britain was starting its program in earnest. Research was conducted on the use of aerosols to disperse bacteria; particular interest was paid to the use of anthrax and botulinum toxin. Animals were exposed to anthrax spores to determine the number necessary to kill them if the spores were inhaled. Field tests were performed to see what concentration of spores could be suspended in an aerosol that had been dispersed from bomblets and how far downwind the cloud would extend. These tests took place on Gruinard Island off the coast of Scotland (the tests targeted sheep).

In 1941, Prime Minister Winston Churchill gave his permission to researchers to build an anthrax biological weapon. Termed "Operation Vegetarian," the project was only intended to produce weapons to be used against cattle and only in the event that the Germans used a biological weapon first.

"Operation Vegetarian" involved the uses of hollowed-out "cakes" made of finely ground linseed meal. Powdered anthrax was sealed into the hollowed area, and the cakes would then be dropped over German agricultural districts whereupon cattle would ingest the cakes, killing the cattle and contaminating part of the food supply. By 1943, more than 5 million cakes had been produced, though they were never used. Contingency plans, also never used, included dropping anthrax over six German cities, including Berlin, Hamburg, and Frankfurt (Barnaby, 2000).

Great Britain also briefly considered using anthrax to assassinate Adolph Hitler during the war. According to documentation released by the Public Record Office, one method explored involved smuggling anthrax into Germany using a common vessel such as a fountain pen, then contaminating his clothes (Carus, 2001). Great Britain ultimately abandoned the plan, and the offensive program ended in 1957.

In May 1943, British researchers reported the Gruinard trials and results to American and Canadian leaders. The three countries developed a working relationship in which the British designed the weapons, the Americans built them, and the Canadians tested them. In early 1944, realizing England did not have the capacity to build the weapons, Churchill ordered half a million bombs from the United States.

Just 1 month before the British reported their findings from the tests at Gruinard, the United States officially started its offensive weapons program at Camp Detrick (later Fort Detrick) in Maryland. Anthrax was produced in large quantities in Building 470. By the end of World War II, there were more than 4,000 people employed in the American offensive weapons program. Most of them were working with anthrax and botulinum toxin.

Approximately one-quarter of a million bombs had been produced that could be filled with anthrax (Barnaby, 2000).

By the 1960s, the United States had stockpiled three lethal agents – anthrax among them. The United States never used its anthrax biological weapon, and the offensive program ended in 1969. The United States also initiated work on the convention to ban biological weapons.

No other country, however, spent as much in terms of human and financial resources on the development of biological weapons in general and on an anthrax weapon in particular as the former Soviet Union. Three facilities in the Soviet Union were dedicated to the production of anthrax – Sverdlovsk , Penza, and Kurgan. All three facilities had ten 50-ton reactors each to manufacture anthrax. Sverdlovsk was the only active facility; the other two were mobilization capacities in the event production needed to be increased. At the height of the program, during the 1980s, the combined estimated capabilities of the anthrax was almost 5,000 tons a year, though actual production was significantly lower.

In March 1979, there was an accidental release of anthrax spores at the Sverdlovsk facility resulting from a neglected organizational maintenance work. Within a week, workers at a ceramics plant downwind from the facility became ill and died. Soon others in the area became sick. The Soviet government did everything possible to cover up the accident. The exact number of people who died is still unkown, but it is estmated that the number is between 66–105 (Alibek, 1999). Soviet officials claimed that the outbreak was the result of people eating contaminated meat. Evaluation of the autopsies of the victims, however, showed that they were infected via the aerosol route.

Following the outbreak at the Sverdlovsk plant, anthrax production was moved to a facility built in the city of Stepnogorsk, which was previously intended to manufacture plague, tularemia, and glanders biological weapons. This facility had a lower capacity – ten 20-ton reactors. However, the facility developed an anthrax weapon based on new production processes that allowed it to produce anthrax on a massive scale. The LD_{50} of this strain for mice was approximately five spores, compared with the Ames strain, with an LD_{50} of 30 spores. Anthrax produced here was based on Anthrax 836, a strain first discovered in Kirov in 1953, which was highly virulent, transportable, and could be reproduced in large quantities. The cover of the facility was that it manufactured pesticides.

The efficacy of the anthrax biological weapon developed at Stepnogorsk was tested in Building 600, which was at the time the largest indoor testing facility in the world. Inside the building were two testing chambers. The chambers were used to test anthrax weapon stability via explosion and to determine infection and mortality characteristics of anthrax weapons when testing aerosol dispersion attributes in monkeys. The Soviet Union considered anthrax as one of a few biological agents that could be used as a strategic weapon in the event that war erupted between the super powers.

While the Soviet Union and the United States had the most developed biological weapons programs, other countries explored the possibility of developing this weapon on a smaller scale. For example, according to Saddam Hussein's son-in-law Kamel, during the 1980s Iraq developed, produced, and stockpiled anthrax. Records from the American Type Culture Collection show that Iraq purchased several strains of *B. anthracis* in 1986 and 1988 (by early 1989, the United States placed a ban on the sale of the anthrax pathogen to Iraq and other hostile nations) (Zimmerman and Zimmerman, 2003).

There is also evidence that South Africa developed an anthrax weapon as well. Dr. Schalk van Rensburg, a member of South Africa's Roodeplaat Research Laboratory (the cover for the country's offensive chemical and biological program), testified as such. Van Rensberg reported to the Truth and Reconciliation Commission, which was established by Bishop Desmond Tutu to investigate human rights abuses – that South African security services attempted to murder three Russian advisers to the African National Council by poisoning their food with anthrax (Barnaby, 2000).

Individual terrorists and terrorist groups have also used anthrax as a weapon for a variety of purposes. In October 1981, a suspicious package was left on the grounds of England's Chemical Defence Establishment. The accompanying letter claimed that the package contained anthrax-contaminated soil from Gruinard Island. Tests indicated that there were low concentrations of anthrax consistent with soil from Gruinard Island (Carus, 2001).

Just a few years later, the Japanese cult Aum Shinrikyo began working on an anthrax weapon. During the mid-1980s, the cult began working on *B. anthracis* in a building in Tokyo (Carus, 2001). Aum Shinrikyo released anthrax up three times; no human deaths resulted from any of the releases. In 1993, the cult released spores from the roof of an office building in Kameido, Japan (near Tokyo). Initially, the release went largely unnoticed, but following testimony of cult members, a retroactive investigation was initiated and spore samples obtained were shown to be consistent with Sterne 34F2, a widely available strain of anthrax used to develop animal prophylaxis. This release was "unsuccessful" because Sterne 34F2 is an attenuated (nonvirulent) strain of anthrax (Klein *et al.*, 2001).

The most infamous and destructive bioterrorist attack involving the use of anthrax occurred in 2001, directly on the heels of September 11th. On September 18, 2001, at least two letters containing anthrax were mailed from the Trenton (New Jersey) area, one to NBC news anchor Tom Brokaw and the other to an editor at the *New York Post*. The letters traveled through the Hamilton (New Jersey) and Morgan (New York City) postal facilities. Approximately 3 weeks later on October 9th, two additional letters were also mailed from the Trenton area, this time destined for Senators Tom Daschle and Patrick Leahy. These letters traveled through the Hamilton facility to the Brentwood postal facility in Washington, D.C., and later to the mail processing unit at the United States Department of State. A probable third letter was sent to the American Media Incorporation (AMI) offices in Florida, but the actual letter was never found.

In total, 22 people were infected with anthrax. These individuals' symptoms presented in two clusters, one lasting from late September to early October and the other from mid to late October (the one case that did not fall into either cluster was that of a 94-year-old woman in Connecticut whose symptoms appeared in mid-November). The first cluster involved nine cases of cutaneous anthrax infection that resulted from the September 18th letters. The second cluster involved seven cases of inhalational anthrax infection and one case of cutaneous anthrax infection that were linked to the October 9th letters. In addition, four cases of inhalational and one case of cutaneous anthrax infection occurred, but were not directly linked to any of the recovered letters. Five people died as a result of this attack – one AMI employee in Florida, two employees of the Brentwood postal facility, an elderly woman in Connecticut, and a hospital employee in New York. To date, the perpetrator(s) have not been caught.

While the loss of even a single human life is exorbitant, the frightening reality is that the number of casualties would have been significantly higher had prophylactic measures (specifically the use of antibiotics as prophylaxis) not been taken in a timely manner. For example, of the 625 people in the Senate Hart Building tested for *B. anthracis*, 28 tested positive.

An epidemiological investigation of the event determined that the average number of days between possible exposure to the spores and the onset of symptoms was 4.5 days. The median age of those infected was 46 years (the ages of the victims ranged from 7 months to 94 years). The average age of people with inhalational anthrax was higher – 56 years – than those with the cutaneous form – 35 years. Additionally, 12 of the 22 victims were mail handlers for either the U.S. Postal Service or the federal government, or in the mail room at AMI.

The anthrax bioterrorist attack of 2001 cost the United States millions of dollars in containment, treatment, decontamination, and associated expenses. Multiple government agencies – including the FBI, CIA, CDC, U.S. Postal Service, and the Environmental Protection Agency – played a role in response to the attack and the subsequent investigation and provided a "trial-run" for a larger bioterrorist event (Jernigan *et al.*, 2002).

More than 100 hoax cases or threats of using anthrax have also been made (Carus, 2001). These threats and hoaxes have been carried out against government agencies, public and private schools, medical clinics, post offices, corporations, religious organizations, court houses, and media organizations. The real and threatened use of anthrax have cost and will continue to cost millions or billions of dollars in worker downtime, civil and criminal investigations, decontamination efforts, and increased attempts to prevent such incidents from occurring in the future. For these reasons, it is obvious that the understanding of inhalational anthrax, which can cost the United States a large number of lives, must be reevaluated and that new therapies and urgent prophylaxes must be developed in the near future.

3. THE ORGANISM

B. anthracis is a Gram-positive, nonmotile, facultative anaerobic, spore-forming, rod-shaped bacterium. Since sporulation requires the presence of free oxygen, bacilli shed by the dying or dead animal will sporulate on contact with air. The proportion of cells that reach the ultimate stage, a dormant spore, is variable. Anthrax spores are highly resistant to heat, ultraviolet and ionizing radiation, pressure, and chemical agents. They are able to survive in the soil for long periods of time, even up to decades or perhaps longer. In a suitable environment (e.g., various tissues and organs), spores start vegetating and multiplying. However, the bacilli are poor survivors (Lindeque and Turnbull, 1994), and it is unlikely that germination, propagation with further resporulation, will occur outside the host in natural conditions. Spores ingested by the host germinate and vegetate within the organism expressing their virulence factors and killing the host. The vegetative form is usually capsulated. The capsule consists of a monotonous linear polymer of D-glutamic acid that is weakly immunogenic (Makino *et al.*, 1989; Wang *et al.*, 2004). The molecular weight of the *poly*-glutamic chains is between 20 and 55 kDa in vitro and is estimated to be 215 kDa in vivo (Goodman and Nitecki, 1967; Wang *et al.*, 2004). In *B. anthracis*, the glutamyl polypeptide of the capsule loosely adheres to the bacterial cell by an as yet unknown mechanism. The capsule contributes to pathogenicity by enabling bacteria to evade host immune defenses. A recent report demonstrated that antibodies against the capsule exhibited a partial protective effect against the virulent Ames strain

(Record and Wallis, 1956). The *B. anthracis* capsule inhibits phagocytosis (Kozel *et al.*, 2004; Zwartouw and Smith, 1956), but is not able to completely protect the bacilli from phagocytes. For example, the histological analysis of livers of Sverdlovsk anthrax victims revealed a large number of bacilli engulfed by Kupffer cells (Grinberg *et al.*, 2001).

The capsule is the outermost element of the cell wall. When *B. anthracis* does not produce its capsule, the exterior of the cell wall appears layered, owing to the S-layer (Mesnage *et al.*, 1998). Various functions have been proposed for S-layers, including shape maintenance and molecular sieving. The capsule and the S-layer seem to have a cumulative effect in virulence, increasing resistance to immune-mediated defenses (Ray *et al.*, 1998).

The complete genome sequences of two virulent anthrax strains, Ames and A2012, have been reported (Read *et al.*, 2003). The chromosome of *B. anthracis* consists of 5.23 megabases and encodes approximately 5,500 genes. Only 2,760 chromosome genes have been assigned specific functions, and there are approximately 2,100 genes with hypothetical functions, as well as 660 unknown genes. An average *B. anthracis* gene spans 800 nucleotides. The *B. anthracis* chromosome portrays a soil-dwelling organism, possessing numerous potential virulence genes. Almost all potential chromosomal virulence-enhancing genes are homologues in *Bacillus cereus*, suggesting that they are not specifically associated with the unique pathogenicity of *B. anthracis*, but are part of the common arsenal of the *B. cereus* group of bacteria (Read *et al.*, 2003). In general, the chromosomal genes among this group of closely related bacteria are very similar.

Key virulence genes of *B. anthracis* are found on plasmids, which represent extrachromosomal, circular, double-stranded DNA molecules. Virulent *B. anthracis* strains typically carry two plasmids, pXO1 and pXO2 (Little and Ivins, 1999); however, the relative virulence of different *B. anthracis* strains varies, depending on their chromosomal sequences, plasmid composition, and animal species (Welkos *et al.*, 1986). The pXO1 and pXO2 plasmids contain genes for three major virulence factors, the lethal and edema toxins (LT and ET, respectively, encoded by pXO1) and a capsule (encoded by pXO2).

The DNA sequence of the pXO1 isolated from the Sterne strain (AF065404) (Okinaka *et al.*, 1999) is 181,654 nucleotides long with 143 open reading frames (ORFs), covering about 61% of the DNA. Among those genes are the structural toxin genes *lef*, *cya*, and *pagA* for three major components of toxins: lethal factor (LF), edema factor (EF), and protective antigen (PA), respectively, along with regulatory elements; a resolvase and a transposase; and gerX, a three-gene germination operon (Guidi-Rontani *et al.*, 1999). The plasmid also carries DNA topoisomerase (Fouet *et al.*, 1994) and genes possibly involved in horizontal transfer. The pXO1 sequence from the Ames strain isolated from letters used in 2001 bioterror attack contains 181,600 nucleotides long with 217 ORFs corresponding to known and hypothetical genes, covering about 77% of the DNA (Read *et al.*, 2003).

pXO2 carries *capB*, *capC*, *capA*, and *dep* genes – all known to encode capsule synthesis and degradation (Thorne, 1993), and *acpA*, which is a regulatory gene. pXO2 from a Pasteur strain is 96,231 bp long and contains 85 ORFs. By comparison, the pXO2 sequence from the Ames strain contains 94,829 nucleotides encoding 113 genes (Read *et al.*, 2003).

LT comprises seven PA molecules tightly bound to each other in a ring structure and of variable number of LF molecules (2–3) associated with the PA heptamer (Mogridge *et al.*, 2002). LF is a zinc-dependent protease that cleaves mitogen-activated protein kinase kinases (Pellizzari *et al.*, 1999; Vitale *et al.*, 1998). Similarly to LT, the ET comprises the PA heptamer and the EF subunit bound to it. EF is a calmodulin-dependent adenylate cyclase.

In addition to plasmid-encoded factors, the chromosome of *B. anthracis* contains a number of genes for proteins that may contribute to pathogenicity, including hemolysins, phospholipases, numerous proteases, iron acquisition proteins, catalases, superoxide dismutases, etc. There is a substantial body of experimental evidence that genes carried on the chromosome contribute directly to virulence (Cole *et al.*, 2001; Eremin *et al.*, 1999; Heidelberg *et al.*, 2000; Ogata *et al.*, 2001; Roux *et al.*, 1997; Stephens *et al.*, 1998). Recently, it was found that chromosomally encoded hemolytic and cytolytic genes of *B. anthracis*, collectively designated as anthralysins (Anls), could be induced in anaerobic conditions (Klichko *et al.*, 2003). It was also demonstrated that Anls genes are expressed at the early stages of infection within macrophages by vegetating bacilli after spore germination. Cooperative and synergistic enhancement of the membrane-perforating and membranolytic activities of the Anls was found in hemolytic tests on human red blood cells. These findings imply Anls as *B. anthracis* pathogenic determinants and highlight oxygen limitation as environmental factor controlling their expression at both early and late stages of infection (Klichko *et al.*, 2003). It has been hypothesized that evolution forced *B. anthracis* to tightly control the Anl genes expression in the presence of oxygen to resolve a conflict between the transcriptional regulation of toxin genes and the genes involved in the process of sporulation, which appeared after it acquired the toxin plasmid pXO1 (Mock and Mignot, 2003).

4. PATHOGENESIS OF ANTHRAX INFECTION

The information on the pathogenesis of anthrax infection in humans is almost nonexistent, and most available knowledge on inhalational anthrax comes from experimental studies in monkeys (Albrink and Goodlow, 1959; Fellows *et al.*, 2001; Friedlander *et al.*, 1993; Hail *et al.*, 1999) and other nonprimate animals (Fellows *et al.*, 2001; Ross, 1957; Zaucha *et al.*, 1998). The disease is initiated by the entry of spores into the host body. This can occur via a minor abrasion, by eating contaminated meat, or inhaling airborne spores. There are four types of human infection: cutaneous, gastrointestinal, oropharyngeal, and inhalational. Each form can progress to fatal systemic anthrax.

Macrophages cells play a central role in the initiation of inhalation anthrax infection. *B. anthracis* spores are usually engulfed by alveolar macrophages, and a certain fraction of the spores – at least in part due to the LT activity, and presumably to superoxide dismutase and catalase activities as well – overcomes this process and germinates within them instead of being killed (Dixon *et al.*, 2000; Guidi-Rontani *et al.*, 2001).

The current model of host-cell intoxication explains how LF or EF secreted by bacteria gain access to the cytosol of target cells by first binding to cell-associated PA. This interaction depends on the proteolytic processing of PA by furin-like proteases, which allows PA to self-oligomerize into a heptamer that contains the binding sites for three EF or LF moieties (Gordon *et al.*, 1995; Milne *et al.*, 1994; Mogridge *et al.*, 2002). The toxin complex is then endocytosed and trafficked to a low pH endosome where the PA heptamer undergoes conformational changes that allow it to insert into the endosomal membrane, forming a channel or pore, and translocating the EF or LF moieties into the host cytosol (Friedlander 1986; Milne and Collier, 1993).

Image cytometry and the technique using sensitive fluorescence-based reporter systems have demonstrated rapid onset of the expression of genes encoding virulence factors such as LF, PA, EF, and the toxin *trans*-activator, AtxA, upon germination of spores inside the phagosomal compartment of spore-infected macrophages (Guidi-Rontani *et al.*, 1999). It is therefore possible that macrophage "disabling" can take place early in the infectious process as a result of intracellular LT and ET expression, but not only from the action of externally produced toxins.

Inside the macrophages, spores germinate and become vegetative bacteria. Macrophages consequently transport them to the regional tracheo-bronchial lymph nodes. After the spore germination, vegetative cells become encapsulated and perhaps express additional virulence factors, such as AnlO, helping them destroy both phagosomal and cellular membranes for the ultimate escape from the macrophage (Klichko *et al.*, 2003).

In the regional lymph nodes, the local production of virulence factors by the multiplying bacilli gives rise to the characteristic pathological picture of mediastinitis, massive hemorrhage, edema, and tissue death in the lymph nodes (Dutz and Kohout, 1971). Organisms or toxins are usually detected in the lymph several hours before appearing in the blood (Klein *et al.*, 1962). Once the clearing ability of the regional nodes is overwhelmed, and the nodes are destroyed opening access into the circulation (most likely through the lymph nodes' blood vessels), bacilli can then spread to the blood, leading to septicemia.

Several lines of evidence indicate that the germination rate has a crucial effect on the outcome of the infection (Guidi-Rontani *et al.*, 1999); germination rates can depend on the distribution of various constituents in different tissues. It is clear that, to establish disease, the vegetative form must emerge rapidly in a niche favorable for survival – *B. anthracis* germinates and multiplies efficiently in the bloodstream. Anthrax bacteria in the blood can reach high concentrations (up to 10^8 per mL or even higher), then lodge in the reticuloendothelial system and also penetrate through the meninges causing hemorrhagic meningitis. Once bacteria appear in the blood, and a septicemia is underway, it progresses at a constant rate until death (Klein *et al.*, 1966). It is interesting that, once bacilli have been released from the macrophages, there is not much evidence that an immune response is initiated against vegetative bacilli, preventing further multiplication. It has been proposed that after this event, the capsule of *B. anthracis* can partially protect bacteria from complement activation and phagocyte-mediated killing on secondary uptake, in a manner similar to that seen in other capsulated pathogenic bacteria (Lindberg, 1999). Some researchers suggest *B. anthracis* is capable of impairing white blood cell function (Ramachandran and Natarajan, 1968) and a relative lack of white blood cells in the necrotized areas (which can be an indication of immune response suppression) was a repeated observation for the anthrax patients' autopsy results, further support for the idea (Abramova and Grinberg, 1993). It has been already shown that the phenomenon of immune suppression takes place early in the course of anthrax infection, helping to initiate the disease. For example, our research indicates that infection of blood mononuclear cells with anthrax spores leads to the appearance of a large population of apoptotic cells with a reduced capacity to eliminate spores and vegetative bacteria (Popov *et al.*, 2002). Moreover, the aminopeptidase inhibitor, bestatin, capable of protecting cells from LT, restored the bactericidal activity of infected cells. These findings explain the role of LT expression within spore-infected phagocytes as an early intracellular virulence factor contributing to bacterial dissemination and disease progression. Concurrently, anthrax LT causes abnormal impairment of dendritic cell function, which breaks the continuity from innate to adaptive immune responses (Fukao, 2004).

Gastrointestinal anthrax occurs as a result of ingesting improperly prepared or cooked meat or meat products from animals infected with anthrax. It is presumed that spores enter the mucosal and submucosal lymphatics via either breaches in the mucosal lining of the intestinal tract or through phagocytic or antigen-processing cells, such as M-cells. The spores are transported by macrophages to mesenteric and other regional lymph nodes, where they germinate, multiply, and invade other tissues, either directly to the peritoneal cavity or indirectly via the blood. In oral-pharyngeal anthrax, spores are acquired by ingestion or perhaps inhalation, and spores spread to the cervical lymph nodes. If the infection generalizes, its further development is the same or similar to what we can see in systemic inhalational anthrax.

Throughout history, there have been many hypotheses regarding the actual cause of death during anthrax infection. For example, capillary blockage by bacilli, kidney shutdown and progressive secondary shock, attachment of toxin to white blood cells (WBC) and destruction of reticulaoendothelial system (RES) cells, and altered capillary permeability and hypotension which perhaps is the basis of pulmonary edema (Walker *et al.*, 1967) have all been suggested. Some researchers have subsequently suggested central nervous system depression and respiratory failure as the cause of death (Bonventre *et al.*, 1967).

Hypoxia has been considered a major contributor in death during anthrax infection (Moayeri *et al.*, 2003). Hansen (1961) and Nordberg *et al.* (1964) reported pathophysiological changes in rabbits after anthrax spore challenge, and Bonventre and Eckert (1963) reported similar observations in rats challenged with sterile anthrax toxin, with the most remarkable observation following both spore and toxin challenge of the extraordinary hypoxia observed immediately prior to death (Klein *et al.*, 1962).

Smith and Stoner (1967) proposed that the main effect of anthrax toxic complex is to increase vascular permeability, which can be seen after either intradermal or intravenous injection. A severe loss of circulating fluid, comparable with that occurring in anthrax infection, was seen when crude anthrax toxin was injected intravenously in guinea pigs. After the intravenous injection of the toxic complex in the rat fluid loss occurs mainly in the lungs as pulmonary edema. Dalldorf and Beall (1967) also showed that the toxic complex could increase vascular permeability at other sites in the rat, such as the peritoneal cavity and skin, but the changes developed much more slowly than in the lungs. The effects of fluid loss could be aggravated in some species and under certain conditions by respiratory embarrassment due to fluid loss into the lungs and by circulatory embarrassment due to pressure from fluid loss into the mediastinum (Smith and Stoner, 1967).

Thrombosis might be considered as one more component contributing to death in anthrax infection (Dalldorf and Beall, 1967). The lungs of rats, rabbits, and guinea pigs that died of anthrax septicemia were studied by light and electronic microscopy. In all three species, the clinical signs of respiratory failure that occurred during the terminal phase of the disease were associated with widespread pulmonary capillary thrombosis. The thrombi occurred in intact capillaries and were composed of platelets and fibrin (Dalldorf and Beall, 1967).

Systemic proinflammatory cytokine release resulting from the oxidative burst of intoxicated macrophages has been previously implicated as a major death-causing factor in anthrax (Hanna *et al.*, 1994). This point of view, however, was later been challenged by a number of publications (Erwin *et al.*, 2001; Liu *et al.*, 2003; Moayeri *et al.*, 2003; Popov *et al.*, 2002a, 2002b) that demonstrated the absence of cytokine induction by LT in macrophages in culture. It is possible that earlier studies used LT preparations produced in *Escherichia coli*, and therefore contained trace amounts of endotoxin responsible for the observed cytokine response. Recently, the issue of cytokine response to LT administration has been studied in mice (Moayeri *et al.*, 2003). The authors confirmed the observation made in cultured cells and further concluded that sensitive macrophages alone do not appear to be the source of the LT-induced death-causing factors.

Smith *et al.* (1955) considered that the major factor contributing to the death in anthrax is LT. However, it is known that the toxicity of the anthrax toxin is not as strong as that of many other bacterial toxins, and it is difficult to explain how this toxin – with a relatively low level of toxicity – could cause death. The only scenario where LT could kill is when *B. anthracis* multiplies rapidly and produces a high amount of the toxin. When this high level of toxemia occurs, the host could be in the state of septicemia, and the real cause of death may be one of many other factors of anthrax infection, but is attributed to the toxin.

Although none of the theories describes the pathogenesis of late-stage anthrax infection in its complexity, manifestations of advanced disease – including shock and sudden death – can be attributed to the combination of multiple factors, which can originate either from systemic bacteremia or local necrotic and hemorrhagic changes in vital organs. The high bacteremia and toxemia in the blood contribute to the systemic increase of vascular permeability causing edema and various effusions, which in turn are capable of causing organs function decrease and failure. For example, respiratory failure can be associated with pulmonary edema, pleural effusions, or both. Direct damage of lung tissues with alveoli lacking their functional capacities additionally promotes the phenomenon. Loss of circulating fluid due to effusions and edema contributes to the development of hypovolemic syndrome, which further increases hypoxia and could lead to hypotonia and circulatory collapse. Shock and death could follow these changes. Another mechanism of injury includes hepatic insufficiency and other damaging changes in the reticuloendothelial system. Sudden death can also be stimulated by bacteremia if the bacilli and their virulence factors cause enough pathological changes in the blood circulation of the brain. Mechanisms can include the development of massive hemorrhages or hematomas, especially near vital centers. Meningitis has been demonstrated to be a very common complication of inhalation anthrax and is almost always fatal.

B. anthracis is unique because instead of having one specific mechanism of killing, it has the capability to act in a variety of ways, expressing various virulence factors and causing different types of damage in different cases. It usually depends on the individual characteristics of the victim, and which system or organ is most vulnerable during the course of infection, and would tend to fail first contributing to the death.

5. CLINICAL MANIFESTATION OF ANTHRAX INFECTION

In the event of an anthrax biological attack, the appearance of inhalational systemic cases, as well as some number of cutaneous forms, should be expected.

5.1. Cutaneous Anthrax

More than 95% of naturally occurring anthrax is the cutaneous form. The incubation period ranges from 1 to 12 days (with an average of approximately 7 days). Anthrax spores are usually introduced at the site of a cut or abrasion, typically on the arms, face, or neck (Swartz, 2001). The resulting lesion of cutaneous anthrax is often described as a *malignant pustule* because of its characteristic appearance. Coagulation necrosis of the center of the pustule results in the formation of a dark-colored *eschar*, which is later surrounded by a ring of vesicles containing serous fluid and an area of edema and induration that may become extensive (Greenwood *et al.*, 1997). The lesion is generally painless, but fever, malaise, headache, and regional lymphadenopathy are present (Brook, 2002). The papule can resemble an insect or spider bite and may itch. Pustules are rarely present in anthrax lesions, and a primary pustular lesion is unlikely to be cutaneous anthrax. Lesions may be solitary or multiple (Carucci *et al.*, 2002). The eschar separates in 1–2 weeks and falls off, leaving a scar in most cases. Diagnosis and treatment of the cutaneous form of anthrax usually does not present serious difficulty for physicians, and mortality rate for it is low (<1% with antibiotic treatment).

5.2. Systemic Anthrax

Current literature has a relatively standardized approach to describing the clinical manifestation of systemic anthrax, which is by now familiar to all physicians as well as scientists working in this area. The incubation period of inhalation anthrax according to current literature may last from 1 to 9 days (very lengthy incubation periods – i.e., up to 45 days – have never been scientifically proven), but the average incubation period for the patients infected in the United States in 2001 was for 4–6 days (Jernigan *et al.*, 2001). Symptoms and physical findings are nonspecific in the beginning of infection.

The clinical presentation is usually biphasic. The initial stage begins with the onset of myalgia, malaise, fatigue, nonproductive cough, occasional sensation of retrosternal pressure, and fever. Unlike those with influenza or other viral respiratory illnesses, people with inhalational anthrax are not contagious. They may experience shortness of breath and vomiting, but in the majority of cases sore throat or rhinorrhea is absent. In some patients, a brief period of apparent recovery follows. Other patients may progress directly to the second, fulminant stage of illness. The second stage develops suddenly with the onset of acute respiratory distress, hypoxemia, and cyanosis. Death sometimes occurs within hours (Inglesby *et al.*, 2002; Jamie, 2002). The patient may have mild fever; alternatively, the patient may have hypothermia and develop shock. Diaphoresis is often present. Enlarged mediastinal lymph nodes may lead to partial tracheal compression and alarming stridor; auscultation of the lungs is remarkable for crackles and signs of pleural effusions. Meningeal symptoms may be present in up to 50% of cases. They may be associated with subarachnoid hemorrhage and reveal themselves through decreased level of consciousness, meningismus, and coma.

Chest radiography typically shows widening of the mediastinum and pleural effusions, whereas the parenchyma may appear normal. In a review of the 11 patients infected by anthrax in October 2001, chest X-ray films from the initial examination showed mediastinal widening, paratracheal and hilar fullness, and pleural effusions or infiltrates. In some patients, the initial findings were subtle and not detected immediately (Cranmer and Martinez, 2003).

To provide physicians with details of the disease that are helpful when trying to understand anthrax infection, we examined more than 40 cases of inhalation anthrax, published over the last 100 years (Albrink *et al.*, 1960; Barakat *et al.*, 2002; Brachman, 1980; Brachman and Pagano, 1961; Brooksher and Briggs, 1920; Bush *et al.*, 2001; Chandramukhi *et al.*, 1987; Cowdery, 1947; Drake and Blair, 1971; Doganay *et al.*, 1986; Enticknap *et al.*, 1968; Gold, 1955; Haight, 1952; Koshi *et al.*, 1981; Krauss, 1943; LaForce *et al.*, 1969; Plotkin *et al.*, 2002; Pluot *et al.*, 1976; Rangel and Gonzalez, 1975; Rozen'er, 1948; Severn, 1976; Shapiro and Galinobskoi, 1923; Suffin, 1978; Teacher and Glasg, 1906; Vessal, 1975). We extensively analyzed both common and rare symptoms, the treatment regimens, and laboratory data and autopsy results when available. Data were then analyzed for possible links and associations, both tenuous and precise, to create a profile of the disease.

The findings were somewhat inconsistent due to the different styles of disease description and different terminology used by the attending physicians trained under different conventions. Therefore, the statistical approach in interpretation of this data was rather difficult, yet it was still possible to obtain important information that is not common knowledge.

It is obvious that the disease does not target a specific age, or range of ages, gender, or locale. The cases ranged in age from 14 to 94 years and were diagnosed on continents and islands, in the developed and developing countries. Ethnicity or race did not appear to be a factor, or more specifically, the disease does not seem to be linked to any particular set of population. The cases that were reported included people who were whites, blacks, and Asians.

The disease progression from the first manifestation of symptoms until death appears to have a considerable range from a few hours (Pluot *et al.*, 1976) to 11 days (Brooksher and Briggs, 1920). In our study of 40 cases (Table 1.1), 7 patients recovered and 33 died. Of the 33 fatal cases, 7 patients expired within 7 days, 3 patients – within 6 days, 5 patients – within 5 days, 4 patients – within 4 days, 6 patients – within 3 days, 3 patients – within 2 days, and 3 patients – within 1 day. There was 1 case each of patients expiring in 8 and 11 days. The average length of terminal episode appeared to be 4.7 days. Histogram of distribution of the length of terminal episode is presented in Figure 1.1.

5.2.1. General Symptoms of Anthrax Infection

Since anthrax symptoms insidiously mimic flu-like symptoms in the beginning and are not characterized by any significant, prominent, or violent abnormalities, they are often underreported by the patients themselves. General symptoms were fever, headache, diaphoresis, chills, myalgias, fatigue, malaise, weakness, backache, and edema. Edema was generally evident in the neck area or axilla and was of the nonpitting variety.

Distribution of general symptoms is presented in Figure 1.2.

A vast majority of the cases (33) reported a fever ranging from "mild" to 105°F (rectal). Two of the cases reported no fever. The complaints of chills were recorded in 13 cases and sweating in 14 cases. Our analysis indicates that profuse diaphoresis is a common symptom in anthrax patients and could serve as one of the signatures of this infection if presented in a proper setting of other signs. The complaints of weakness (3 cases), fatigue (11 cases), and

Table 1.1
Inhalational Anthrax Cases 1–40[a]

Case no.	Year	Age	Gender	Outcome	Country	Reference
1	1905	36	Male	d	UK	Teacher and Glasg (1906)
2	1913	25	Female	d	Russia	Shapiro and Galinobskoi (1923)
3	1920	36	Male	d	US	Brooksher and Briggs (1920)
4	1941	71	Male	d	US	Krauss (1943)
5	1942	42	Male	d	US	Gold (1955)
6	1947	46	Male	d	US	Cowdery (1947)
7	1948	50	Female	d	US	Brachman and Pagano (1961)
8	1948	25	Female	d	Russia	Rozen'er (1948)
9	1948	55	Male	d	Russia	Rozen'er (1948)
10	1952	29	Female	d	US	Haight (1952)
11	1957	60	Male	d	US	Albrink *et al.* (1960), Plotkin *et al.* (2002)
12	1957	49	Male	d	US	Albrink *et al.* (1960), Plotkin *et al.* (2002)
13	1957	33	Male	d	US	Albrink *et al.* (1960), Plotkin *et al.* (2002)
14	1957	65	Female	d	US	Plotkin *et al.* (2002)
15	1957	46	Male	R	UK	Plotkin *et al.* (2002)
16	1957	31	Male	d	US	Brachman and Pagano (1961)
17	1958	53	Male	d	US	Koshi *et al.* (1981)
18	1961	51	Female	d	US	Koshi *et al.* (1981)
19	1965	54	Male	d	UK	Enticknap *et al.* (1968)
20	1966	46	Male	d	Turkey	LaForce *et al.* (1969)
21	1971	35	Male	d	Zimbabwe	Drake and Blair (1971)
22	1975	16	Female	d	Iran	Vessal (1975)
23	1975	34	Male	d	Iran	Vessal (1975)
24	1975	14	Male	d	South America	Rangel and Gonzalez (1975)
25	1976	53	Male	d	UK	Severn (1976)
26	1976	39	Male	d	France	Pluot *et al.* (1976)
27	1978	32	Male	d	US	Suffin (1978)
28	1981	63	Male	d	India	Koshi *et al.* (1981)
29	1987	45	Female	d	India	Chandramukhi *et al.* (1987)
30	2001	63	Male	d	US	Jernigan *et al.* (2001), LaForce *et al.* (1969)
31	2001	73	Male	R	US	Jernigan *et al.* (2001)
32	2001	56	Male	R	US	Jernigan *et al.* (2001)
33	2001	56	Male	R	US	Jernigan *et al.* (2001)
34	2001	55	Male	d	US	Jernigan *et al.* (2001)
35	2001	47	Male	d	US	Jernigan *et al.* (2001)
36	2001	59	Male	R	US	Jernigan *et al.* (2001)
37	2001	56	Female	R	US	Jernigan *et al.* (2001)
38	2001	43	Female	R	US	Jernigan *et al.* (2001)
39	2001	61	Female	d	US	Jernigan *et al.* (2001)
40	2001	94	Female	d	US	Barakat *et al.* (2002)

d, death; R, recovery.
[a]*Note*: These references will not be repeated throughout the rest of the chapter since all symptoms, findings, therapies, and pathologic changes are taken from these cases. However, information from other sources will be referenced.

malaise (10 cases) could be systemic indications of severe infection, but could also be considered to represent a neurological component, depending on which phase of the disease the patient was in at the time of observation. Headaches and myalgias were also among the frequently reported complaints (17 and 12 cases, respectively), although the appearance of headaches could have a neurological origin. Sore throat was seen in 6 patients, and nasal

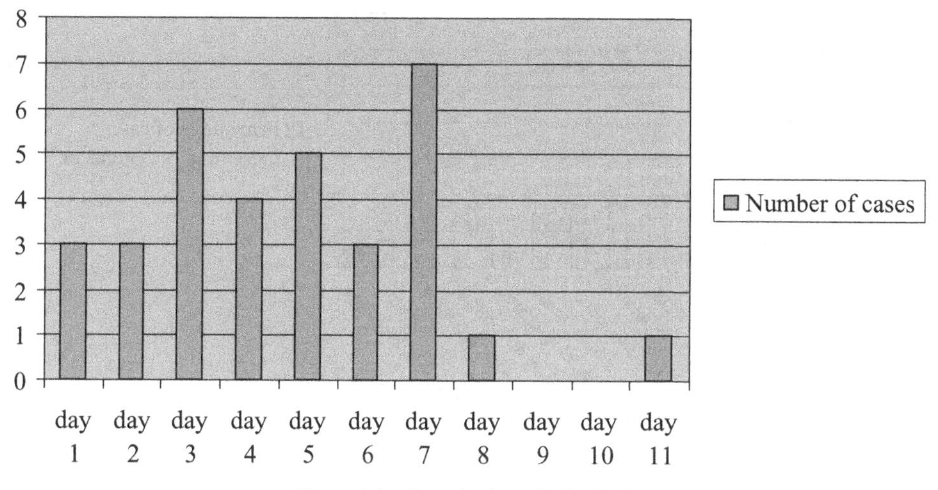

Figure 1.1. Length of terminal episode.

discharge was seen in 5 patients (technically these are the signs of upper respiratory system disorder, but are also included in flu-like symptoms).

5.2.2. *Respiratory Symptoms and Findings*

Respiratory symptoms were characterized as cough, chest pain, sore throat, shortness of breath, dyspnea, orthopnea, and presence of sputum or nasal discharge. Auscultation provided findings of bronchial breathing and rales, and the picture was rounded out by documentation of the mediastinal widening on X-ray examination.

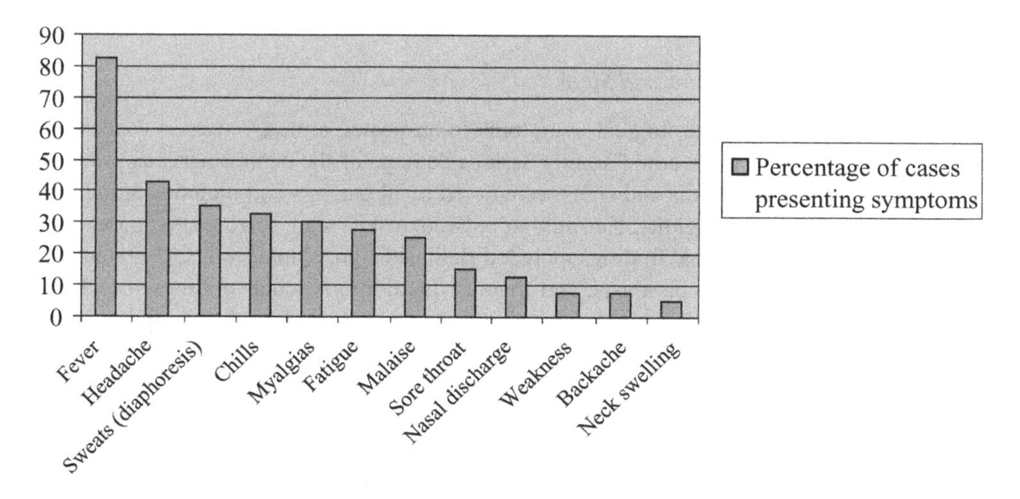

Figure 1.2. General symptoms of anthrax infection.

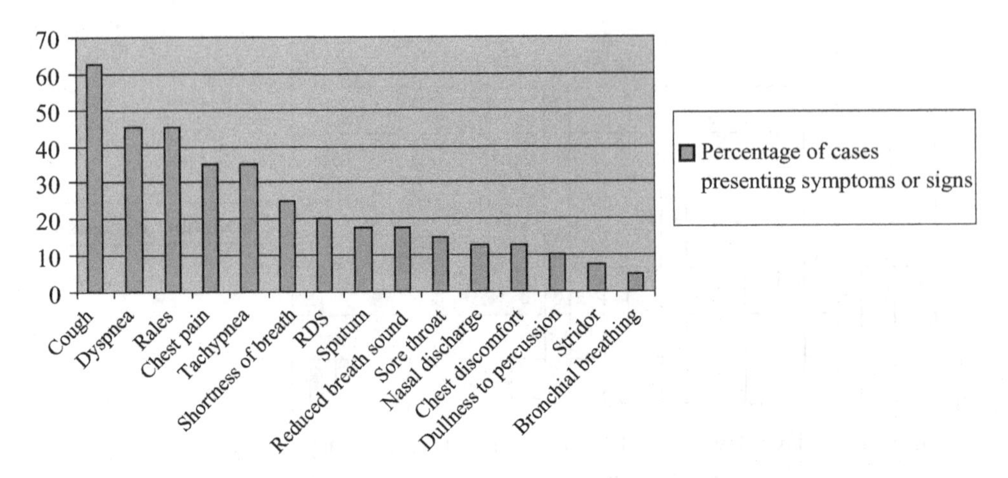

Figure 1.3. Respiratory symptoms and signs. RDS, respiratory distress syndrome.

Thirty-nine cases recorded a variety of respiratory symptoms. Of these, two-thirds (or 25 cases) presented with a cough, and 7 cases had discharge (sputum) that ranged from hemoptysis to green to white. Fourteen patients complained of chest pain, and 5 patients complained of chest discomfort. Breathing difficulties were evident in the rate of respirations, which ranged from 20 to 50 per minute, although they were recorded in only 14 patients. Shortness of breath was reported in 10 patients, and dyspnea and orthopnea was reported in 17 and 6 patients, respectively. Eight patients were diagnosed with respiratory distress syndrome.

Abnormal chest sounds were discovered in the majority of patients and were further subclassified as rales (18), egophony (1) and rhonchi (1), decreased breath sounds (7), and bronchial breathing (2 patients). Stridor was noted in 3 patients, and 1 patient was described as "groaning with each breath." Dullness to percussion was seen in 4 patients. Respiratory symptoms and findings are presented in Figure 1.3.

5.2.3. Neurological Symptoms and Signs

The appearance of neurological symptoms in suspected anthrax cases is especially alerting, since some of them could signify serious damage of the central nervous system/ brain. Neurological symptoms and signs were noted in 34 patients and included agitation, confusion, delirium, neck rigidity, Kernig's sign, Brudzinski's sign, seizures, syncope, and coma. Abnormal corticospinal findings included Babinski sign, plantar reflex, deep tendon reflexes, flaccid and spastic paralysis, and trismus. In addition, ocular findings – such as conjugate and dysconjugate deviation of the eyes, nystagmus, dilated and fixed pupils, visual distortion, and photophobia – suggested brainstem and cranial nerves involvement.

With this wide variety of signs of neurological origin, no clear picture can be seen as the frequency the signs and their presence had a wide statistical dispersion. The majority of patients had confusion and/or anxiety at some point during the course of their anthrax, making these the most constant neurological symptoms. The Babinski sign was present bilaterally in one case, hypertonia was seen in 2 cases, deep tendon reflexes were symmet-

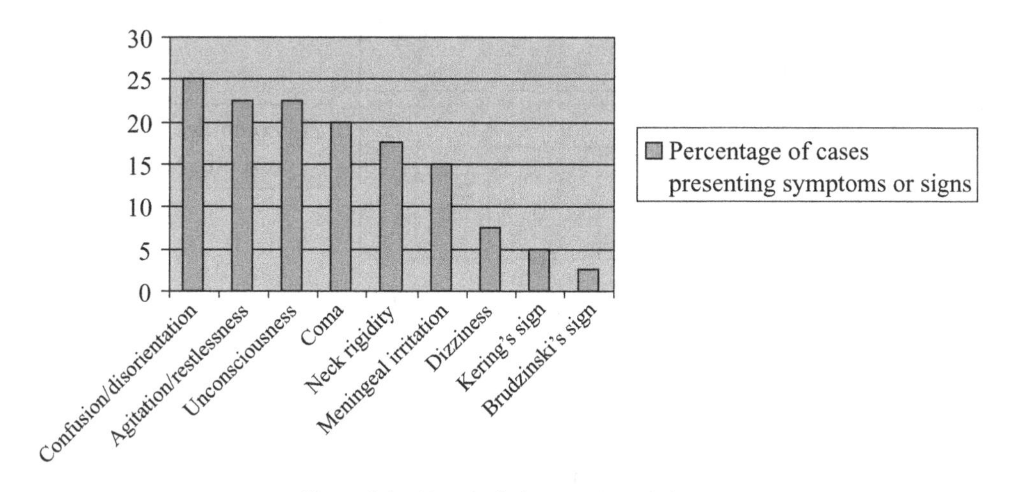

Figure 1.4. Neurological symptoms and signs.

rically hyperactive in 3 cases, plantar reflexes were extensor in 3 cases, flaccid paralysis was present in 2 cases, ineffective response to painful stimuli was present in 2 cases, and spasticity of extremities was present in 1 case. Unintentional hand tremor and trismus were also observed in one patient each. Dysphagia was noted in two patients, but it was not clear if this was of neurological or oropharyngeal origin.

Vomiting was reported in 13 cases, and nausea was reported in 12 cases, 8 of which were concurrent. Both symptoms could have their origins in disturbance of either the neurological or gastrointestinal systems. Although vomiting and nausea can be considered of neurological origin, at least in some patients, we included them only in the group of gastrointestinal symptoms since they manifest themselves primarily as signs of gastrointestinal tract involvement. Neurological symptoms and signs are presented in Figure 1.4.

5.2.4. Cardiovascular Symptoms and Signs

Symptoms of the cardiovascular system were recorded as abnormal blood pressure, pulse/heart rate, and cyanosis. Twenty-nine patients reported tachycardia; pulse rates were significantly elevated and ranged from 92 to 180. Twelve patients had hypotension of various degrees. Cyanosis, which is generally considered a sign of hypoxia, was noted in 15 cases. Cardiovascular signs, although quite consistently present in the cases studied, gave a significant indication of generalized stress to the body, but careful detailed analysis of more cases is clearly warranted to formalize a better picture. Cardiovascular system signs are presented in Figure 1.5.

5.2.5. Gastrointestinal Symptoms and Signs

Gastrointestinal symptoms fell into the general categories of abdominal pain (8 cases), anorexia (4 cases), and nausea and vomiting, although the latter two can possibly be cross-linked to the neurological disorders. Vomiting was seen in 13 cases and was accompanied by

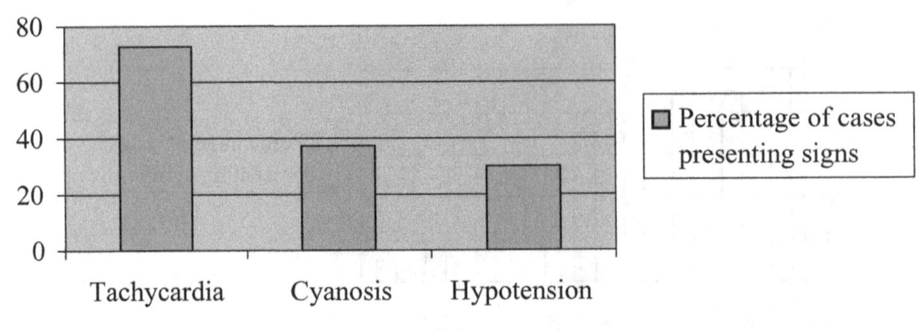

Figure 1.5. Cardiovascular signs.

nausea in only 8 of those cases. Nausea by itself was seen in 12 patients. Absence of bowel sounds was noted in 1 patient, and diarrhea was noted in 1 patient, although the majority of the cases did not mention any intestinal motility abnormalities. Gastrointestinal symptoms and signs are presented in Figure 1.6.

5.2.6. Miscellaneous

Renal disturbances were rare. One case of renal failure was reported and two cases of urinary incontinence, although one case of urinary incontinence episodes improved and lasted for only 1 day. Dehydration was noted in two cases and hypothetically can be attributed to the fluid loss as the result of tissue edema and effusion.

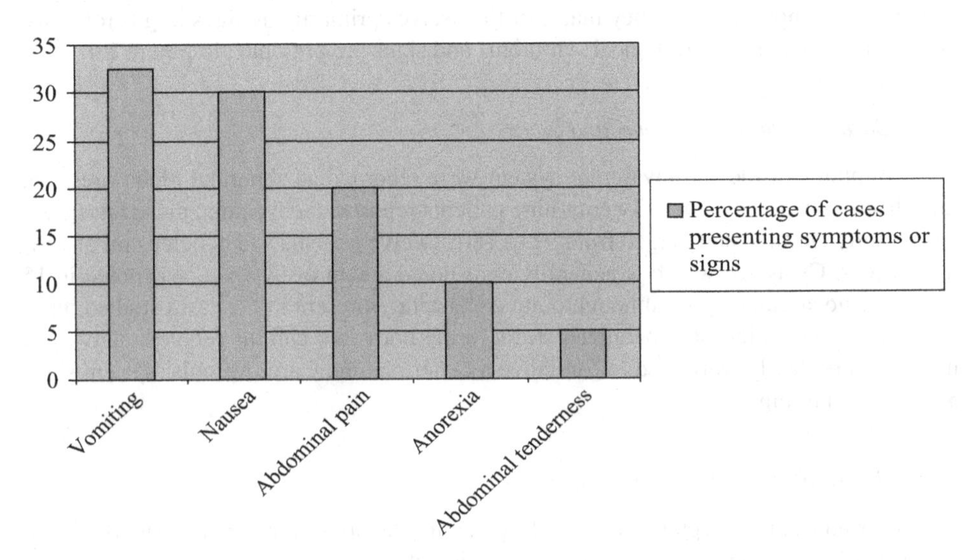

Figure 1.6. Gastrointestinal symptoms and signs.

5.2.7. *Findings – Autopsies*

Of the 40 cases examined (33 lethal cases), 29 reported autopsies were conducted in varying degrees. The principal sites of pathological alteration seem to be in the thorax and the central nervous system, with some involvement of the gastrointestinal system, the liver, and the spleen. The most prominent findings reported appear to be those of hemorrhage and edema and congestion in organs along with the presence of the pathogen. Our data correlate with the results of 42 necropsies from the Sverdlovsk anthrax outbreak of 1979 (Abramova *et al.*, 1993; Grinberg *et al.*, 2001). The most prominent findings consisted of hemorrhagic thoracic lymphadenitis and hemorrhagic mediastinitis, with various manifestations of hematogenous spread of *B. anthracis* infection, including serohemorrhagic and hemorrhagic leptomeningitis. Gastrointestinal lesions were observed in most of the cases and appeared in particular in the small intestine, stomach, and colon.

A notable effect of the disease was edema and other evidence of increased vascular permeability, including gelatinous edema of the mediastinum, pleural effusions, leptomeningeal edema, and pulmonary edema. Ulceration, pseudomembrane formation, fibrin-rich edema, and hemorrhage were present in the bronchi and trachea of many patients. Acute bronchopneumonia was reported in nearly half of the patients with microscopic observations of intraalveolar exudate, interstitial exudate, edema, hemorrhage, and capillaritis (Abramova *et al.*, 1993; Grinberg *et al.*, 2001).

In our cases, hemorrhage was also the most widely reported finding, being present in the lungs (8 of 29 cases), mediastinum (8 cases), liver (8 cases), spleen (8 cases), stomach (8 cases), and small intestine (8 cases). The lymph nodes and brain had repeated reports of hemorrhage; however, the brain was omitted from examination in some cases where autopsies were performed, so the ratios are somewhat misleading. Hemorrhage was also reported in the kidneys (2 cases), heart (2 cases), and the thyroid (1 case). In some cases, the subcutaneous tissues were edematous, and petechial hemorrhages were reported in two cases.

Some interesting details were forthcoming in describing the hemorrhage of the brain. Meningitis and even leptomeningitis were present in five autopsy reports. In one case, the leptomeninges covering the superior lateral surface of the cerebral hemispheres were described as being hazy and evidence of bilateral hippocampal uncal herniation was present. In five other cases, subarachnoid hemorrhage was observed. In three cases, it was present over the entire cerebrum, along with hemorrhage in the Virchow-Robin space and in the lateral ventricle, and two other cases had hemorrhage extending to the cerebellum. One autopsy report of an anthrax patient who died within a few hours of disease progression included a massive left cerebromeningeal hematoma, with extensive involvement of the frontal lobe of white matter.

Hemorrhage of the lungs was recorded for almost every autopsy report. The inability of the lung to expand was noted as "patchy atelectasis," "atelectasis," "atelectic," "collapsed lung covered with thick tenacious mucus," or "partial collapse." Reports of pleural effusion included descriptions in various forms ("straw," "bloody," "cloudy,"and "fibrinous") and with the amounts ranging from 500 mL in both lungs to 2,000 mL in each.

Hemorrhage was widely present in the mediastinum, and the mediastinal lymph nodes have been described as "anthracotic," "black," "completely destroyed," and "necrotic." Both edema and hemorrhage were seen in these. The hilar lymph nodes were also hemorrhagic – with some having consolidation areas – and were "black," had "necrotic liquefaction," and some were "destroyed."

The gastrointestinal tract was also reported to have hemorrhages. The stomach has been described in different cases as "swollen" and hemorrhaging with "multiple acute ulcers in antrum of stomach," "circumscribed mucosal erosions," and "submucosal hemorrhage." The small intestine has been variously described as having "hemorrhage on tips of the villi," "superficial ulcers covered with gelatinous exudate," "acute ulcers," "swollen loop," and "swollen with ulcers." Ulcerated mucosal areas were seen in the ileum.

Hemorrhage was reported in the spleen, and the spleen has been described as being "mushy," "enlarged," "swollen," "splenomegaly," "splenitis," with "ill-defined architecture," or "complete destruction of cytoarchitecture." The liver had reports of hemorrhage, as well as congestion, edema, and parenchymatous degeneration.

There were several reports of subendocardial hemorrhage, with one case reporting considerable amount of blood-staining under the endocardium. Effusion into the pericardium was reported in three cases, with volume varying from 10 to 100 mL. Hemorrhage was described as having occurred in the thyroid, but there were no extra markers for this description.

Edema was reported in the mediastinum in eight cases. Examination revealed that the mediastinum was greatly enlarged due to edema, including one case where the entire mediastinum was occupied by hemorrhagic edema with scattered, enlarged, and hemorrhagic lymph nodes.

Edema was also seen in the spleen (5 cases), small intestine (3 cases), ileum (1 case), with two instances each of the brain, stomach, and liver, and one report of the kidney. In the brain, narrowing of sulci and flattening of the gyri (in two cases) evidenced edema, and one case had a prominent cerebellar pressure cone. In the spleen, edema resulted in obliteration of the cytoarchitecture in five cases. Indistinct cytoarchitecture was also reported for the liver. The kidneys reported glomerulitis in one case, but in many other cases it was grossly normal. In the nine reported cases of edematous appearance of the lungs, some of the researchers further classified the edema as being "subpleural" or being an "edematous upper lobe."

Passive congestion was reported in the liver and spleen in 7 cases and in 2 cases for the kidney. Congestion was noted in 9 cases in the lung and 2 cases each for the kidney and small intestine. A large number of macrophages containing black pigment granules were seen inside the alveolar spaces in the lungs.

B. anthracis was found in various tissues that were tested at the time of autopsy. The presence of *B. anthracis* was found in the lungs (9 cases); mediastinum (7 cases); spleen (6 cases); brain and liver (5 cases each); lymph nodes (4 cases); small intestine, kidneys, and cerebrospinal fluid (3 cases each); and stomach, heart, and thyroid (1 case each). It was also found in the meninges, cerebrospinal fluid, and brain in different descriptions.

Death has been attributed to a combination of causes, including profound hemorrhagic syndrome, disruption of the respiratory system function (due to pleural effusion, atelectasis in the lungs, accumulation of mucous in the alveoli and bronchioles, increased permeability, and vasculitis in the lung vessels), and shock.

6. CONSIDERATIONS ON CLINICAL MANIFESTATIONS OF SYSTEMIC ANTHRAX

Diagnosis of inhalation anthrax, especially during the first stage, is difficult. Analyzing literature on inhalational anthrax cases, we made a conclusion that to provide physicians with a useful diagnostic tool, it is important to divide clinical presentations of systemic anthrax into certain provisional groups. Within all 40 cases, at least four clinical groups could be distinguished with some level of certainty:

- patients with prevalent symptoms of respiratory system disorder and collapse
- patients with prevalent symptoms of general neurological (or meningial) origin
- patients with septicemic syndrome, and
- apoplectic (a rare form) anthrax where death occurs suddenly within few hours after the disease onset (in terms of pathogenesis, this form is a lightening meningial/ neurological form; it has significant clinical distinctions).

The initial stage of anthrax with prevalent symptoms of respiratory system affection consists of the insidious onset of mild fever, malaise, fatigue, myalgia, nonproductive cough, and at times chest pain or discomfort. This initial phase typically lasts for several days, following which there may be some improvement in the clinical condition. The second stage develops suddenly with the onset of acute dyspnea and subsequent cyanosis. The patient may appear moribund with accelerated pulse and respiration rate. The body temperature may be mildly elevated, or it may be subnormal because of shock. Stridor may occur perhaps as a result of partial extrinsic obstruction of the trachea by enlarged mediastinum nodes. Profuse perspiration is frequently present. Physical examination of the chest reveals moist, crepitant rales and signs of pleural effusion. A chest X-ray typically shows widening of the mediastinum.

Septicemic anthrax resembles that of respiratory anthrax, but differs by less prominent symptoms of respiratory failure and more prominent features of bacterial sepsis. The difference is relatively small, yet this type of systemic anthrax will be more difficult to diagnose because chest pain or mediastinal widening would not be evident. Patients experience fever, weakness, and might be moribund or semiconscious. There is usually a profuse diaphoresis and there might be dehydration. Marked cyanosis of the skin is common, but in some cases patients may have a cyanotic face with paleness of the rest of the body (Litusov *et al.*, 2002). Pulse is usually weak and rapid, with tachycardia reaching up to 150 beats a minute. Without successful treatment, the patient's condition deteriorates, blood pressure drops, and septic shock develops. At this stage, respiratory symptoms – including cough, dyspnea, and tachypnea – usually appear, making this type of anthrax clinically undistinguishable from the form with prevalent respiratory symptoms.

Anthrax with a prevalent neurological (meningial) syndrome usually starts with a sudden onset of malaise, intense headache, dizziness, chilliness or frank chills, myalgia, nausea and vomiting, and general restlessness. Delirium, stupor, or coma may develop within 24 hours. In the cases with rapid progression of the disease, seizures may be one of the first signs of progression to the manifestation phase. Cerebrospinal fluid analyses usually demonstrates hemorrhagic meningitis with positive cultures. The average duration of the illness is usually from 2 to 4 days. In every instance, hemorrhagic meningitis would be noted.

Apoplectic anthrax is characterized by extremely sudden onset of the disease. Patients might expire within hours after they appear to be in good health. Usually, there is no treatment implemented, and the diagnosis is made through the autopsy results. It is almost certain that subarachnoid hemorhage due to *B. anthracis* damage of the brain is the cause of death in this type of anthrax infection.

7. LABORATORY DIAGNOSIS

Current methodologies of anthrax diagnosis are performed focusing on several levels of specificity. Specimens of sputum, cerebrospinal fluid, blood, or pustules for cutaneous anthrax may be collected and cultured on appropriate laboratory media. According to the CDC, genus-level presumptive identification of *Bacillus* infection is performed through positive Gram staining, along with colony identification. Species-level presumptive identification requires motility tests, positive capsule visualization, γ-phage lysis, hemolysis, wet mount, and malachite green staining (for spores). Confirmatory identification is done by direct flourescent antibody staining against both capsular antigens, and cell wall polysacharhides (CDC Emergency Preparedness and Response Website, 2004).

In addition to these procedures, the polymerase chain reaction may also be used to amplify genes specific for PA (a toxin gene) and the capsule. Each of these virulence factors are encoded on separate plasmids harbored by pathogenic anthrax strains, and their presence in samples is indicative of an active disease. ELISA assays against the PA toxin itself are also available (WHO Model G, 2004). Additionally, the inhibition of bacterial growth by Penicillin G can also be measured on culture plates and compared with known standards.

Delayed confirmation of the diagnosis can be provided by detection of antibodies to *B. anthracis* in the blood. Ouchterlony gel diffusion, microhemagglutination, and ELISA procedures can demonstrate antibodies to the organism. Additionally, acute and convalescent sera of suspected cases are typically submitted to the CDC for confirmation of infection.

8. VACCINATION

A Food and Drug Administration-licensed vaccine – derived from the supernatant fluid of an attenuated, nonencapsulated *B. anthracis* strain (Sterne) – is available and has been used in hundreds of thousands of military troops and at-risk civilians. The vaccine, designated "anthrax vaccine adsorbed" (AVA) is nonliving, and the primary antigen is a protein called protective antigen. The vaccination series, as currently licensed, consists of six doses (0, 2, and 4 weeks and 6, 12, and 18 months) followed by annual boosters. AVA is administered subcutaneously as a 0.5-mL dose. Efforts are underway to reduce the number of doses now required by the package insert. There is not enough data from exposure of humans to determine protective efficacy of the vaccine against aerosol challenge, but studies in rhesus monkeys indicate that the vaccine is effective, even when as few as two doses are administered (Fellows *et al.*, 2001; Ivins *et al.*, 1996). The U.S. Army has also developed a next-generation anthrax vaccine based on the same antigen (PA) produced through recombinant technologies. Although there is no reason to believe that the new vaccine will be more protective, it will be more easily produced in available production facilities and may be slightly less reactogenic and possibly less costly if large lots are needed.

The safety of AVA has been the subject of numerous studies. Manufacturer labeling cites a 30% rate of mild local reactions for the current U.S. vaccine (Pile *et al.*, 1998). The Institute of Medicine recently published a report that concluded that AVA is effective against inhalational anthrax and may help prevent onset of disease postexposure if given with appropriate antibiotics. The Institute of Medicine committee also concluded that the vaccine was acceptably safe (Committee to Assess the Safety and Efficacy of the Anthrax Vaccine, 2002). However, it is important to remember that it is impossible to test the actual efficacy of any vaccine in case of aerosol attack; therefore, it remains unclear.

9. POSTEXPOSURE PROPHYLAXIS

Postexposure prophylaxis against anthrax may be achieved with oral ciprofloxacin (500 mg orally every 12 hours) or doxycycline (100 mg orally every 12 hours), and all persons exposed to a bioterrorist incident involving anthrax should be administered one of these regimens at the earliest possible opportunity. Adherence to the antibiotic prophylaxis program must be strict, as disease can result at any point within 30–60 days after exposure if antibiotics are stopped. In case of threatened or suspected release of anthrax, chemoprophylaxis can be delayed 24–48 hours until the threat is verified. Chemoprophylaxis can be discontinued if the threat is found false. Levofloxacin and ofloxacin would be acceptable alternatives to ciprofloxacin. In addition to receiving chemoprophylaxis, exposed persons should be immunized. If the vaccine is unavailable, chemoprophylaxis should be continued for 8 weeks.

10. TREATMENT OF ANTHRAX INFECTION

According to CDC recommendations, cutaneous anthrax associated with a bioterrorism attack should be treated with ciprofloxacin or doxycycline as the first-line therapy. Cutaneous anthrax with signs of systemic involvement, extensive edema, or lesions on the head and neck require intravenous therapy, and a multidrug approach is reccommended (CDC Treatment Protocols, 2004).

Limited clinical experience is available, and no controlled trials in humans have been performed to validate current treatment recommendations for inhalational anthrax (CDC Treatment Protocols, 2004). In the cases we analyzed, physicians used a variety of drugs that encompassed diverse categories – antibiotics of various groups, anticonvulsants, anti-inflammatory drugs, analgesics, and stimulants.

Available anthrax treatment in the beginning and during a large portion of the 20th century was of course different from today's treatments; but, in each case, the physician was

Table 1.2
Cases Receiving Penicillin and/or Other Drugs

Case no.	Outcome	Penicillin dosage
7	Death	Penicillin, 300,000 units, intramuscularly
12	Death	Aqueous procaine penicillin, 800,000 units
13	Death	Oral penicillin , 8–10 tablets (200,000 units)
14	Death	Aqueous procaine penicillin, 400,000 units, intramuscularly
15	Recovery	Aqueous procaine penicillin in combination with dihydrostreptomycin
16	Death	Penicillin, 600,000 units, intramuscularly
20	Death	Penicillin, dosage not available
22	Death	Penicillin, dosage not available
24	Death	Aqueous penicillin 20×10^6 units, intravenously in 1 L of 5% glucose and water
27	Death	Penicillin, 5 million units/4 hrs, intravenous
28	Death	Penicillin, dosage not available
29	Death	Parenteral crystalline penicillin 2×10^6 units/2 hrs
30	Death	Penicillin G in combination with other antibiotics
35	Death	Penicillin in combination with other antibiotics
36	Recovery	Penicillin IV added (for 24 hrs) to ciprofloxacin, rifampin, and vancomycin; ciprofloxacin started on day 2 of disease

aware of a severe infection and was trying to treat it with conventionally available drugs and practices. Much information on old therapies is available now since very early treatments have been carefully and meticulously documented in the old-fashioned way of written patient history notes. Attending physicians made tremendous efforts to save their patients by treating them with a variety of drugs given either in combination or sequentially.

Since the discovery of penicillin in 1928, the "wonder drug" was used in the course of anthrax infection and administered in various doses from several hundred thousand units to millions of units. Of the 40 studied cases, 15 patients listed in Table 1.2 were treated with penicillin alone or in combination with another antibiotic.

The majority of patients who received penicillin as their treatment for anthrax infection died (13 of 15). It should not be considered, however, that the other two patients were successfully treated with penicillin. One of them (case 36) had received penicillin just for a limited time and among three other antibiotics, including ciprofloxacin, rifampin, and vancomycin; and another patient (case 15) who was originally considered to have inhalational anthrax, had a very mild course of infection and serological reaction results were questionable. In the last case, a low-level agglutination in the patient's samples with "antilethal toxin serum" was shown. However, in 1957 (when the case occurred), pure anthrax antilethal toxin antibodies simply did not exist. It is likely that the serum used contained antibodies to other components of *B. anthracis* (e.g., proteases, hemolysins, cell wall components, etc.) that as we know now may cross-react with antigens of some other bacilli from the *Bacillus* genus. Moreover, according to a microbiological analysis, the isolated culture of pleural fluid was characterized as "*Bacillus subtilis*." Even though it is impossible to obtain more precise information half a century after the case has taken place, it is possible to assume that, with high level of certainty, it was not anthrax or at least not a "classical inhalational anthrax." There is no doubt that these two cases cannot be considered successful treatments of inhalational anthrax. If we exclude these two cases, *there is no evidence that penicillin has any efficacy in the treatment of inhalational anthrax.*

According to our analysis, prior to the 2001 cases, inhalational anthrax patients did not survive often. Among the medications used without apparent success were sulfonamides, penicillin, chloramphenicol, chloromycetin, amoxycillin, streptomycin, tetracycline, aureomycin, and even anti-anthrax serum. In addition to the antimicrobials, there were a number of other medications used in the course of anthrax treatment that provided no evident therapeutic effect. For example, 4 patients were given narcotic analgesics and all 4 died (3 of them within 1½ hours of being given morphine or Demerol, and the 4th within 6 hours of being given morphine). Five patients received cortisone during the course of illness. Of these, only one survived (the patient was on a multiple-antibiotic/thoracentesis combination therapy). Anticonvulsants were given to 4 patients – all 4 died. Similarly, vasopressor agents were given to 5 patients – all 5 died. Digitalis was given to 3 patients – all of them died.

Even though the patients who were given the abovementioned medications died, it is impossible to say whether the medications used were simply ineffective or they worsened the course of disease. We do not want to make any final conclusion since the number of patients was limited and treatment often started just prior to the lethal outcome. However, it is obvious that the use of some medications in the case of anthrax infection (e.g., narcotic drugs) must be reevaluated.

Modern treatment regimens are most comprehensively presented through the first 11 cases of inhalational anthrax caused by intentional release of *B. anthracis* in the United States in 2001. Patients started to get care on average of 3.5 days (range 1–7 days) after onset of symptoms. Five patients were in the initial stage when the antibiotic treatment was started (days 1–3). All 5 patients received a combination of antibiotics and thoracentesis, 4 survived, and 1 died. One of the patients who survived has also received plasmapheresis as a supplemental treatment. The other six patients started receiving antibiotic therapy between days 4 and 7 after disease initiation. Four of these six patients died. One of the four patients who died received just a single antibiotic – levofloxacin – but died the very same day. Three of four patients who died received combination of antibiotics and 1 of them had therapeutic thoracentesis. Both patients who survived received a combination of antibiotics plus therapeutic thoracentesis.

All patients who survived had ciprofloxacin in their therapeutic regimens, some in combination with other antibiotics: rifampin was used in 4, clindamycin in 3, azithromycin in 2, vancomycin in 2, and levofloxacin in 2 cases. The combination of ciprofloxacin with rifampin seems to be most efficient, but available information is not sufficient to make a final conclusion.

Pleural effusion was a remarkably consistent clinical feature of inhalational anthrax in this series, occurring in all patients. Drainage of the pleural cavity was performed in 8 contemporary cases and also in 1 patient in 1957. It is important to note that all 7 patients who survived inhalation anthrax (6 in 2001 and 1 in 1957) received therapeutic thoracentesis among the conducted therapeutic efforts, where only 2 of 33 lethal cases have this procedure listed. One of these two lethal cases was a 94-year-old female patient from Connecticut who had additional pathology in anamnesis and developed kidney insufficiency through the course of the infection. Another patient who was provided with thoracentesis had the course of disease complicated by a pericardial effusion that progressed to tamponade. The latter was originally misdiagnosed and treated for congestive heart failure by administering cardiostimulating drugs.

The apparent improvement in survival compared with previous cases suggests that early treatment with a fluoroquinolone antibiotic and at least one or two other effective antibiotics combined with thoracentesis when necessary may significantly improve survival of anthrax patients. Though we cannot say that ciprofloxacin is the best antibiotic for anthrax therapy, it is at least one of the best among the antibiotics that have been relatively widely used for this infection. It is warranted, however, to preclinically and clinically test other antibiotics that could give higher survival and less toxicity, especially those having protease-inhibiting capability. However, at this point in time, it is possible to make some general conclusions and recommendations on anthrax therapy:

1. The highest rate of survival in cases of inhalation anthrax was provided by a combination of ciprofloxacin and rifampin plus therapeutic thoracentesis.
2. Penicillin was ineffective for the treatment of inhalational anthrax, but could be still used for cutaneous anthrax infection since there is evidence (from other publications) of its efficacy for the mild (cutaneous) forms of infection.
3. Penicillin should be reconsidered as a medication for the treatment of inhalational anthrax in pediatric practices, and possibly replaced with combination of ciprofloxacin and rifampin as a treatment of choice. Regardless of these antibiotics' relative toxicity, it is imperative to save lives of children at the time of infection rather than use ineffective drugs while treating inhalation anthrax.
4. Artificial ventilation/oxygenation is not recommended as a sole procedure due to its possible negative effect (the lungs are atelectatic and compressed by effusion, and the respiratory tract could be full of mucous). But it might be effective if it follows initial thoracentesis and a gentle bronchopulmonary suction to remove excessive mucus from trachea and bronchi.
5. The use of narcotic analgesics should be avoided because of their possible suppressive effect on breathing activity that could worsen the course of inhalational anthrax (all four patients who received them died within a short period of time).

11. PROTECTION

According to the *CDC Biosafety in Microbiological and Biomedical Laboratories Manual*, 4th edition, Biosafety Level 2 practices, containment equipment, and facilities are recommended for activities using clinical materials and diagnostic quantities of infectious cultures. Animal Biosafety Level 2 practices, containment equipment, and facilities are recommended for studies utilizing experimentally infected laboratory rodents. Biosafety Level 3 practices, containment equipment, and facilities are recommended for work involving production quantities or concentrations of cultures and for activities – such as centrifugation – with a high potential for aerosol production.

12. ISOLATION

Anthrax infection is not contagious and cannot be transmitted by human-to-human contact. Therefore, no isolation is necessary. (*Note*: Standard precautions are still indicated for patient care as part of routine infection control practice.) Cadavers should not be opened within 30 minutes of death, as the bacteria sporulate immediately on contact with ambient air.

13. AFTERWARD

There is no doubt that anthrax is still one of the most threatening pathogens that could be used in bioterrorist attacks. That is why, in the course of the work on this chapter, we tried to highlight the most important areas related to this still devastating disease, hoping that our work will help scientists and physicians remove this threat from the table. However, there are many unanswered questions in practically all areas related to this pathogen and disease. But the most devastating conclusion we made was that there is still a huge gap between scientists and physicians: if you take a look at the areas where we scientists work and the actual therapeutic regimens that physicians use, you would see what we mean. If this chapter helps to fill even a part of this gap, we will consider our work successful.

ACKNOWLEDGMENTS

The authors wish to acknowledge Ms. Kathryn Crockett, Dr. Alison Andrews, and Mrs. Muffarah Jahangeer for their assistance in preparing this manuscript.

References

Abramova, A.A., and Grinberg, L.M. (1993). [Pathology of anthrax sepsis according to materials of the infectious outbreak in 1979 in Sverdlovsk (microscopic changes)] *Arkh. Patol.* 55:18–23 (in Russian).

Abramova, F.A., Grinberg, L.M., Yampolskaya, O.V., and Walker, D.H. (1993). Pathology of inhalational anthrax in 42 cases from the Sverdlovsk outbreak of 1979. *Proc. Natl. Acad. Sci. U.S.A.* 90:2291–2294.

ABS, a Subsidiary of Analex. (2004). Report to the Department of State, August.

Albrink, W., Brooks, S., Biron, R., and Kopel, M. (1960). Human inhalation anthrax: a report of three fatal cases. *Am. J. Pathol.* 36:457–468.

Albrink, W.S., and Goodlow, R.J. (1959). Experimental inhalation anthrax in chimpanzees. *Am. J. Pathol.* 35:1055–1065.

Alibek, K. (1999). *Biohazard: The Chilling True Story of the Largest Covert Biological Weapons Program in the World – Told from Inside by the Man Who Ran It.* Random House, New York, p. 74.

Barakat, L.A., Quentzel, H.L., Jernigan, J.A., Kirschke, D.L., Griffith, K., Spear, S.M., Kelley, K., Barden, D., Mayo, D., Stephens, D.S., Popovic, T., Marston, C., Zaki, S.R., Guarner, J., Shieh,

W.J., Carver, H.W. 2nd, Meyer, R.F., Swerdlow, D.L., Mast, E.E., and Hadler, J.L. [Anthrax Bioterrorism Investigation Team]. (2002). Fatal inhalational anthrax in a 94-year-old Connecticut woman. *J.A.M.A.*, 287(7):863–868.

Barnaby, W. (2000). *The Plague Makers: The Secret World of Biological Weapons*. Continuum, New York, 123.

Belluck, P. (2001). Anthrax Outbreak of '57 Felled a Mill but Yielded Answers. *New York Times*, October 27, 2001.

Bonventre, P.F., and Eckert, N.J. (1963). The biologic activities of *Bacillus anthracis* and *Bacillus cereus* culture filtrates. *Am. J. Pathol.* 43:201–211.

Bonventre, P.F., Sueoka, W., True, C.W., Klein, F., and Lincoln, R. (1967). Attempts to implicate the central nervous system as a primary site of action for *Bacillus anthracis* lethal toxin. *Fed. Proc.* 26:1549–1553.

Brachman, P., and Pagano, J. (1961). Two cases of fatal inhalation anthrax, one associated with sarcoidosis. *N. Engl. J. Med.* 265:203–208.

Brachman, P.S. (1980). Inhalation anthrax. *Ann N.Y. Acad. Sci.* 353:83–93.

Brook, I. (2002). The prophylaxis and treatment of anthrax. *Int. J. Antimicrob. Agents* 20(5):320–325 [review].

Brooksher, W., and Briggs, J. (1920). Pulmonary anthrax: report of a case. *J.A.M.A.* 74:323–324.

Bush, L.M., Abrams, B.H., Beall, A., and Johnson, C.C. (2001). Index case of fatal inhalational anthrax due to bioterrorism in the United States. *N. Engl. J. Med.* 345:1607–1610.

Carucci, J.A., McGovern, T.W., Norton, S.A., Daniel, C.R., Elewski, B.E., Fallon-Friedlander, S., Lushniak, B.D., Taylor, J.S., Warschaw, K., and Wheeland, R.G. (2002). Cutaneous anthrax management algorithm. *J. Am. Acad. Dermatol.* 47(5):766–769.

Carus, W.S. (2001). *Bioterrorism and Biocrimes: The Illicit Use of Biological Agents Since 1900*. Washington, D.C.: National Defense University Center for Counterproliferation Research, pp. 108–109.

CDC Emergency Preparedness and Response Website. (2001). Anthrax Q & A. http://www.bt.cdc.gov/agent/anthrax/faq/diagnosis.asp.

CDC Treatment Protocols for Cases of Inhalational and Cutaneous Anthrax Associated with Bioterrorist Attack Website. (2001). Anthrax Q & A. http://www.cdc.gov/mmwr/preview/mmwrhtml/mm5042a1.htm.

Chandramukhi, A., Shankar, P., Rao, T., Sundararajan, S., and Swamy, H.S. (1987). Acute leptomeningitis due to *Bacillus anthracis*. *Trop. Geol. Med.* 35:79–82.

CNN.com (2001). Anthrax through the ages. http://www.cnn.com/2001/HEALTH/conditions/10/16/anthrax.timeline/

Cole, S.T., Eiglmeier, K., Parkhill, J., James, K.D., Thomson, N.R., Wheeler, P.R., Honore, N., Garnier, T., Churcher, C., Harris, D., Mungall, K., Basham, D., Brown, D., Chillingworth, T., Connor, R., Davies, R.M., Devlin, K., Duthoy, S., Feltwell, T., Fraser, A., Hamlin, N., Holroyd, S., Hornsby, T., Jagels, K., Lacroix, C., Maclean, J., Moule, S., Murphy, L., Oliver, K., Quail, M.A., Rajandream, M.A., Rutherford, K.M., Rutter, S., Seeger, K., Simon, S., Simmonds, M., Skelton, J., Squares, R., Squares, S., Stevens, K., Taylor, K., Whitehead, S., Woodward, J.R., and Barrell, B.G. (2001). Massive gene decay in the leprosy bacillus. *Nature* 409:1007–1011.

Committee to Assess the Safety and Efficacy of the Anthrax Vaccine, Medical Follow-Up Agency. (2002). *The Anthrax Vaccine: Is It Safe? Does It Work?* Institute of Medicine, National Academy Press, Washington, D.C., http://www.iom.edu/iom/iomhome.nsf/WFiles/Anthrax-8-pager1FIN AL/$file/Anthrax-8-pager1FINAL.pdf.

Cowdery, J. (1947). Primary pulmonary anthrax with septicemia. *Arch. Pathol.* 43:396–399.

Cranmer, H., and Martinez, M. (2003). CBRNE – anthrax infection. http://www.emedicine.com/emerg/topic864.htm/.

Dalldorf, F.G., and Beall, F.A. (1967). Capillary thrombosis as a cause of death in experimental anthrax. *Arch. Pathol.* 83:154–161.

Dixon, T.C., Fadl, A.A., Koehler, T.M., Swanson, J.A., and Hanna, P.C. (2000). Early *Bacillus anthracis*-macrophage interactions: intracellular survival survival and escape. *Cell Microbiol.* 2:453–463.

Doganay, M., Almac, A., and Hanagasi, R. (1986). Primary throat anthrax. *Scand. J. Infect. Dis.* 18:415–419.

Drake, D.J., and Blair, A.W. (1971). Meningitic anthrax. *Cent. Afr. J. Med.* 17:97–98.

Dutz, W., and Kohout, E. (1971). Anthrax. *Pathol. Annu.* 6:209–248.

Enticknap, J.B., Galbraith, N.S., Tomlinson, A.J.H., and Elias-Jones, T.F. (1968). Pulmonary anthrax caused by contaminated sacks. *Br. J. Industr. Med.* 25:72–74.

Eremin, S.A., Nikiforov, A.K., Kostiukova, T.A., and Mikshis, N.I. (1999). The capacity of the causative agent of anthrax to reduce methylene blue. *Zh. Mikrobiol. Epidemiol. Immunobiol.* 4:91–92 (in Russian).

Erwin, J.L., DaSilva, L.M., Bavari, S., Little, S.F., Friedlander, A.M., and Chanh, T.C. (2001). Macrophage-derived cell lines do not express proinflammatory cytokines after exposure to *Bacillus anthracis* lethal toxin. *Infect. Immun.* 69:1175–1177.

Fellows, P., Linscott, M., Ivins, B., *et al.* (2001). Efficacy of a human anthrax vaccine in guinea pigs, rabbits, and rhesus macaques against challenge by *Bacillus anthracis* isolates of diverse geographical origin. *Vaccine* 19:3241–3247.

Fouet, A., Sirard, J.C., and Mock, M. (1994). *Bacillus anthracis* pXO1 virulence plasmid encodes a type 1 DNA topoisomerase. *Mol. Microbiol.* 11:471–479.

Friedlander, A.M. (1986). Macrophages are sensitive to anthrax lethal toxin through an acid-dependent process. *J. Biol. Chem.* 261:7123–7126.

Friedlander, A.M., Welkos, S.L., Pitt, M.L.M., Ezzell, J.W., Worsham, P.L., Rose, K.J., Ivins, B.E., Lowe, J.R., Howe, G.B., Mikesell, P., and Lawrence, W.B. (1993). Postexposure prophylaxis against experimental inhalation anthrax. *J. Infect. Dis.* 167:1239–1243.

Fukao, T. (2004). Immune system paralysis by anthrax lethal toxin: the roles of innate and adaptive immunity. *Lancet Infect. Dis.* 4:166–170 [review].

Gold, H. (1955). Anthrax: a report of one hundred seventeen cases. *AMA Arch. Intern. Med.* 96:387–396.

Goodman, J.W., and Nitecki, D.E. (1967). Studies on the relation of a prior immune response to immunogenicity. *Immunology* 13:577–583.

Gordon, V.M., Klimpel, K.R., Arora, N., Henderson, M.A., and Leppla, S.H. (1995). Proteolytic activation of bacterial toxins by eukaryotic cells is performed by furin and by additional cellular proteases. *Infect. Immun.* 63:82–87.

Greenwood, D., Slack, R., and Peutherer, J. (1997). *Medical Microbiology. A Guide to Microbial Infections: Pathogenesis, Immunity, Laboratory Diagnosis and Control.* Churchill Livingstone, New York.

Grinberg, L.M., Abramova, F.A., Yampolskaya, O.V., Walker, D.H., and Smith, J.H. (2001). Quantitative pathology of inhalational anthrax I: quantitative microscopic findings. *Mod. Pathol.* 14:482–495.

Guidi-Rontani, C., Levy, M., Ohayon, H., and Mock, M. (2001). Fate of germinated *Bacillus anthracis* spores in primary murine macrophages. *Mol. Microbiol.* 42:931–938 [erratum in *Mol. Microbiol.* 2002;44:297].

Guidi-Rontani, C., Pereira ,Y., Ruffie, S., Sirard, J.C., Weber-Levy, M., and Mock, M. (1999). Identification and characterization of a germination operon on the virulence plasmid pXO1 of *Bacillus anthracis*. *Mol. Microbiol.* 33:407–414.

Guidi-Rontani C., Weber-Levy, M., Labruyère, E., and Mock, M. (1999). Germination of *Bacillus anthracis* spores within alveolar macrophages. *Mol. Microbiol.* 31:9–17.

Haight, T.H. (1952). Anthrax meningitis: review of literature and report of two cases with autopsies. *Am. J. Med. Sci.* 224:57–69.

Hail, A.S., Rossi, C.A., Ludwig, G.V., Ivins, B.E., Tammariello, R.F., and Henchal, E.A. (1999). Comparison of non-invasive sampling sites for early detection of *Bacillus anthracis* spores from rhesus monkeys after aerosol exposure. *Mil. Med.* 164:833–837.

Hanna, P.C., Kruskal, B.A., Ezekowitz, R.A., Bloom, B.R., and Collier, R.J. (1994). Role of macrophage oxidative burst in the action of anthrax lethal toxin. *Mol. Med.* 1:7–18.
Hansen, H.J., Norberg, B.K., and Schmiterloew, C.G. (1961). Pathophysiological studies on anthrax. *Acta Pathol. Microbiol. Scand.* 51(Suppl 144):307–308.
Heidelberg, J.F., Eisen, J.A., Nelson, W.C., Clayton, R.A., Gwinn, M.L., Dodson, R.J., Haft, D.H., Hickey, E.K., Peterson, J.D., Umayam, L., Gill, S.R., Nelson, K.E., Read, T.D., Tettelin, H., Richardson, D., Ermolaeva, M.D., Vamathevan, J., Bass, S., Qin, H., Dragoi, I., Sellers, P., McDonald, L., Utterback, T., Fleishmann, R.D., Nierman, W.C., and White, O. (2000). DNA sequence of both chromosomes of the cholera pathogen *Vibrio cholerae*. *Nature* 406:477–483.
Inglesby, T.V., O'Toole, T., Henderson, D.A., Bartlett, J.G., Ascher, M.S., Eitzen, E., Friedlander, A.M., Gerberding, J., Hauer, J., Hughes, J., McDade, J., Osterholm, M.T., Parker, G., Perl, T.M., Russell, T.K., and Tonat, K. (2002). Anthrax as a biological weapon, 2002: updated recommendations for management. *J.A.M.A.* 287:2236–2252 [review; erratum in *J.A.M.A.* 2002;288:1849].
Ivins, B.E., Fellows, P., and Pitt, M.L., *et al.* (1996). Efficacy of standard human anthrax vaccine against *Bacillus anthracis* aerosol spore challenge in rhesus monkeys. *Salisbury Med. Bull.* 87:125–126.
Jamie, W.E. (2002). Anthrax: diagnosis, treatment, prevention. *Primary Care Update for OB/GYNs* 9:117–121.
Jernigan, D.B., Raghunathan, P.L., Bell, B.P., Brechner, R., Bresnitz, E.A., Butler, J.C., Cetron, M., Cohen, M., Doyle, T., Fischer, M., Greene, C., Griffith, K.S., Guarner, J., Hadler, J.L., Hayslett, J.A., Meyer, R., Petersen, L.R., Phillips, M., Pinner, R., Popovic, T., Quinn, C.P., Peefhuis, J., Reissman, D., Rosenstein, N., Schuchat, A., Shieh, W.J., Siegal, L., Swerdlow, D.L., Tenover, F.C., Traegar, M., Ward, J.W., Weisfuse, I., Wiersma, S., Yeskey, K., Zaki, S., Ashford, D.A., Perkins, B.A., Ostroff, S., Hughes, J., Fleming, D., Koplan, J.P., Gerberding, and the National Anthrax Investigation Team. (2002). Investigation of bioterrorism-related anthrax, United States, 2001: epidemiologic findings. *Emerg. Infect. Dis.* 8:1019–1028.
Jernigan, J.A., Stephens, D.S., Ashford, D.A., Omenaca, C., Topiel, M.S., Galbraith, M., Tapper, M., Fisk, T.L., Zaki, S., Popovic, T., Meyer, R.F., Quinn, C.P., Harper, S.A., Fridkin, S.K., Sejvar, J.J., Shepard, C.W., McConnell, M., Guarner, J., Shieh, W.J., Malecki, J.M., Gerberding, J.L., Hughes, J.M., Perkins, B.A., and members of the Anthrax Bioterrorism Investigation Team. (2001). Bioterrorism-related inhalational anthrax: the first 10 cases reported in the United States. *Emerg. Infect. Dis.* 7:933–944.
Keim, P., Smith, K.L., Keys, C., Takahashi, H., Kurata, T., and Kaufmann, A. (2001). Molecular investigation of the Aum Shinrikyo anthrax release in Kameido, Japan. *J. Clin. Microbiol.* 39(12):4566–4567.
Klein, F., Hodges, D.R., Mahlandt, B.G., Jones, W.I., Haines, B.W., and Lincoln, R.E. (1962). Anthrax toxin: causative agent in the death of rhesus monkeys. *Science.* 138:1331–1333.
Klein, F., Walker, J.S., Fitzpatrick, D.F., Lincoln, R.E., Mahlandt, B.G., Jones, W.I. Jr, Dobbs J.P., and Hendrix, K.J. (1966). Pathophysiology of anthrax. *J. Infect. Dis.* 116:123–138.
Klichko, V.I., Miller, J., Wu, A., Popov, S.G., and Alibek, K. (2003). Anaerobic induction of *Bacillus anthracis* hemolytic activity. *Biochem. Biophys. Res. Commun.* 303:855–862.
Knights, E.M. (0000). Anthrax. *History Magazine*. Accessed online at: http://www.historymagazine.com/anthrax.html.
Koshi, G.M., Lalitha, D.J., Chacko, A., and Pulimood, B. (1981). Anthrax meningitis, a rare clinical entity. *J Assoc. Phys. Ind.* 29:59–62.
Kozel, T.R., Murphy, W.J., Brandt, S., Blazar, B.R., Lovchik, J.A., Thorkildson, P., Percival, A., and Lyons, C.R. (2004). mAbs to *Bacillus anthracis* capsular antigen for immunoprotection in anthrax and detection of antigenemia. *Proc. Natl. Acad. Sci. U.S.A.* 101:5042–5047
Krauss, F. (1943). Anthrax appearing primarily in the nose. *Arch. Otolaryngol.* 37:238–241.
LaForce, F.M., Bumford, F.H., Feeley, J.C., Stokes, S.L., and Snow, D.B. (1969). Epidemiologic study of a fatal case of inhalation anthrax. *Arch. Environ. Health.* 18:798–805.

Lindberg, A.A. (1999). Polyosides (encapsulated bacteria). *C.R. Acad. Sci.* 322:925–932.
Lindeque, P.M., and Turnbull, P.C. (1994). Ecology and epidemiology of anthrax in the Etosha National Park, Namibia. *Onderstepoort J. Vet. Res.* 61:71–83.
Little, S.F., and Ivins, B.E. (1999). Molecular pathogenesis of *Bacillus anthracis* infection. *Microb. Infect.* 2:131–139.
Litusov, N.V., Vasilyev, N.T., Vasilyev, P.G., and Burgasov, P.N. (2002). Pathomorphology of anthrax. *Meditsina.*
Liu, S., Schubert, R.L., Bugge, T.H., and Leppla, S.H. (2003). Anthrax toxin: structures, functions and tumour targeting. *Expert Opin. Biol. Ther.* 3:843–853 [review].
Macher, A. (2002). An industry-related outbreak of human anthrax, Massachusetts, 1868. *Emerg. Infect. Dis.* 8:1182.
Makino, S.I., Uchida, I., Terakado, N., Sasakawa, C., and Yoshikawa, M. (1989). Molecular characterization and protein analysis of the *cap* region, which is essential for encapsulation in *Bacillus anthracis. J. Bacteriol.* 171:722–730.
Mangold, T., and Goldberg, J. (1999). *Plague Wars: The Terrifying Reality of Biological Warfare.* St. Martins Griffin, New York, pp. 19–24.
Mesnage, S., Tosi-Couture, E., Gounon, P., Mock, M., and Fouet, A. (1998). The capsule and S-layer: two independent and yet compatible macromolecular structures in *Bacillus anthracis. J. Bacteriol.* 180:52–58.
Mikesell, P., Ivins, B.E., Ristroph, J.D., Vodkin, M.H., Dreier, T.M., and Leppla, S.H. (1983). Plasmids, Pasteur, and Anthrax. *ASM News* 49:320–322.
Milne, J.C., and Collier, R.J. (1993). pH-dependent permeabilization of the plasma membrane of mammalian cells by anthrax protective antigen. *Mol. Microbiol.* 10:647–653.
Milne, J.C., Furlong, D., Hanna, P.C., Wall, J.S., and Collier, R.J. (1994). Anthrax protective antigen forms oligomers during intoxication of mammalian cells. *J. Biol. Chem.* 269: 20607–20612.
Moayeri, M., Haines, D., Young, H.A., and Leppla, S.H. (2003). *Bacillus anthracis* lethal toxin induces TNF-alpha-independent hypoxia-mediated toxicity in mice. *J. Clin. Invest.* 112:670–682.
Mock, M., and Mignot, T. (2003). Anthrax toxins and the host: a story of intimacy. *Cell Microbiol.* 5:15–23 [review].
Mogridge, J., Cunningham, K., and Collier, R.J. (2002). Stoichiometry of anthrax toxin complexes. *Biochemistry* 41:1079–1082.
Nobel Lectures, Physiology or Medicine 1901–1921, Elsevier Publishing Company, Amsterdam, 1967.
Norberg, B.K., Schmiterloew, C.G., Bergrahm, B., and Lundstroem, H. (1964). Further pathophysiological investigations into the terminal course of experimental anthrax in the rabbit. *Acta. Pathol. Microbiol. Scand.* 60:108–116.
Office of Public Health Service Historian – National Library of Medicine. (2001). A Brief History of Anthrax. http://lhncbs.nlm.nih.gov/apdb/phsHistory/resources/anthrax/anthrax02.html.
Ogata, H., Audic, S., Renesto-Audiffren, P., Fournier, P.E., Barbe, V., Samson, D., Roux, V., Cossart, P., Weissenbach, J., Claverie, J.M., and Raoult, D. (2001). Mechanisms of evolution in *Rickettsia conorii* and *R. prowazekii. Science* 293:2093–2098.
Okinaka, R.T., Cloud, K., Hampton, O., Hoffmaster, A.R., Hill, K.K., Keim, P., Koehler, T.M., Lamke, G., Kumano, S., Mahillon, J., Manter, D., Martinez, Y., Ricke, D., Svensson, R., and Jackson, P.J. (1999). Sequence and organization of pXO1, the large *Bacillus anthracis* plasmid harboring the anthrax toxin genes. *J. Bacteriol.* 181:6509–6515.
Pellizzari, R., Guidi-Rontani, C., Vitale, G., Mock, M., and Montecucco, C. (1999). Anthrax lethal factor cleaves MKK3 in macrophages and inhibits the LPS/IFNgamma-induced release of NO and TNFalpha. *FEBS Lett.* 462:199–204.
Pile, J.C., Malone J.D., and Etizen, E.M., *et al.* (1998). Anthrax as a potential biological warfare agent. *Arch. Intern. Med.* 158:429–434.

Plotkin, S.A., Brachman, P.S., Utell, M., Bumford, F.H., and Atchison, M.M. (2002). An epidemic of inhalation anthrax, the first in the twentieth century: clinical features. *Am. J. Med.* 112:4–12.
Pluot, M., Vital, C., Aubertin, J., Croix, J.C., Pire, J.C., and Poisot D. (1976). Anthrax meningitis: report of two cases with autopsies. *Acta Neuropathol.* 36:339–345.
Popov, S.G., Villasmil, R., Bernardi, J., Grene, E., Cardwell, J., Popova, T., Wu, A., Alibek, D., Bailey, C., and Alibek, K. (2002a). Effect of *Bacillus anthracis* lethal toxin on human peripheral blood mononuclear cells. *FEBS Lett.* 527:211–215.
Popov, S.G., Villasmil, R., Bernardi, J., Grene, E., Cardwell, J., Wu, A., Alibek, D., Bailey, C., and Alibek, K. (2002b). Lethal toxin of *Bacillus anthracis* causes apoptosis of macrophages. *Biochem. Biophys. Res. Commun.* 293:349–55.
Ramachandran, S., and Natarajan, M. (1968). Tissue response to experimental anthrax of sheep and goats. *Indian Vet. J.* 45:381–387.
Rangel, R.A., and Gonzalez, D.A. (1975). *Bacillus anthracis* meningitis. *Neurology* 25:525–530.
Ray, K.C., Mesnage, S., Washburn, R., Mock, M., Fouet, A., and Blaser, M. (1998). Complement binding to *Bacillus anthracis* mutants lacking surface structures. Presented at the 98th Annual ASM General Meeting, Atlanta, Georgia.
Read, T.D., Peterson, S.N., Tourasse, N., Baillie, L.W., Paulsen, I.T., Nelson, K.E., Tettelin, H., Fouts, D.E., Eisen, J.A., Gill, S.R., Holtzapple, E.K., Okstad, O.A., Helgason, E., Rilstone, J., Wu, M., Kolonay, J.F., Beanan, M.J., Dodson, R.J., Brinkac, L.M., Gwinn, M., DeBoy, R.T., Madpu, R., Daugherty, S.C., Durkin, A.S., Haft, D.H., Nelson, W.C., Peterson, J.D., Pop, M., Khouri, H.M., Radune, D., Benton, J.L., Mahamoud, Y., Jiang, L., Hance, I.R., Weidman, J.F., Berry, K.J., Plaut, R.D., Wolf, A.M., Watkins, K.L., Nierman, W.C., Hazen, A., Cline, R., Redmond, C., Thwaite, J.E., White, O., Salzberg, S.L., Thomason, B., Friedlander, A.M., Koehler, T.M., Hanna, P.C., Kolsto, A.B., and Fraser, C.M. (2003). The genome sequence of *Bacillus anthracis* Ames and comparison to closely related bacteria. *Nature* 423:81–86.
Record, B.R., and Wallis, R.G. (1956). Physicochemical examination of polyglutamic acid from *Bacillus anthracis* grown *in vivo*. *Biochem. J.* 63:443–447.
Ross, J.M. (1957). The pathogenesis of anthrax following the administration of spores by the respiratory route. *J. Pathol. Bacteriol.* 73:485–494.
Roux, V., Bergoin, M., Lamaze, N., and Raoult, D. (1997). Reassessment of the taxonomic position of *Rickettsiella grylli*. *Int. J. Syst. Bacteriol.* 47:1255–1257.
Rozen'er, L.A. (1948). *Anthrax in Humans*. State publisher, Moldova, Kishinev.
Severn, M. (1976). A fatal case of pulmonary anthrax. *Br. Med. J.* 27:748.
Shapiro, L.B., and Galinobskoi, C.E. (1923). Anthrax case via inhalation. *clin. med.* 4:13–14.
Smith, H., Keppie, J., and Stanley, J.L. (1955). The chemical basis of the virulence of *B. anthracis*. *Br. J. Exp. Pathol.* 36:460–472.
Smith, H., and Stoner, H.B. (1967). Anthrax toxic complex. *Fed Proc.* 26:1554–1557.
Stephens, R.S., Kalman, S., Lammel, C., Fan, J., Marathe, R., Aravind, L., Mitchell, W., Olinger, L., Tatusov, R.L., Zhao, Q., Koonin, E.V., and Davis, R.W. (1998). Genome sequence of an obligate intracellular pathogen of humans: *Chlamydia trachomatis*. *Science* 282:754–759.
Suffin, S. (1978). Inhalation anthrax in a home craftsman. *Human Pathol.* 9:594–597.
Swartz, M.N. (2001). Recognition and management of anthrax—an update. *N. Engl. J. Med.* 345(22): 1621–1626.
Teacher, J.H., and Glasg, M.D. (1906). Primary intestinal anthrax in man; septicemia; haemorrhagic lepto-meningitis. *Lancet* 12:1306–1311.
Thorne, C.B. (1993). *Bacillus anthracis*. In: Sonenshein, A.L., Hoch, J.A., and Losick, R. (eds.), *Bacillus subtilis and Other Gram-positive Bacteria*. Washington, D.C., ASM, pp. 113–124.
Vessal, K. (1975). Radiological changes in inhalational anthrax. A report of radiological and patological corellations in two cases. *Clin. Radiol.* 26:471–474.

Vitale, G., Pellizzari, R., Recchi, C., Napolitani, G., Mock, M., and Montecucco, C. (1998). Anthrax lethal factor cleaves the N-terminus of MAPKKs and induces tyrosine/threonine phosphorylation of MAPKs in cultured macrophages. *Biochem. Biophys. Res. Commun.* 248:706–711.

Walker, J.S., Lincoln, R.E., and Klein, F. (1967). Pathophysiological and biochemical changes in anthrax. *Fed Proc.* 26:1539–1544.

Wang, T.T., Fellows, P.F., Leighton, T.J., and Lucas, A.H. (2004). Induction of opsonic antibodies to the gamma-D-glutamic acid capsule of *Bacillus anthracis* by immunization with a synthetic peptide-carrier protein conjugate. *FEMS Immunol. Med. Microbiol.* 40:231–237.

Welkos, S.L., Keener, T.J., and Gibbs, P.H. (1986). Differences in susceptibility of inbred mice to *Bacillus anthracis*. *Infect. Immun.* 51:795–800.

WHO Model G for PCR detection of *Bacillus anthracis*, Standard Operating Procedure. WHO Website. (2004). http://w3.whosea.org/bct/anthrax/modelgf.htm.

Witkowski, J.A. (2002). The story of anthrax from antiquity to present: a biological weapon of nature and humans. *Clin. Dermatol.* 20:336–242.

Zaucha, G.M., Pitt, L.M., Estep, J., Ivins, B.E., and Friedlander, A.M. (1998). The pathology of experimental anthrax in rabbits exposed by inhalation and subcutaneous inoculation. *Arch. Pathol. Lab. Med.* 122:982–992.

Zimmerman, B.E., and Zimmerman, D.J. (2003). *Killer Germs: Microbes and Diseases that Threaten Humanity.* Contemporary Books, New York, p. 219.

Zwartouw, H.T., and Smith, H. (1956). Polyglutamic acid from *Bacillus anthracis* grown *in vivo*: structure and aggressin activity. *Biochem. J.* 63:437–454.

2

Plague as a Biological Weapon

David T. Dennis

Plague [L. plaga] A stroke, blow or wound; an affliction, calamity or scourge, especially a visitation of divine justice or anger.

(Anonymous, 1998)

1. HISTORY OF PLAGUE AND ITS POTENTIAL AS A WEAPON OF BIOTERRORISM

1.1. Pandemic History and Epidemic Potential

Three well-documented plague pandemics have occurred in the past two millennia, resulting in more than 200 million deaths and great social and economic chaos (Perry and Fetherston, 1997; Pollitzer, 1954). The Justinian pandemic arose in northern Africa in the mid-6th century, and by the 7th century had spread throughout the Mediterranean and near-eastern regions – severely impacting both the Roman and Byzantine empires. The second pandemic, the Black Death or great pestilence, originated in Central Asia, was carried to Sicily in 1347 via ships from the Crimea, and rapidly swept through medieval Europe. By 1352, it had killed 30% or more of afflicted populations, slowly playing itself out in successive epidemics, including the Great Plague of London in 1665 (Perry and Fetherston, 1997). The third (Modern) pandemic began in southwestern China in the mid-19th, struck Hong Kong in 1894, and was soon carried by rat-infested steamships to port cities on all inhabited continents, including several in the United States (US) (Link, 1955; Pollitzer, 1954). By 1930, the third pandemic had caused more than 26 million cases and 12 million deaths. Plague in these three pandemics was predominantly the bubonic form, emanating

Bioterrorism and Infectious Agents
Edited by Fong and Alibek, Springer Science+Business Media, Inc., New York, 2005

from *Yersinia pestis*-infected rats and fleas, although terrifying outbreaks of the more virulent person-to-person spreading pneumonic form were recorded during the course of each. The explosive contagiousness and severity of pneumonic plague was most completely documented in Manchurian epidemics in the early 20th century, which involved tens of thousands of cases, virtually all of them fatal (Wu, 1926).

Improved sanitation, hygiene, and modern disease control methods have, since the early 20th century, steadily diminished the impact of plague on public health, to the point that an average of 2,500 cases is now reported annually (World Health Organization, 2003). The plague bacillus is, however, entrenched in rodent populations in scattered foci on all inhabited continents except Australia (Gage, 1998; Gratz, 1999b), and eliminating these natural transmission cycles is unfeasible. Furthermore, although treatment with antimicrobials has reduced the case fatality ratio of bubonic plague to 10% or less, the fatality ratio for pneumonic plague remains high. A review of 420 reported plague cases in the US in the period 1949–2000 identified a total of 55 cases of plague pneumonia, of which 22 (40.0%) were fatal (Centers for Disease Control and Prevention, unpublished data); of 7 primary pneumonic cases, the fatality ratio was 57.1% (Centers for Disease Control and Prevention, 1997). Although pandemics are unlikely to recur, plague – including the pneumonic form – holds considerable outbreak potential (Boisier *et al.*, 2002; Campbell and Hughes, 1995; Chanteau *et al.*, 1998, 2000; Gabastou *et al.*, 2000; Ratsitorahina, 2000a). This potential could be exploited for purposes of terrorism or warfare.

1.2. Plague as a Weapon of Biological Warfare

The idea of using plague as a weapon is not new. Anecdotal reports describe catapulting of plague cadavers into enemy fortifications in 14th and 18th century warfare (Derbes, 1996; Gasquet, 1908; Marty, 2001). In World War II, the Japanese military experimented with plague in human subjects at their clandestine biological research facilities in Manchuria, and on several occasions dropped *Y. pestis*-infested fleas from low-flying planes on Chinese civilian populations, causing limited outbreaks of bubonic plague and initiating cycles of infection in rats (Bellamy and Freedman, 2001; Harris, 1992; Kahn, 2002). Biological warfare research programs begun by the Soviet Union (USSR) and the US during the Second World War intensified during the Cold War, and in the 1960s both nations had active programs to "weaponize" *Y. pestis*. In 1970, a World Health Organization (WHO) expert committee on biological warfare warned of the dangers of plague as a weapon, noting that the causative agent was highly infective, that it could be easily grown in large quantities and stored for later use, and that it could be dispersed in a form relatively resistant to desiccation and other adverse environmental conditions (World Health Organization, 1970). Models developed by this expert committee predicted that the intentional release of 50 kg of aerosolized *Y. pestis* over a city of 5 million would, in its primary effects, cause 150,000 cases of pneumonic plague and 36,000 deaths. It was further postulated that, without adequate precautions, an initial outbreak of pneumonic plague involving 50% of a population could result in infection of 90% of the rest of the population in 20–30 days and could cause a case fatality ratio of 60–70%. The work of this committee provided a basis for the 1972 international Biological Weapons and Toxins Convention prohibiting biological weapons development and maintenance, and that went into effect in 1975 (Marty *et al.*, 2001). It is now known that, despite signing this accord, the USSR continued an aggressive clandestine program of research and development that had begun decades earlier, stockpiling battle-ready plague weapons (Alibek, 1999). The Soviets prepared *Y. pestis* in liquid and dry forms as aerosols to be released by bomblets, and plague was considered by them as one of the most important strategic weapons in their arsenal. They also developed virulent fraction 1 (F1) capsular antigen-deficient and antimicrobial-resistant strains of *Y. pestis* and performed experiments to create an agent that could evade vaccine-induced immunity, be unresponsive to standard antibiotic treatment, and be difficult to identify. Moreover, the USSR was capable through a number of industrial plants to manufacture a plague weapon in hundreds of tons (Alibek, 1999).The US military biowarfare program also recognized that aerosolized *Y. pestis* had the basic attributes suitable for a large-scale attack (Martin and Marty, 2001; Marty, 2001), but US military scientists failed in their attempts to weaponize plague, apparently because they were unable to produce sufficient quantities of *Y. pestis* in stable form. Offensive biological weapons research was halted by the US in 1970, but Soviet efforts continued at least until 1990. Although Russia converted its civilian biological weapons to legitimate ends, it is unclear whether their military program has ceased development work and eliminated all of its stores (Alibek, 1999). Many nations maintain biological weapons defense programs that adhere to the ban on the development of offensive weapons; however, there is an obvious potential offensive value of studies to better understand candidate agents, and their possible modes of delivery, dispersal, and effectiveness under varying conditions.

1.3. US Countermeasures to Plague as a Weapon of Terrorism

No longer is the capacity for biological weapons development limited to the most technologically advanced states, but has expanded to include small rogue states, terrorist groups, cults, and even individuals. Because of the gathering terrorist threat, the US Congress passed the Biological Weapons Act of 1989 and the Chemical and Biological Weapons Control and Warfare Elimination Act of 1991 (Ferguson, 1997). In the wake of terrorist bombings of the World Trade Center in New York and the Alfred P. Murruh Federal Office Building in Oklahoma City, the arrest in the US of a microbiologist for illegally acquiring *Y. pestis* for suspicious purposes, and numerous bioweapons hoaxes, Congress passed the Anti-Terrorism Act of 1996 (Anonymous, 1996; Ferguson, 1997). Under this Act, the federal Centers for Disease Control and Prevention (CDC) was directed to identify dangerous biological agents that could be used by terrorists (select agents), and to establish a regulatory system for governing their acquisition, use, and transfer (U.S. Department of Health and Human Services, 1996). *Y. pestis* was classified by CDC as 1 of 6 Category A select biological agents that posed the highest risk to national security and was placed under these regulations (Centers for Disease Control and Prevention, 2000, 2002; Rotz *et al.*, 2002). Since 1997, federal law (42 FR 72.6) (Anonymous, 1997) has strengthened regulations on the transfer of *Y. pestis* and other restricted agents from one facility to another, requiring that the shipping and receiving facilities each complete an official transfer form prior to shipment. These regulations were supplemented by 42 CFR part 73

(Possession, Use, Transfer of Select Agents and Toxins), as specified in the Public Health Security and Bioterrorism Preparedness and Response Act of 2002 (effective 2003). 42 CFR part 73 details requirements for laboratories that handle plague and other select agents, including registration, security risk assessments, safety plans, security plans, emergency response plans, training, transfers, record keeping, inspections, and notifications (Centers for Disease Control and Prevention, 2002). These and other regulations have raised questions about protection of freedom of scientific research and civil liberties while combating potential terrorist actions (Annas, 2002; Fidler, 2001).

1.4. Preparedness and Response to a Possible Plague Attack

The CDC has developed a strategic plan to address the deliberate dissemination of plague and other select agents (Centers for Disease Control and Prevention, 2000; Khan *et al.*, 2000; Rotz *et al.*, 2002). The Johns Hopkins University Schools of Medicine and Public Health formed a Working Group on Civilian Biodefense Strategies to draw up consensus recommendations for measures to be taken following use of plague or other select agents as biological weapons against a civilian population (Johns Hopkins Center for Civilian Biodefense Strategies, 2003; Inglesby *et al.*, 2000). A critical advance was the development by the Johns Hopkins group of a Model State Emergency Health Powers Act that provides a model for state officials to follow in assuming extraordinary powers necessary to better detect and contain a potentially catastrophic disease outbreak. These powers range from pre-emergency planning to compensation for private property, and include tracking and reporting of certain diseases, the management of property, and the protection of persons (Johns Hopkins Center for Civilian Biodefense Strategies, 2003; Mair *et al.*, 2002).

It is assumed that a terrorist attack would most likely use a *Y. pestis* aerosol, possibly resulting in large numbers of severe and fatal primary and secondary pneumonic plague cases. Especially given plague's notoriety, even a limited event would likely cause public panic, create large numbers of the "worried-well," foster irrational evasive behavior, and quickly place an overwhelming stress on medical and other emergency response elements working to save lives and bring about control of its spread (Glass and Schoch-Spana, 2002; Osterholm and Schwartz, 2000; O'Toole and Inglesby, 2001). This was the situation that arose in the large industrial city of Surat, India, during the plague emergency there in 1994 (Dennis, 1994; Ramalingaswami, 2001).

In the US, emergency preparedness to counter bioterrorism includes the stockpiling of drugs, vaccines, and medical equipment for rapid deployment; the establishment of emergency operations centers with advanced communications systems; and a permanently staffed terrorism response unit established at CDC that works closely with the Department of Homeland Security (Centers for Disease Control and Prevention, 2000; Centers for Disease Control and Prevention, Bioterrorism Program, 2003; Rotz *et al.*, 2002). Several simulations of a plague attack have been conducted in the US as learning exercises leading to strengthened national and local preparedness and response; these have involved all levels of government, numerous agencies, and a wide range of first responders, including public safety and law enforcement personnel, hazardous materials teams, emergency medical and public health staff, and information specialists. Two of these, TOPOFF I (Denver, 2001) and TOPOFF II (Chicago, 2003), so-named because they involved top government officials, were based on coordinated national and local responses to simulated plague attacks. During these simulations, critical deficiencies in emergency response became obvious, including the following: problems in leadership, authority, and decision-making; difficulties in prioritization and distribution of scarce resources; failures to share information; and overwhelmed health care facilities and staff. The need to formulate in advance sound principles of disease containment, and the administrative and legal authority to carry them out without creating confusing new government procedures were glaringly obvious (Block, 2003; Hoffman, 2003; Hoffman and Norton, 2000; Khan and Ashford, 2001; Inglesby *et al.*, 2001).

2. PLAGUE MICROBIOLOGY AND PATHOGENESIS

2.1. The Agent

2.1.1. General Characteristics

Y. pestis is a Gram-negative, microaerophilic, pleomorphic coccobacillus (1.0–2.0 μm × 0.5 μm) belonging to the family Enterobacteriacae (Perry and Fetherston, 1997). In direct smears, *Y. pestis* presents as single cells or short chains of cells, characteristically appearing as plump bacilli exhibiting a bipolar staining (closed safety-pin) appearance when treated with Wayson, Giemsa, or Wright stains. The bacillus is nonmotile and nonsporulating. It does not ferment lactose and is catalase-positive, and urease-, oxidase-, and indole-negative. *Y. pestis* is relatively nonreactive, and automated biochemical systems may lead to misidentifications unless correctly programmed (Wilmoth *et al.*, 1996). Growth occurs in a variety of media at a wide range of temperatures (4–40°C; optimal 28–37°C) and pH values (5.0–9.6; optimal 6.8–7.6). However, *Y. pestis* grown at 28–30°C and with pH over 7.2 is more stable under natural conditions and in an aerosol form (K. Alibek, personal communication). *Y. pestis* is relatively slow growing in culture, with pinpoint colonies usually visible after 24 hours of growth at 28°C on sheep blood agar and later on MacConkey agar. The colonies are raised and opalescent in appearance, developing a hammered copper-appearing surface and irregular borders as they grow larger. In broth culture, a stalactite pattern of growth occurs along the sides of the vessel and settles to the bottom in clumps if disturbed. Almost all naturally occurring *Y. pestis* strains have been found to be susceptible in vitro to tetracyclines, chloramphenicol, sulfonamides, aminoglycosides, and fluoroquinolones (Frean *et al.*, 1996, 2003; Smith *et al.*, 1995). Rarely, isolates from several areas of the world have shown incomplete susceptibility to one or more antimicrobials recommended for treating plague (Rasoamanana *et al.*, 1989); these have occurred in isolated instances, have not been followed by a recognized wider emergence of these strains, and have not required modifications of standard protocols for treatment and control. However, of greater concern was the identification in 1995 of a strain of *Y. pestis* from a patient in Madagascar who was multiply resistant to the principal recommended antimicrobials used in treating plague, and the resistance was plasmid-mediated and transferable (Galimand *et al.*, 1997). This finding elicited calls for intensified surveillance of patients and the environment (Dennis and Hughes, 1997), which fortunately have not led to identification of other such strains in Madagascar or elsewhere.

2.1.2. Molecular Genetics

Gene sequencing comparisons of multiple strains of *Y. pestis*, *Y. pseudotuberculosis*, and *Y. enterocolitica* show that *Y. pestis*, a blood-borne organism, only recently (1,500–20,000 years ago) evolved from *Y. pseudotuberculosis*, an enteric pathogen (Achtman *et al.*, 1999). Decoding of the entire genome of *Y. pestis* (consisting of 4.65 Mb chromosome and three plasmids of 96.2 kb, 70.3 kb, and 9.6 kb) disclosed that the evolution of *Y. pestis* was made possible by the acquisition of virulence determinants suitable for systemic invasion of mammalian hosts and replication in the flea, and by the inactivation of genes required for enteric survival (Parkhill *et al.*, 2001). These genomic studies suggest that *Y. pestis* is a pathogen that has undergone large-scale genetic flux and provide a unique insight into how new and highly virulent pathogens evolve. Three classic biovars of *Y. pestis* have been identified, including biovar Antiqua, Medievalis, and Orientalis, linked respectively to the three historical pandemics. Results of typing by restriction fragment-length polymorphism analysis of rRNA genes (ribotyping) support these distinctions and have shown chromosomal rearrangements in the Orientalis biotype that occurred following its spread around the world about 100 years ago (Guiyoule *et al.*, 1994). Further studies of polymorphisms show considerable genome plasticity, even within strains from one geographic area (Guiyoule *et al.*, 1997; Radnedge *et al.*, 2002).

2.2. Pathogenicity of *Y. pestis*

2.2.1. Virulence Factors

Y. pestis is among the most pathogenic bacteria known. Although it is a facultative intracellular pathogen that normally grows in extracellular environments, virulence is in part dependent on invasion and multiplication within cells, including phagocytes that transport the bacterium in the initial phases of infection (Hinnebusch, 1997; Perry and Fetherston, 1997). Both chromosomal and plasmid-encoded gene products are associated with adaptability to its various hosts and to virulence (Carniel, 2003; Hinnebusch, 1997; Hinnebusch *et al.*, 2002; Koornhof *et al.*, 1999; Perry and Fetherston, 1997; Smego *et al.*, 1999).

Chromosomal genes of *Y. pestis* express a potent lipopolysaccharide endotoxin and a factor that controls the absorption of exogeneous iron. Of the three major plasmids, the pesticin (*pst*) plasmid (~9.5 kb) has genes that encode for a plasminogen activator (Pla) and a bacteriocin or pesticin (Pst). The low calcium response or *Lcr* plasmid (~70 kb), which is shared with other yersiniae, encodes products that activate the V and W antigens and outer surface proteins (Yops) under low calcium conditions. The *caf* operon of the *pFra* (Tox) plasmid (~110 kb) encodes the F1 glycoprotein envelope antigen (Caf1) and a murine toxin (Ymt) unique to *Y. pestis*. F1 antigen is produced only when *Y. pestis* grows at 30°C or greater; strains expressing F1 antigen are able to resist phagocytosis in the absence of opsonizing antibodies. In summary, *Y. pestis* virulence factors are thought to mediate the following responses between invading organism and the human host:

- The lipopolysaccharide endotoxin activates complement and triggers the release of kinins and other proinflammatory mediators;
- The chromosomally mediated hemin storage molecule (Hms) enhances *Y. pestis* survival in phagocytes and facilitates uptake of the bacillus into eukaryotic cells;
- Another chromosomally encoded product, the pH 6 antigen (Psa), inhibits *Y. pestis* phagocytosis;
- The plasminogen activator (Pla) is a single surface protease that degrades fibrin and other extracellular proteins, facilitating systemic spread of *Y. pestis*;
- The V and W antigens block phagocytosis of *Y. pestis*, and the V antigen promotes survival of *Y. pestis* in macrophages;
- Yops expressed by the 70-kb plasmid inhibit phagocytosis and platelet aggregation, and block an effective inflammatory response;
- The 110-kb plasmid-associated 17-kDa polypeptide F1 antigen (produced optimally at 37°C) is antiphagocytic and also elicits a strong humoral immune response.

Several factors have been identified that are selectively expressed by *Y. pestis* in the gut of fleas. For example, hemin storage locus (*hms*) products expressed at <30°C enable the bacteria to form blockages of the flea gut necessary for efficient transmission. Expression in the midgut of murine toxin (recently identified as phospholipidase D) protects *Y. pestis* from cytotoxic digestion by blood plasma products (Perry, 2003).

2.2.2. Pathology of Infection

The virulence of *Y. pestis* is expressed in a wide spectrum of disease that reflects the portal of pathogen entry and the organ systems targeted (Butler, 1972, 1983; Crook and Tempest, 1992; Dennis and Gage, 2004; Dennis and Meier, 1997; Hull *et al.*, 1987; Welty *et al.*, 1985; Wu, 1926). Plague organisms inoculated through the skin or mucous membranes typically are transported within lymphatic vessels to afferent, regional nodes, where they multiply. In the early stages of infection, affected nodes show vascular congestion, edema, and minimal inflammatory infiltrates or vascular injury; later, however, nodes may contain enormous numbers of infectious plague organisms and demonstrate vascular breaks, hemorrhagic necrosis, and infiltration by neutrophilic leukocytes. These affected nodes (buboes) are typically surrounded by a collection of serous fluid, often blood-tinged. When several adjacent lymph nodes are involved, a boggy, edematous mass can result. In later stages, abscess formation and spontaneous rupture of buboes may occur.

Y. pestis can invade and cause disease in almost any organ, and untreated infection usually results in widespread and massive tissue destruction. Diffuse interstitial myocarditis with cardiac dilatation, multifocal necrosis of the liver, diffuse hemorrhagic necrosis of the spleen and involved lymph nodes, and fibrin thrombi in renal glomeruli are commonly found in fatal cases (Butler, 1972; Dennis and Meier, 1997; Finegold, 1968). Pneumonitis, pleuritis, and meningitis occur less frequently. Abcesses may form in affected organs. Disseminated intravascular coagulation (DIC) is associated with generation of microthrombi, thrombocytopenia, necrosis, and bleeding in affected tissues (Butler, 1972). Petechiae and ecchymoses commonly appear in the skin, and on mucosal and serosal surfaces. Ischemia and gangrene of acral parts, such as fingers and toes, may occur in the late stages of this process (Dennis and Meier, 1997; Pollitzer, 1954; Wu, 1926).

Primary plague pneumonia resulting from inhalation of infective respiratory particles usually begins as a lobular process and then extends by confluence, becoming lobar and then multilobar. Typically, plague organisms are numerous in the alveoli and in pulmonary secretions. Secondary plague pneumonia arising from hematogeneous seeding of the lungs may begin more diffusely as an interstitial process. In untreated cases of both primary and secondary plague pneumonia, disease progresses to diffuse pulmonary congestion, hemorrhagic necrosis of pulmonary parenchyma, and infiltration by neutrophilic leukocytes (Wu, 1926). In advanced untreated stages, the alveolae are filled with fluid containing massive numbers of plague bacilli.

3. CLINICAL SPECTRUM

3.1. Bubonic Plague

Bubonic plague is characterized by the development of one or more swollen, tender, inflamed lymph nodes termed *buboes*, from the Greek *bubon*, meaning groin. Bubonic plague has a usual incubation period of 2–6 days, occasionally longer. Typically, bubonic plague is heralded by the sudden onset of chills, fever that rises within hours to 100.4°F (38°C) or higher, accompanied by headache, myalgias, arthralgias, and a profound lethargy. Soon, usually within a few hours of symptom onset, increasing swelling, tenderness, and pain occur in one or more regional lymph nodes proximal to the portal of entry. The femoral and inguinal groups of nodes are most commonly involved, axillary and cervical nodes less frequently, varying with the site of inoculation. Buboes occur at a single site in about 90% of cases; sometimes, more than one regional gland grouping may be affected, and bacteremic spread can result in a generalized lymphadenopathy. Typically, the patient guards against palpation and limits movement, pressure, and stretching around the bubo. The surrounding tissue often becomes edematous, sometimes markedly; and the overlying skin is typically reddened, warm, tense, and occasionally desquamated. The bubo of plague differs from lymphadenitis of most other causes by its rapid onset, extreme tenderness, surrounding edema, accompanying signs of toxemia, and usual absence of cellulitis or obvious ascending lymphangitis. Inspection of the skin surrounding the bubo or distal to it may reveal the site of bacterial inoculation marked by a small papule, pustule, scab, or ulcer (phlyctenule). A large furuncular lesion at site of entry, resulting in an ulcer that may be covered by an eschar, occurs occasionally (Figure 2.1). Presenting manifestations in a series of 40 Vietnamese bubonic plague patients were as follows (Butler, 1972):

- Fever (100%; mean of 39.4°C in 32 patients)
- Bubo (100%): groin (88%); axilla (15%); cervical (5%); and epitrochlear (3%)
- Headache (85%)

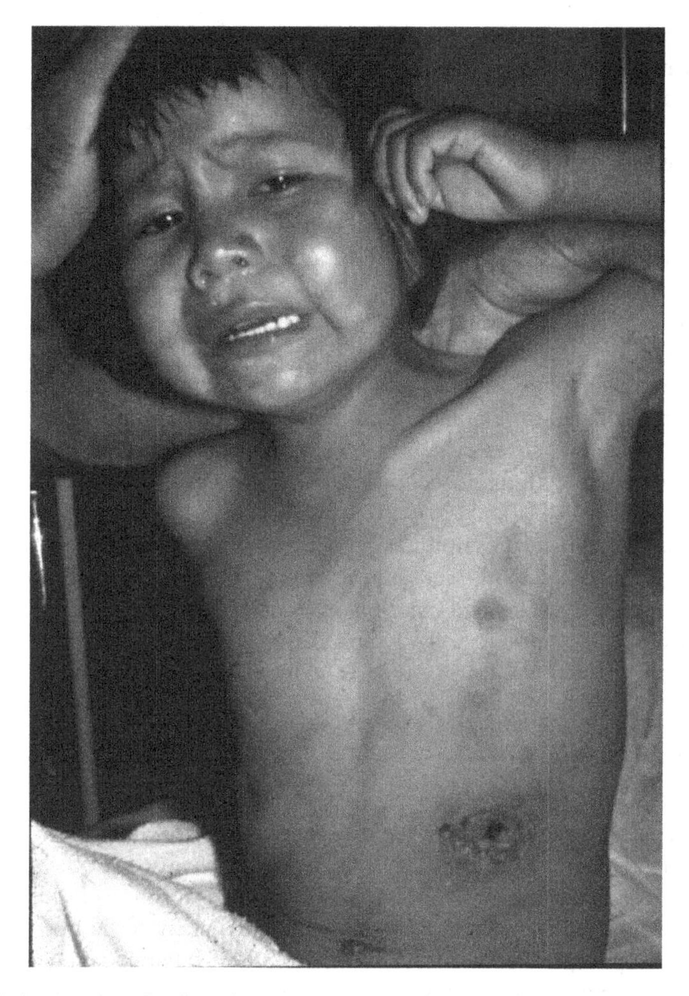

Figure 2.1. Native American showing primary cutaneous plague ulcer and eschar on anterior chest wall at site of *Yersinia pestis* inoculation, ascending lymphangitis, and axillary bubo.

- Prostration (75%)
- Chills (40%)
- Anorexia (33%)
- Vomiting (25%)
- Cough (25%)
- Skin rash, including petechiae, purpura, and papular eruptions (23%)
- Abdominal pain (18%)
- Chest pain (13%)

Altered brain function – manifest as lethargy, confusion, delirium, seizures – was also common in patients in the Vietnam series.

If treated with an appropriate antimicrobial agent, uncomplicated plague responds quickly, with resolution of fever and other systemic manifestations over a 2- to 5-day period. Buboes often remain enlarged and tender for a week or more after treatment has begun, and infrequently become purulent and fluctuant, and may require incision and drainage. Untreated, they may spontaneously rupture and drain. Without effective antimicrobial treatment, bubonic plague patients typically become increasingly toxic, with fever, tachycardia, lethargy leading to prostration, agitation and confusion, and, occasionally, convulsions and delirium. In the preantibiotic era, the case fatality ratio for bubonic plague was greater than 50%, and it is now about 5%. Mild forms of bubonic plague, called pestis minor, have been described in South America and elsewhere; in these cases, the patients are ambulatory and only mildly febrile and have subacute buboes (Legters *et al.*, 1970). The epidemiology and pathophysiology of these mild cases are poorly described, and the syndrome has been attributed to immunological tolerance rather than to a lesser virulence of infecting strains. There is some evidence from serological studies in endemic areas that subclinical *Y. pestis* infections do occur in endemic populations (Ratsitorahina *et al.*, 2000a, 2000b).

Differential diagnostic possibilities for bubonic plague include streptococcal or staphylococcal adenitis, tularemia, cat scratch disease, mycobacterial infection, acute filarial lymphadenitis, aspergillosis and other fungal conditions, chancroid and other sexually transmitted diseases that cause regional lymphadenitis, and strangulated inguinal hernia.

3.2. Septicemic Plague

Plague sepsis is manifest as a rapidly progressive, overwhelming endotoxemia (Butler *et al.*, 1976; Hull *et al.*, 1987). Plague sepsis in the absence of signs of localized infection, such as a bubo, is termed primary septic plague. It can result from direct entry of *Y. pestis* through broken skin or mucous membranes, or from the bite of an infective flea. Secondary septic plague can occur in the course of bubonic or pneumonic plague when lymphatic or pulmonary defenses are breached, and the plague bacillus enters and multiplies within the bloodstream. Bacteremia is common in all forms of plague; septicemia is less common and is immediately life threatening. A diagnosis of primary plague sepsis is often not made until results of blood culture are reported by the laboratory, since there is little to clinically distinguish plague sepsis from other causes of sepsis. Occasionally, plague organisms are visible in stained peripheral blood smears, indicating a poor prognosis (Butler *et al.*, 1976; Hull *et al.*, 1987; Mann *et al.*, 1984). The clinical diagnosis of plague sepsis may be obscured by prominent gastrointestinal symptoms, such as nausea, vomiting, diarrhea, and abdominal pain (Hull *et al.*, 1986). If not treated early with appropriate antibiotics and aggressive supportive care, septic plague is usually fulminating and fatal. Petechiae, ecchymoses, bleeding from puncture wounds and orifices, and subsequent ischemia and gangrene of acral parts are some manifestations of DIC (Figure 2.2). Refractory hypotension, renal shut down, stupor, and other signs of shock are preterminal events. Acute respiratory distress syndrome, which can occur at any stage of septic plague, may be confused with other conditions, such as the hantavirus pulmonary syndrome.

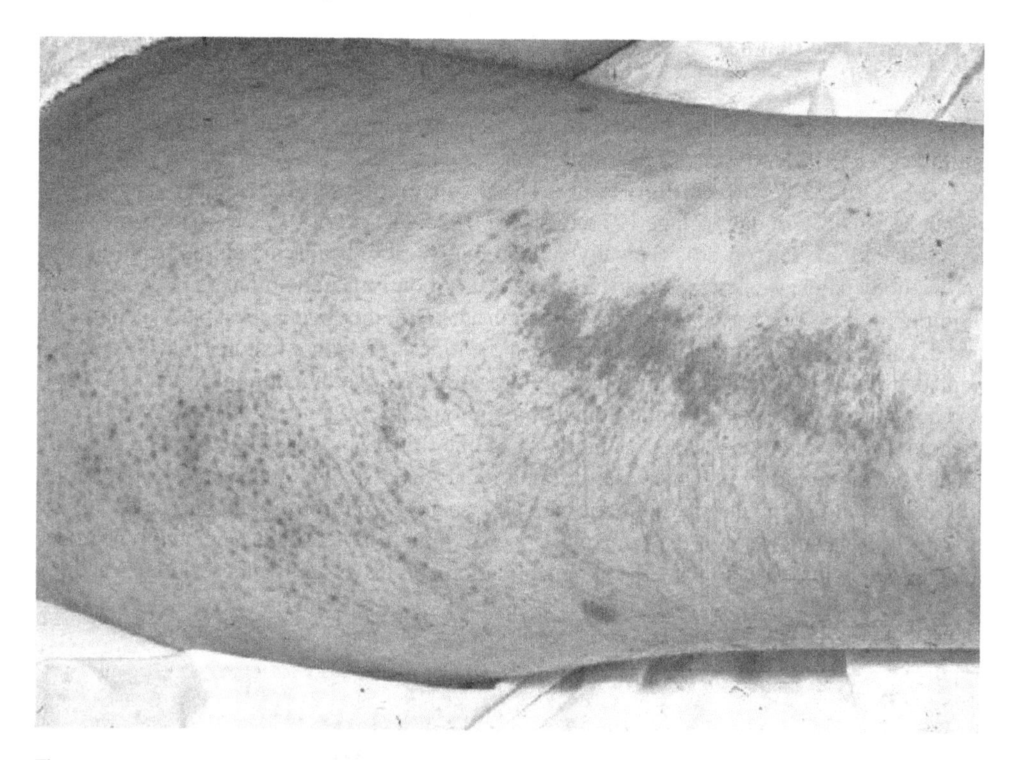

Figure 2.2. Petechiae and ecchymoses on lateral thigh of a patient with plague pneumonia, septicemia, and intravascular coagulation. This patient subsequently developed ischemia and gangrene of acral parts.

Presenting manifestations of septicemic plague in a case series of 18 patients in New Mexico (Hull *et al.*, 1987) include:

- Fever (100%); mean temperature 38.5°C, range 35.4–40.4°C
- Any gastrointestinal symptom (72%)
- Chills (61%)
- Vomiting (50%)
- Nausea (44%)
- Headache (44%)
- Diarrhea (39%)
- Abdominal pain (39%)

Because the diagnosis of plague is often made late, the case fatality ratio is 25% or greater among septicemic patients treated in the US (Crook and Tempest, 1992; Dennis and Chow, 2004; Hull *et al.*, 1987) and approaches 100% in those not receiving appropriate antibiotics.

Differential diagnostic possibilities include any other overwhelming systemic infection, including Gram-negative sepsis with agents other than the plague bacterium, meningococcemia, and bacterial endocarditis.

3.3. Pneumonic Plague

Pneumonic plague is the most rapidly developing and fatal form of plague (Doll, 1994; Laforce *et al.*, 1971; Meyer, 1961; Ratsitorahina *et al.*, 2000a; Tieh *et al.*, 1948; Wu, 1926; Wu *et al.*, 1922; Wynne-Griffith, 1948). The incubation period for primary pneumonic plague is usually 3–5 days (range 1–7 days) (Wu, 1926; Wu *et al.*, 1922; K. Alibek, personal communication). The onset is typically sudden, with severe headache, chills, fever, tachycardia, body pains, weakness, dizziness, and chest discomfort. Abdominal pain, nausea, and vomiting may also be present. Cough, sputum production, increasing chest pain, tachypnea, and dyspnea typically predominate on day 2 of the illness, and these features may be accompanied by bloody sputum, increasing respiratory distress, cardiopulmonary insufficiency, and circulatory collapse. In primary plague pneumonia, the sputum is most often watery or mucoid, frothy, and blood-tinged, but it may become frankly bloody. Chest signs in primary plague pneumonia may indicate localized pulmonary involvement in the early stage; a rapidly developing segmental consolidation may be seen before bronchopneumonia occurs in other segments and lobes of the same and opposite lung (Figure 2.3). Liquefaction necrosis and cavitation may develop at sites of consolidation and leave significant residual scarring.

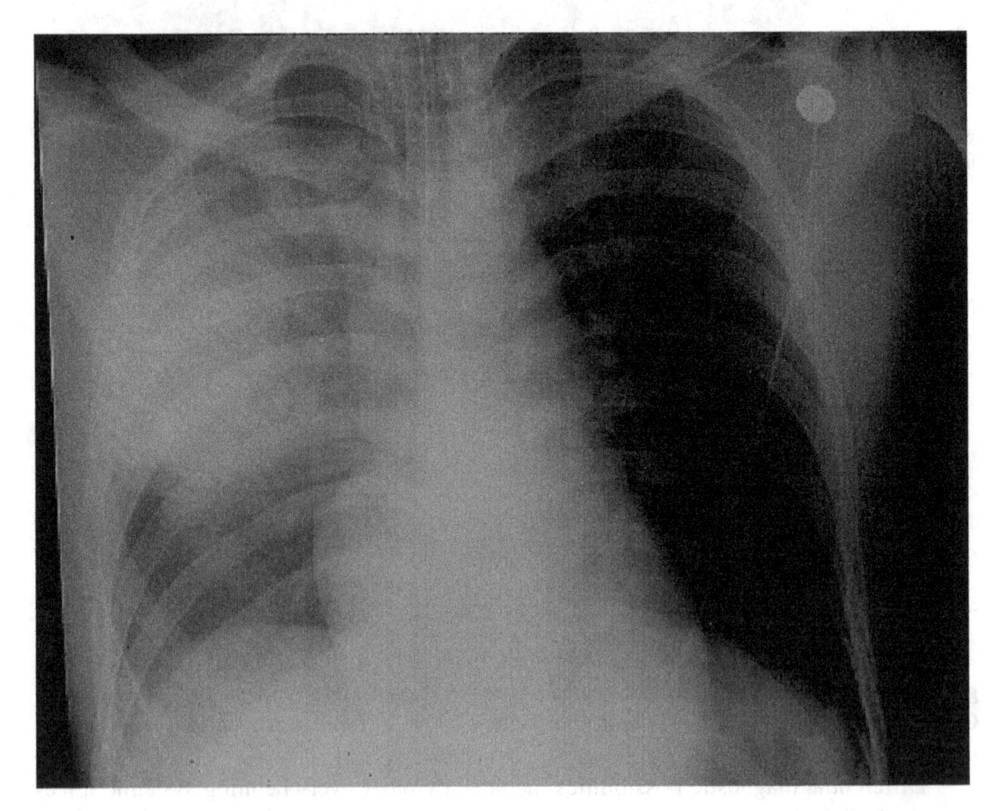

Figure 2.3. Chest radiographs showing rapid progression of primary plague pneumonia on days 3 and 4 of fatal illness.

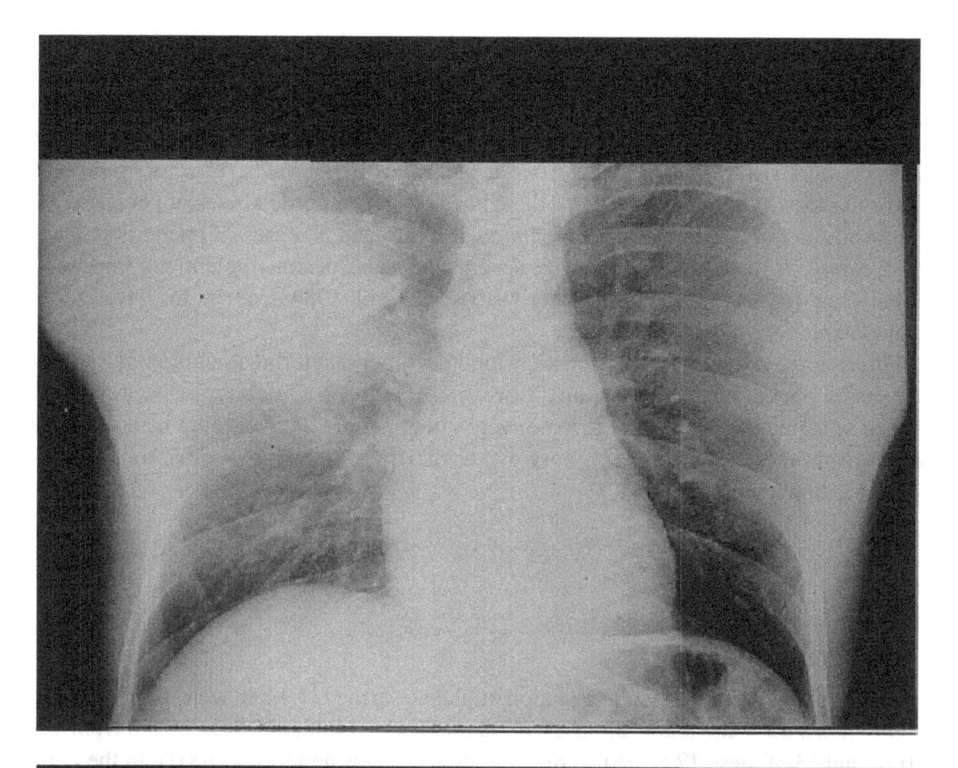

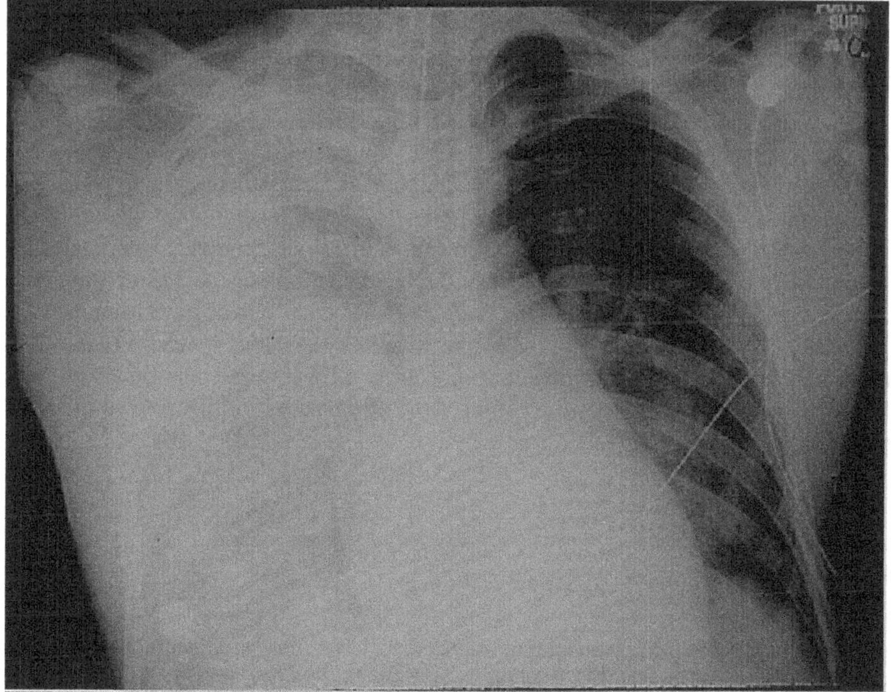

Figure 2.3. Continued

Plague pneumonia arising from metastatic spread is more likely to present in early stages as an interstitial pneumonitis in which sputum production is at first scant. The disease progresses rapidly, however, and chest radiographs described for nine cases of secondary plague pneumonia showed alveolar infiltrates in all cases and pleural effusions in more than half of patients (Alsofrom *et al.*, 1981). Advanced cases often develop refractory pulmonary edema and sepsis syndrome. In the US, there have been no recorded cases of person-to-person transmission of plague since 1924, although more than 50 cases of pneumonic plague have occurred in that time period, with several thousand persons potentially exposed to infection from these patients (Centers for Disease Control, 1984; Centers for Disease Control and Prevention, unpublished data).

Differential diagnostic possibilities include other bacterial pneumonias, such as mycoplasma pneumonia, Legionnaire's disease, staphylococcal or streptococcal pneumonia, tularemia pneumonia, and Q fever. Severe viral pneumonia, including hantavirus pulmonary syndrome and acute respiratory syndrome from coronavirus infection, could be confused with plague.

3.4. Other Clinical Syndromes

Meningitis is an unusual manifestation of plague. In the US, there were 17 (0.4%) cases of meningitis among the total 421 cases of plague reported in the 45-year period from 1947 to 2001, and 13 of these 17 (77%) were in children (Dennis and Chow, 2003). In the majority of these cases, meningitis was a late-arising occurrence in treated bubonic plague and nearly all patients survived (Becker *et al.*, 1987; Centers for Disease Control and Prevention, 1997; Mann *et al.*, 1982). Less commonly, infection of the meninges can occur in the apparent absence of a bubo. The patient typically has fever, meningismus, altered mental status, and a purulent cerebrospinal fluid with early polymorphonuclear cells predominating; endotoxin has been identified within the cerebrospinal fluid of these patients (Butler *et al.*, 1976).

Occasionally, plague presents as pharyngitis, following inhalation of infectious particles or ingestion of *Y. pestis*, such as may occur by eating undercooked, contaminated meat (Christie *et al.*, 1980). Pharyngeal colonization without symptoms also sometimes occurs (Marshall *et al.*, 1967; Tieh *et al.*, 1948). Rare cases of ocular plague have been described, which could arise by inoculation of the conjunctiva or by systemic spread (Poland, 1989). Primary ocular and pharyngeal plague might arise following an intentional release of *Y. pestis*, by direct inoculation of the conjunctivae or inhalation of aerosolized organisms, respectively.

3.5. Pediatric Plague

The clinical presentation of plague in children is similar to that in adults. Based on reported case series (Burkle, 1973; Mann *et al.*, 1982), children with bubonic plague are somewhat more likely than adults to develop complications of systemic spread (septicemia, pneumonia, and meningitis) and to have a higher case fatality ratio. In a review of 183 childhood cases reported in the US during the period 1947–2001 (Dennis and Chow, 2003), 167 (91%) were classified as bubonic plague; of these, 54 (32%) developed secondary complications, including 41(25%) with sepsis (23 with sepsis alone, 9 combined with meningitis, 7 with pneumonia, and 2 with both pneumonia and meningitis). There were 13 secondary pneumonic cases without mention of sepsis, two of which were complicated by meningitis. In the total series, there were 31 (17%) deaths; mortality was highest in patients with primary sepsis or pneumonia (39%), less in patients with bubonic plague and secondary spread (30%), and least in patients with bubonic plague without recorded spread (9%). Comparing findings with adult cases reported over the same time period, children were more likely to develop complications of dissemination (32% vs. 27%), especially meningitis (13/183 vs. 4/238). Children were also more likely to have a fatal outcome (17% vs. 14%). These findings are similar to those reported in a series of 38 children from New Mexico (Mann *et al.*, 1982). Delay in diagnosis of plague in children is common; in the New Mexico series, the correct diagnosis was considered in less than 10% of patients on first medical encounter and in only about 40% at time of hospital admission. Presenting manifestations of plague in the New Mexico series included:

- Fever (95%)
- Lethargy, malaise, and anorexia (40%)
- Vomiting (50%)
- Chills (29%)
- Headache (29%)
- Abdominal discomfort (26%)
- Diarrhea (8%)

3.6. Plague in Pregnancy

Plague in the pregnant woman can result in infection of the fetus, with stillbirth, abortion, and perinatal infection as outcomes (Mann and Moskowitz, 1977; Welty *et al.*, 1985).

4. DIAGNOSIS

4.1. Laboratory Diagnosis

4.1.1. Laboratory Response Capabilities

Laboratory tests for plague are highly reliable when conducted by persons trained and experienced in the microbiology of *Y. pestis*; until recently, this was limited to a very few reference laboratories. In response to concerns with biological terrorism, the US initiated in 1999 a national Laboratory Response Network to provide upgraded, standardized diagnostic testing for plague and other select agents (Centers for Disease Control and Prevention, 2000; Morse *et al.*, 2003). This network links state and local public health laboratories with other advanced-capacity laboratories, including those at CDC, the National Institutes of Health, the Food and Drug Administration (FDA), Department of Defense, Federal Bureau of Investigation, and the Department of Agriculture. Member laboratories operate either as sentinel laboratories (Level A) or as reference Levels B–D, representing progressively stringent safety, containment, and technical proficiency capabilities.

Sentinel (Level A) laboratories include hospital and other community clinical laboratories that practice BSL-2 safety procedures and perform initial tests to presumptively identify or rule out *Y. pestis* infection, using such procedures as direct staining, bacterial culture, and biochemical screening tests. Clinical specimens from most victims of a bioterrorist attack would probably be first handled for routine diagnostic procedures in Level A laboratories.

Suspicious isolates and source materials would be forwarded to Level B and C laboratories (federal, state, and local public health laboratories with BSL-2 and BSL-3 capabilities), which are prepared to perform advanced rapid diagnostic tests, to confirm the microbiological identification, and to characterize strain attributes. Level B and C laboratories are also prepared to carry out standardized antimicrobial susceptibility studies, and Level C laboratories routinely perform molecular subtyping tests, such as multiple locus variable number tandem-repeat assay, restriction fragment-length polymorphism, and pulsed-field gel electrophoresis (Henchal *et al.*, 2001; Morse *et al.*, 2003). These latter assays may be useful for tracing epidemiological links and for forensic purposes.

4.1.2. Collection and Processing of Specimens

Clinical Specimens. When plague is suspected, clinical specimens should be obtained promptly for microbiologic studies (Henchal *et al.*, 2001; Miller, 2001), chest radiographs taken, and antimicrobial therapy initiated pending confirmation of diagnosis. Blood and other clinical materials – such as bubo aspirates, sputum, tracheal-bronchial washes, swabs of skin lesions or pharyngeal mucosal, and cerebrospinal fluid – should be inoculated onto suitable media (e.g., brain–heart infusion broth, sheep blood agar, chocolate agar, or MacConkey agar). Blood for culture should be collected prior to administration of antibiotics. Blood culture counts typically range from fewer than 10, to 4×10^7 colony-forming units per mL. Direct bubo aspirates typically yield only small amounts of serosanguinous fluid, and 1–2 mL of saline should be injected into the bubo and withdrawn to ensure adequate aspirate for diagnosis. Smears of each specimen should be stained with Gram stain and a polychromatic stain, such as Wayson or Giemsa stain. Direct fluorescent antibody testing is a useful presumptive diagnostic procedure available at specialized laboratories. An acute-phase serum specimen should be collected for *Y. pestis* antibody testing, followed by a convalescent-phase specimen collected 3–4 weeks later. In the absence of an isolated organism, plague can be diagnosed by demonstrating a four-fold or greater change in serum antibodies to *Y. pestis* antigen using passive hemagglutination testing. A serum antibody titer of 1:128 or greater in a single serum sample from a patient who has a compatible illness and who has not received plague vaccine is also diagnostic. A few plague patients will develop detectable antibodies as soon as 5 days after the onset of illness, most seroconvert 1–2 weeks after onset, a few seroconvert three or more weeks after onset, and a few (<5%) fail to seroconvert (Butler and Hudson, 1977; Centers for Disease Control and Prevention, unpublished data). Early specific antibiotic treatment may delay seroconversion by several weeks. Positive serologic titers diminish gradually from months to years. Enzyme-linked immunosorbent assays for detecting IgM and IgG antibodies to *Y. pestis* have been found to be useful in identifying antibodies in early infection and in differentiating them from antibodies developed in response to previous vaccination. Presumptive identification of *Y. pestis* can be made by polymerase chain reaction (PCR) or antigen-capture ELISA (Henchal *et al.*, 2001; Loiez *et al.*, 2003; Radnedge *et al.*, 2001; Rahalison *et al.*, 2000). A recently developed rapid immunogold dipstick assay designed to detect *Y. pestis* antigens in patient samples also appears highly promising for rapid presumptive diagnosis at the bedside, even under primitive field conditions (Chanteau *et al.*, 2003). Protocols and algorithms have been developed for clinical laboratories to follow in diagnosing plague and other select agent diseases in the event of a terrorist event (American Society of Microbiology, Biological Weapons Resources Center, 2003; Centers for Disease Control and Prevention, Bioweapons Laboratory Issues, 2001; Henchal *et al.*, 2001).

Nonspecific laboratory findings typically include elevation of various enzymes resulting from damage to the liver, heart, and other organs and tissues, thrombocytopenia, and white cell counts of 10,000–25,000/mm^3 with a predominance of early-stage polymorphonuclear leukocytes. Leukemoid reactions, with white cell counts of 50,000/mm^3 or higher, sometimes occur (Butler *et al.*, 1974; Welty *et al.*, 1985).

Autopsy Specimens. For diagnosis in fatal cases, tissues, samples of lymph nodes, liver, spleen, lungs, bone marrow, and other affected tissues should be collected at necropsy for culture, fluorescent antibody testing, and histological studies, including possible immunohistochemical staining (Guarner *et al.*, 2002). Cary Blair medium or a similar holding medium can be used to transport *Y. pestis*-infected tissues for later isolation of *Y. pestis*.

4.2. Recognizing a Plague Outbreak Resulting from Intentional Release

The identification of plague resulting from an intentional release must be made quickly to prevent excessive mortality among those persons initially exposed and to interrupt secondary person-to-person transmission. This requires a high level of expertise in clinical, laboratory, and public health services and a smooth integration of response. It is recognized that the most critical components for bioterrorism outbreak detection and reporting are the frontline health care professionals and the local health departments. (Ashford *et al.*, 2003). Principal expected features of an outbreak arising from an aerosol exposure are outlined in Table 2.1.

4.3. Detection of *Y. pestis* in the Environment

Microbiological culture of *Y. pestis* is the standard method for confirming its presence. However, growth of the bacillus is relatively time-consuming and insensitive, and rapid and

Table 2.1
Diagnosis of Plague following Release of an Aerosol of *Y. pestis*

Epidemiology and Symptoms	Sudden outbreak of geographically linked persons with fever, cough, shortness of breath, hemoptysis, and chest pain; point source exposure pattern, incubation period 1–7 days, peaking 3–5 days postexposure; gastrointestinal symptoms common (e.g., nausea, vomiting, abdominal pain, and diarrhea); rapidly developing toxemia; fulminant and fatal course common.
Clinical Signs	Tachypnea, dyspnea, and cyanosis; pneumonic consolidation on chest examination; septic shock and organ failure; disseminated intravascular coagulation, petechiae, ecchymoses; occasional cases of plague pharyngitis and cervical adenitis, conjunctivitis, possible.
Laboratory Studies	Sputum, throat swab, tracheal-bronchial washes for Gram, and Wayson or Giemsa staining; culture at 28°C and 37°C on suitable media (e.g., sheep's blood agar, McConkey's, chocolate agar, BHI broth); if suspicion high, send above to LRN reference laboratory for DFA, advanced rapid testing; standard blood culture as per institution protocol; if suspicion high, culture subset at 28°C; acute-phase serum specimen to be held at 4°C until plague ruled out; antimicrobial susceptibility testing of *Y. pestis* isolates; subtyping studies to characterize isolates, e.g., PFGE, PCR, MLVA; virulence testing; chest radiographs.
Pathology	Lobular exudation, bacillary aggregation, hemorrhagic necrosis of pulmonary parenchyma; tissues processed for direct detection (DFA, IHC) and isolation of *Y. pestis*

BHI, brain–heart infusion; DFA, direct fluorescence antibody testing; IHC, immunohistochemical; LRN, Laboratory Response Netwrok; MLVA, multiple locus variable number tandem-repeat assay; PCR, polymerase chain reaction; PFGE, pulsed-field gel electrophoresis.

reliable tests are needed for early warning systems, for rapid identification of a contaminated environment, and for epidemiological and forensic investigations. Gene amplification systems have been developed that use PCR technology to amplify and characterize specific DNA sequences of *Y. pestis*, such as Pst, Cafl, YopM, and Pla targets (Henchal *et al.*, 2001; Radnedge *et al.*, 2001; Rahalison *et al.*, 2000). Most such rapid sequencing approaches use fluorogenic 5′ nuclease chemistry. PCR assay coupled to probe hydrolysis (e.g., TaqMan^R^) allows "real-time" detection of PCR products. Rapid thermocycling instruments, such as the Lightcycler™ (Roche Molecular Systems), the Ruggedized Advanced Pathogen Identification Device (RAPID™, Idaho Technologies), or the SmartCycler™ (Cepheid) allow identification of the agent in 20–40 minutes after nucleic acid purification (Henchal *et al.*, 2001). In the US, several "sniffing devices" to detect aerosolized microbial pathogens have been developed and tested. The Department of Homeland Security and the Environmental Detection Agency have deployed a PCR-based detection system named BioWatch to continuously monitor filtered air in major cities for *Y. pestis* and other select agents. Other monitoring systems include the Interim Biological Agent Detector System (IBADS™) that uses immunoassay to detect particles captured on a flowthrough membrane, and the Biological Integrated Detection System (BIDS™) that uses a light addressable potentiometric device as the detector (Henchal *et al.*, 2001). Such technologies, designed by the US military for the battlefield, have been used experimentally to monitor specific high-profile civilian events and sensitive sites that could be targets of terrorism. In the event of a real or suspected aerosol release by terrorists, immediate environmental sampling could be performed using PCR, hand-held immunochromatographic assay, and cultural isolation techniques to identify the agent used and to determine the extent of contamination. Only cultural isolation determines viability of the agent involved, and allows characterization of important attributes, such as virulence and antimicrobial susceptibility.

In some circumstances, there may be a concern that release of live *Y. pestis* could result in infection of rodents or other susceptible animals, and potentially pose a risk to humans from of animal- or flea-borne plague. In this circumstance, field teams would be deployed and standard procedures used to collect and process susceptible animals and their fleas for *Y. pestis* infection, including rapid detection methods and cultural isolation. If required, standard procedures for flea and rodent control would be implemented (Gage, 1998; Gratz, 1999). Special surveillance for plague in domestic cats might be indicated, since these animals are susceptible and can transmit infection – including respiratory infection – to humans (Gage *et al.*, 2000).

5. MEDICAL MANAGEMENT OF PLAGUE PATIENTS

5.1. Antimicrobial Treatment of Acute Illness in Naturally Occurring Plague

Untreated, plague is fatal in more than 50% of bubonic plague patients and in nearly all patients suffering from septic or pneumonic plague. The overall mortality in plague cases in the US in the past 50 years has been approximately 15% (Centers for Disease Control and Prevention, 1997; Craven *et al.*, 1993; Dennis and Campbell, 2004). Fatalities are almost always due to delays in seeking treatment, misdiagnosis and delayed or incorrect treatment (Centers for Disease Control and Prevention, 1997; Crook and Tempest, 1992). Rapid diagnosis and treatment with an efficacious antimicrobial agent are essential, and initiation of treatment within 24 hours of onset of pulmonary symptoms is often cited as the critical period in successful treatment of pneumonic plague (Butler, 1994; Butler and Dennis, 2004). With rare exceptions, antimicrobial susceptibility studies of human and animal strains of *Y. pestis* have shown low minimum inhibitory concentrations for the standard agents recommended for treating plague (i.e., streptomycin, gentamicin, doxycyline, chloramphenicol, and trimethoprim-sulfamethoxazole), as well as for several amino-quinolones, such as ciprofloxacin, ofloxacin, and trovafloxacin (Frean *et al.*, 1996; Smith *et al.*, 1995; Wong *et al.*, 2000). In vitro activity of ceftriaxone has also been found to be high, although clinical experience with cephalosporins has not been favorable. Studies in Russia have shown that doxycycline or ciprofloxacin was not inferior to streptomycin or tetracycline in treating bubonic plague in baboons, and that gentamicin or streptomycin were highly efficacious in treating aerosol-induced pneumonic plague in these animals (Romanov *et al.*, 2001a, 2001b). Antimicrobials that have been shown to have poor or only modest efficacy in experimental pneumonic plague in mice include rifampin, ampicillin, aztreonam, ceftazidime, cefotetan, and cefazolin, as compared with high efficacy of streptomycin and gentamicin (Byrne *et al.*, 1998).

Streptomycin has long been considered the drug of choice for treating plague (Butler and Dennis, 2004), and is FDA-approved for this use; however, there is no longer any manufacture of streptomycin in the US, availability is not widespread, and the drug must be obtained by special request. Although not FDA-approved for treating plague, gentamicin has increasingly been used in the US in place of streptomycin because of its ready availability and intravenous administration. Gentamicin has been anecdotally reported to be efficacious in treating plague in the US (Crook and Tempest, 1992; Welty *et al.*, 1985). A recent restrospective analysis of 50 plague patients treated with gentamycin in New Mexico since 1970 revealed that gentamicin – or a combination of gentamicin and doxycycline – is at least as efficacious as streptomycin, and was more often used than streptomycin in treating plague in that state (Boulanger *et al.*, 2004). Tetracyclines or chloramphenicol are suitable alternatives to the aminoglycosides. Doxycycline has – because of its ease of administration, rapid gastrointestinal absorption, and superior ability to achieve peak serum concentrations – become the tetracycline of choice for treating plague. Doxycycline treatment should be initiated with a loading dose, either intravenously or orally, depending on the severity of illness. In adults, a loading dose of 200 mg every 12 hours rapidly achieves a peak serum concentration of ~8 μg/mL (Cunha, 2003). Chloramphenicol is indicated for conditions in which high tissue penetration is important, such as plague meningitis, pleuritis, endophthalmitis, or myocarditis. It may be used separately or in combination with an aminoglycoside. Trimethoprim-sulfamethoxazole (co-trimoxazole) has been used successfully to treat bubonic plague but response may be delayed and incomplete, and it is not considered a first-line choice. Penicillins, cephalosporins, and macrolides have a suboptimal clinical effect and are not recommended for use in treating plague. In general, antimicrobial treatment should be continued for 7–10 days or for at least 3 days after the patient has become afebrile and has made a clinical recovery (Butler and Dennis, 2004; Dennis, 2001; Inglesby *et al.*, 2000). Patients begun on intravenous antibiotics may be switched to oral regimens, as indicated by clinical response; in uncomplicated cases, this can usually be made on the 4th or 5th day of treatment. Clear signs of improvement are usually evident 2–3 days from the start of treatment, even though fever of lessened amplitude may continue for several more days. Strains of *Y. pestis* that are resistant to antimicrobials have only rarely been isolated from humans. Such resistant strains have usually involved partial resistance to a single agent only and have not been associated with treatment failures. Fortunately, the recent single plasmid-mediated multidrug-resistant strain of *Y. pestis* isolated from a bubonic plague patient in Madagascar (Galimand *et al.*, 1997) appears to be a unique finding. Antimicrobial guidelines for treating plague in non-bioterrorism settings are given in Table 2.2.

Common complications of delayed treatment include DIC, acute respiratory distress syndrome, and other consequences of bacterial sepsis and endotoxemia. Patients with these disorders require intensive monitoring and close physiologic support, often taxing personnel and facilities resources. In addition to prompt initiation of antibiotics and measures to countershock, a number of investigational treatments have been proposed for treating sepsis, including recombinant-activated protein C (Dellinger, 2003; Wheeler and Gordon, 1999). A recent report describes the successful treatment of plague sepsis and peripheral gangrene with ciprofloxacin and sympathetic blockade (Kuberski *et al.*, 2003). Abscessed nodes can

Table 2.2
Plague Treatment Guidelines in Usual (Nonbioterrorism) Circumstances

Drug	Dosage	Route of administration
Streptomycin[a]		
Adults	1 g q 12 hr	IM
Children	15 mg/kg q 12 hr[a]	IM
Gentamicin[b]		
Adults	1–1.5 mg/kg q 8 hr[c]	IV or IM
Children	2.0–2.5 mg/kg q 8 hr[c]	IV or IM
Infants/neonates	2.5 mg/kg q 8 hr[c]	IV or IM
Tetracycline[d]		
Adults	0.5 g q 6 hr	PO
Children >8 yr	6.25–12.5 mg/kg q 6 hr	PO
Doxycycline[d]		
Adults	100 mg q 12 hr	IV or PO
Children >8 yr and >45 kg	100 mg q 12 hr	IV or PO
Children >8 yr and <45 kg	2.2 mg/kg q 12 hr	IV or PO
Chloramphenicol[d]		
Adults	12.5 mg/kg q 6 hr[e]	IV or PO
Children >1 yr	12.5 mg/kg q 6 hr[e]	IV or PO

IM, intramuscularly; IV, intravenously; PO, orally.
[a] Not to exceed 2 g/day.
[b] Not FDA-approved for use in treating plague.
[c] Daily dose should be reduced to 3 mg/kg as soon as clinically indicated.
[d] An initial loading dose is usually indicated.
[e] Up to 100 mg/kg/day initially. Dosage should be adjusted to maintain plasma concentrations at 5–20 μg/mL. Hematological values should be monitored closely.

be a cause of recurrent fever in patients who have otherwise made a satisfactory recovery, and the cause may be occult if intrathoracic or intraabdominal nodes are involved.

5.2. Postexposure Prophylaxis

Antimicrobial prophylactic treatment of persons having unprotected direct respiratory exposure to a patient with plague pneumonia is recommended as a public health control measure (Centers for Disease Control and Prevention, 1996; Inglesby *et al.*, 2000). Treatment is indicated if exposure has occurred in the previous 7 days, and the recommended duration of treatment is 7 days. Tetracycline, doxycyline, sulfonamides, and chloramphenicol have been recommended for postexposure prophylaxis (Centers for Disease Control and Prevention, 1996; Poland and Dennis, 1999), and tetracycline and doxycycline are approved by FDA for this purpose. Studies in mice have suggested that fluoroquinolones might also be effective in prophylaxis (Russell *et al.*, 1996), and their use has been advocated for use in responding to a plague terrorist event (Inglesby *et al.*, 2000). Short courses of antimicrobial prophylaxis are sometimes recommended for household members of bubonic plague patients because of the possibility of common rodent flea exposures. Prophylaxis is only rarely warranted for people who visit or reside in an area where plague is occurring. Respiratory plague patients are generally considered to be noncontagious following 48 hours of antibiotic treatment (Anonymous, 2000; Inglesby *et al.*, 2000).

5.3. Treatment of Cases and Case Contacts in a Bioterrorism Event

A plague aerosol attack would require rapid identification and treatment of pneumonic plague cases and case contacts. The Department of Homeland Security is prepared to dispatch immediately shipments of antimicrobials and other materials essential in the management of plague patients, as contained in the National Pharmaceutical Stockpile (NPS). The Johns Hopkins Working Group on Civilian Biodefense has developed consensus-based recommendations for treatment and postexposure prophylaxis of terrorist-caused plague utilizing antimicrobials contained in the NPS (Inglesby *et al.*, 2000). These guidelines consider both a *contained casualty* situation (a situation where a modest number of patients require treatment and can be given maximum individual care, and parenteral antimicrobial treatment is recommended) and a *mass casualty* situation (large numbers of patients require treatment, not allowing full individual care, and oral administration of antimicrobials is recommended). In the contained situation, streptomycin or gentamicin is considered the preferred initial treatment for adults and children, whereas parenteral doxycycline, ciprofloxacin, and chloramphenicol are considered alternatives. Although full courses of treatment are for 10 days, it is recommended that oral therapy should be substituted when the patient's condition improves. In the mass casualty situation, oral doxycycline and ciprofloxacin are the preferred choices, and chloramphenicol is an alternative choice (Table 2.3). The regimen recommended for postexposure prophylaxis is the same as that for persons being treated in the mass casualty situation (i.e., oral doxycycline or ciprofloxacin), except that the duration of recommended treatment is 7 days rather than 10 days (Table 2.4).

Criteria for treatment include the following:

- Potentially exposed persons in an affected community who develop fever of 38.5°C or greater or a new cough should be evaluated and treated for presumptive plague following contained casualty or mass casualty protocols, as appropriate. Young children who develop tachypnea should also be considered for treatment.
- Persons who develop fever or cough while receiving antimicrobial prophylaxis should be evaluated for *Y. pestis* infection and the possibility of antimicrobial resistance or noncompliance with treatment and managed appropriately.

Categories of persons to be considered for postexposure antibiotic prophylaxis within 7 days of an infectious exposure are the following:

- Persons exposed to aerosol or other potentially infective *Y. pestis* release
- Household members of cases with respiratory plague
- Health care workers at facilities that screen or care for suspected plague patients
- Emergency workers who have responded to calls for assistance related to the event
- Persons who have transported suspected plague patients to health care facilities

Table 2.3
Plague Treatment Guidelines Using National Pharmaceutical Stockpile Components

	Initial therapy	Duration
ADULTS	Gentamicin 5 mg/kg IM or IV once daily (or in 3 divided doses)[a–e] OR	7–10 days (switch to oral doxycycline when clinically appropriate to complete10-day therapy)
Parenteral	Doxycycline 100 mg IV twice daily OR	7–10 days
Oral	Doxycycline 100 mg PO twice daily	7–10 days
CHILDREN	Gentamicin 7.5 mg/kg IM or IV once daily (or, in 3 divided doses)[a–e] OR	(switch to oral doxycycline when clinically appropriate to complete 10-day therapy)
Parenteral	Doxycycline[f] ≥45 kg: 100 mg IV twice daily ≤45 kg: 2.2 mg/kg IV twice daily OR	
Oral	Doxycycline[f] ≥45 kg: 100 mg PO twice daily ≤45 kg: 2.2 mg/kg PO twice daily	7–10 days
PREGNANCY	Same as for nonpregnant adults[g,h]	
IMMUNOCOMPROMISED	Same as for nonimmunocompromised adults and children	

IM, intramuscularly; IV, intravenously; PO, orally.

[a] Historic treatment of choice for plague is streptomycin. Streptomycin can be difficult to obtain; therefore, gentamicin is recommended and is included in the National Pharmaceutical Stockpile. Oral therapy should be substituted when clinically indicated.

[b] The frequency of administration is left up to the discretion of the clinician. The manufacturers recommend that the daily dose be given in equally divided doses at 8-hr intervals; however, current evidence suggests that once-daily dosing of aminoglycosides is at least as effective as, and may be less toxic than, conventional dosing regimens using multiple daily doses of the drugs.

[c] An initial loading dose of 2 mg/kg body weight is standard medical practice when gentamicin is given as three doses per day.

[d] Not a US Food and Drug Administration-approved use.

[e] Refer to package insert to adjust dose in the event of renal insufficiency.

[f] In 1991, the American Academy of Pediatrics amended its recommendation to allow treatment of young children with tetracyclines for serious infections for which doxycycline may be indicated. Doxycycline is preferred for its twice daily dosing and low incidence of gastrointestinal side effects.

[g] Aminoglycosides can cause fetal ototoxicity when administered to pregnant women; benefits may, however, outweigh risks when treating serious infections.

[h] Tetracyclines can cause damage to fetal teeth and bones when administered to pregnant women.

- Co-workers, friends, and other associates who have had close contact with symptomatic respiratory plague cases
- Persons who have investigated or performed remediation at the release site
- Public health or other personnel who have interviewed patients or investigated sites of possible environmental contamination

Some high-risk categories may also be considered for pre-exposure prophylaxis (effectiveness of this measure is unknown) and the use of close-fitting face masks designed to block respiratory droplets.

Table 2.4
Plague Prophylaxis Guidelines Using National Pharmaceutical Stockpile Components[a]

	Therapy	Duration
ADULTS	Doxycycline 100 mg PO twice daily OR Ciprofloxacin 500 mg PO twice daily[b,c]	7 days
CHILDREN	Doxycycline[d] ≥45 kg: 100 mg PO twice daily ≤45 kg: 2.2 mg/kg PO twice daily OR Ciprofloxacin 20 mg/kg PO twice daily[b,c,e]	7 days
PREGNANCY	Same as for nonpregnant adults[f]	
IMMUNOCOMPROMISED	Same as for nonimmunosuppressed adults and children	

PO, orally.
[a]One antibiotic regimen, appropriate for patient age, should be chosen from among alternatives. The duration of all recommended therapies is 7 days.
[b]Not a US Food and Drug Administration-approved use.
[c]Refer to package insert to adjust dose in the event of renal insufficiency.
[d]In 1991, the American Academy of Pediatrics amended its recommendation to allow treatment of young children with tetracyclines for serious infections, for which doxycycline may be indicated. Doxycycline is preferred for its twice-a-day dosing and low incidence of gastrointestinal side effects.
[e]The American Academy of Pediatrics states that the use of quinolones in children younger than 18 yr of age may be justified in special circumstances. For the treatment of plague, the assessment of the risks and benefits indicates that administration of ciprofloxacin to pediatric patients is appropriate. Ciprofloxacin dose should not exceed 1 g/day in children.
[f]Tetracyclines can cause damage to fetal teeth and bones when administered to pregnant women.

6. INFECTION CONTROL

6.1. Hospital Infection Control

Some previous public health authorities have recommended strict isolation of untreated pneumonic plague patients (Anonymous, 2000). However, the current consensus of scientific opinion is that transmission of infection from a pneumonic plague patient is via respiratory droplets (droplets larger than 5 μm in diameter) rather than true airborne transmission (droplet nuclei of suspended evaporated droplets, or dust particles, less than 5 μm in diameter). Respiratory droplets are generated primarily during coughing, sneezing, and talking; and during the performance of certain patient care procedures, such as suctioning and bronchoscopy. Respiratory droplets are typically transmitted across short distances only (<2 meters) from the source patient. Because they do not remain suspended in air, special air handling and filtered ventilation are not required in isolation spaces. Accordingly, in addition to applying Standard Precautions recommended for the care of all patients, patients with suspected pneumonic plague should be isolated in a private room, or cohorted with other suspect plague patients and managed under Respiratory Droplet Precautions as promulgated by the CDC and the Hospital Infection Control Practices Advisory

Committee (Centers for Disease Control and Prevention, Hospital Infections Program, 1997; Garner, 1996). In addition to Respiratory Droplet Precautions, Standard Precautions, including eye protection, use of gloves, mask, and gowns should be followed at all times when working within 2 meters of the contagious patient (Grow and Rubinson, 2003). Patients should be moved from isolation for essential purposes only, avoiding contact with others and, if possible, masked. Environmental control guidelines can be obtained at the CDC website for hospital infection control practices (Centers for Disease Control and Prevention, Hospital Infections Program, 2003).

6.2. The Role of Isolation and Quarantine

Isolation is defined as the separation of a person or group of persons, for the period of communicability, from others to prevent the spread of infection (Last, 1983). All confirmed, probable, or suspect cases of plague pneumonia should be isolated under respiratory droplet precautions during the first 48 hours of antimicrobial treatment and until clinical improvement occurs or the diagnosis of plague is ruled out. Although the isolation of a limited number of confirmed or presumptive plague cases may be initially accomplished in a hospital setting, health authorities should be prepared to use alternative facilities if the capacity is exceeded. These facilities should be secured to prevent unwanted movement of persons in or out; be supported with water, heat, electricity, and sanitation; have provision for adequate medical care; and good communications.

Asymptomatic persons with possible infective exposures must be quickly identified, given antimicrobial prophylaxis, and monitored for the development of fever and cough. Although isolation would not likely be required for asymptomatic contacts receiving prophylactic antimicrobials, these persons should be advised that they might be infected and a potential risk to others in close contact with them, and that it would be prudent to avoid intimate interpersonal contacts with others and to restrict activities that could spread infection, at least for the first 48 hours of prophylactic treatment.

Quarantine is defined as the restriction of activities or limitation of freedom of movement, for a period not longer than the longest incubation period, of well persons exposed to a communicable disease in order to prevent transmission of the disease to others should they become contagious (Last, 1983; World Health Oganization, 1983). The term is usually applied to groups of persons or populations. The legal basis for quarantine rests both with state and federal powers. The draft of a model law for state emergency powers that might be required in the event of a release of a biological weapon, including quarantine, was recently made available by under the Model State Emergency Health Powers Act (Johns Hopkins Center for Civilian Biodefense Strategies, 2003; Mair *et al.*, 2002). This Act addresses the legal basis for states to close buildings, take over hospitals, and order quarantine during a biological attack. Some quarantine measures used in response to an intentional release of *Y. pestis* might include the suspension of public gatherings, closing of public places, restriction of travel (air, rail, motor vehicle, pedestrian), and/or *cordon sanitaire* (literally a sanitary cord or line around a quarantined area guarded to prevent spread of disease by restricting passage in and out of an area) (World Health Organization, 1983). There should be a high threshold for instituting restrictions on travel and imposition of a *cordon sanitaire*, since these are seen by most civilians as a potential infringement on basic rights and may conflict with an individual's actions to protect self or family. These restrictions could also lead to irrational fear and civil disobedience.

7. PREVENTION

7.1. Prevention and Control of Naturally Occurring Plague

7.1.1. General Guidelines

The WHO recommends a four-phased system of plague prevention and control. The first two phases address emergency measures to be implemented whenever a human plague case occurs; Phases 3 and 4 outline the establishment of a surveillance system and development of long-term prevention and control measures (Gage, 1999; World Health Organization, 1980, 1983). The four phases are:

- Case recognition and medical intervention
- Epidemiological/epizootiological investigation and emergency control
- Predictive surveillance and preventive control
- Management

In endemic areas, public health services should provide a continuing system of human and animal plague surveillance, epidemiologic investigations, and control actions. The principal environmental remediation measures during outbreaks of human plague or dangerous epizootics are insecticidal flea control, rodent control, and sanitation to remove food and harborage for rodents. If killing of rodents is considered, flea control should be carried out before or simultaneously with the killing to reduce the chances that infective fleas will feed on humans (Gage, 1998; Gratz, 1999a).

In the event of an outbreak of human plague, measures should be taken to rapidly control spread, as described in international regulations and plague control manuals (Gage, 1999; Gratz, 1999a; World Health Organization, 1983). These measures include:

- Establishing the source
- Defining the geographic limits of activity
- Instituting active surveillance
- Laboratory confirmation of cases and isolation of pneumonic cases
- Rapid treatment of cases and close contacts of infectious pneumonic plague cases
- Control of fleas and rodents in plague-infected areas, with attention to port facilities, ships, and other conveyances
- In the event of a pneumonic plague outbreak, measures may be instituted to screen travelers departing from the epidemic area and to quickly identify and isolate suspect cases in travelers arriving from an outbreak area (Fritz *et al.*, 1996).

7.2. Plague Vaccine

A killed, whole-cell plague vaccine has limited availability and utility. Although no longer manufactured in the US, a comparable killed vaccine is manufactured by the Commonwealth Serum Laboratories (CSL Ltd., 45 Poplar Road, Parkville, 3052, Australia). The vaccine is given subcutaneously at a recommended initial course for adults of two 0.5-mL doses at an interval of 1 to 4 weeks, followed by 6-monthly booster doses (Titball *et al.*, 2004). Recommendations for use of the killed vaccine have been limited to certain groups at high risk, including research laboratory workers handling virulent *Y. pestis* strains, biologists working with susceptible animal populations in plague enzootic areas, and some military personnel. The efficacy of killed plague vaccines has never been evaluated in controlled clinical trials, and evidence for protection has been based on animal experiments, immunogenicity studies in humans, and observations on its use in US servicemen during the Vietnam conflict. Killed vaccines are thought to be protective against flea-borne exposures but to be only partially protective, if at all, against respiratory exposures (Centers for Disease Control and Prevention, 1996; Titball *et al.*, 2004).

Vaccines made of live attenuated strains have been used extensively in the past by the former Soviet Union, India, and former French colonies. These vaccines require only a single immunizing dose and provide early (albeit incomplete) protection against either flea-borne or respiratory routes of transmission. Revaccination is required after a year to maintain protection. Their use is accompanied by a high incidence of adverse reactions, especially fever, aching pain, and malaise in the first several days after administration. Reactions may sometimes be severe. Live vaccines are not commercially available, and the only country currently using a live plague vaccine (EV strain of *Y. pestis*) is Russia, where it is applied as an annual vaccination to persons at high risk (K. Alibek, personal communication).

Research is underway to develop improved plague vaccines that are likely to be protective against airborne routes of exposure (Titball *et al.*, 2004; Williamson, 2001). At present, the most promising candidates are recombinant subunit vaccines that express both the F1 and V antigens of *Y. pestis* (Titball and Williamson, 2003). These recombinant vaccines have been prepared both as combination and fusion products, and appear to protect animals against infective aerosol exposures. Experiments are being conducted to develop vaccine formulations that can be delivered as an inhaled aerosol (Eyles *et al.*, 2000). Interest in developing effective plague vaccines has increased greatly in recent years because of concerns with protection against biological weapons.

8. RESEARCH DIRECTIONS

Preparing for the emergency of a release of plague places a high imperative on new and improved technologies for early detection of the agent in the environment, such as B-cell lines engineered to express bioluminescent protein detectors (Rider *et al.*, 2003) and other technologies to improve performance of biological sniffing devices; new and improved rapid diagnostic markers of infection, such as sensitive, specific, and easy-to-use hand-held assays (Chanteau *et al.*, 2003); rapid characterization of the organism and its virulence and pathogenic potential through better understanding of the *Y. pestis* genome (Parkhill *et al.*, 2001) and resultant proteomic advances; improved and standardized antimicrobial treatments of infection, using agents that are effective, simple to use, and that circumvent engineered resistance, such as bacteriocidal/permeability increasing proteins (Beamer, 2002); the development of materials that can combat shock and other serious consequences of endotoxemia, such as recombinant-activated protein C (Dellinger, 2003; Wheeler and Gordon, 1999); and the manufacture of recombinant inhalant vaccines (Eyles *et al.*, 2000), DNA vaccines (McDonnell and Askari, 1996), and postexposure immunoprotective agents, such as monoclonal antibodies directed against *Y. pestis* and its major virulence factors (Casadevall, 2002).

References

Achtman, M., Zurth, K., Morelli, G., Torrea, G., Guiyole, A., and Carniel, E. (1999). *Yersinia pestis*, the cause of plague, is a recently emerged clone of *Yersinia pseudotuberculosis*. *Proc. Natl. Acad. Sci. U.S.A.* 24:14043–14048.

Alibek, K. (1999). *Biohazard*. Random House, Inc., New York.

Alsofrom, D.J., Mettler, F.A., Jr., and Mann, J.M. (1981). Radiographic manifestations of plague in New Mexico, 1975–1980: a review of 42 proved cases. *Radiology* 139:561–565.

American Society of Microbiology, Biological Weapons Resources Center. (2003). http://www.asmusa.org.

Annas, G.J. (2002). Bioterrorism, public health, and civil liberties. *N. Engl. J. Med.* 346:1337–1342.

Anonymous. (1996). Anti-Terrorism and Effective Death Penalty Act of 1996, Pub. L No. 104–132, April 24, 1996.

Anonymous. (1997). Code of Federal Regulations: Additional requirements for facilities transferring or receiving select agents, Title 42, vol. 1, Part 72, section 72.6. US Government Printing Office, Denver, CO.

Anonymous. (1998). Plague. In: Simpson, J.A., and Weiner, E.S.C. (eds.) *The Oxford English Dictionary*, vol. XI, 2nd ed. Oxford University Press, Oxford, p. 948.

Anonymous. (2000). Plague (pestis). In: Chin, J. (ed.), *Control of Communicable Diseases Manual*, 17th ed. American Public Health Association, Washington, D.C., pp. 381–387.

Ashford, D.A., Kaiser, R.M., Bales, M.E., Shutt, K., Patrawalla, A., McShan, A., Tappero, J.W., Perkins, B.A., and Dannenberg, A.L. (2003). Planning against biological terrorism: lessons from outbreak investigations. *Emerg. Infect. Dis.* 9:515–519.

Beamer, L. (2002). Human BPI: one protein's journey from laboratory into clinical trials. *ASM News.* 11:543–548.

Becker, T.M., Poland, J.D., and Quan, T.J. (1987). Plague meningitis: a retrospective analysis of cases reported in the United States, 1970–1979. *West. J. Med.* 147:554–557.

Bellamy R.J., and Freedman, A.R. (2001). Bioterrorism. *Q. J. Med.* 94:227–234.

Block, R. (2003). FEMA points to flaws, flubs in terror drill. *Wall Street Journal,* Friday, October 31, 2003, p B1.

Boisier, P., Rahalison, L., Rasolomaharo, M., Ratsitorahina, M., Mahafaly, M., Razafimahefa, M., Duplantier, J.M., Ratsifasoamanana, L., and Chanteau, S. (2002). Epidemiologic features of four successive annual outbreaks of bubonic plague in Mahajanga, Madagascar. *Emerg. Infect. Dis.* 8:311–316.

Boulanger, L., Ettestad, P., Fogarty, J., Dennis, D.T., Romig, D., and Mertz, G. (2004). Gentamicin and tetracyclines for the treatment of human plague: a review of 75 cases on New Mexico from 1985–1999. *Clin. Infect. Dis.* 38:663–669.

Burkle, F.M., Jr. (1973). Plague as seen in South Vietnamese children. *Clin Pediatr.* 12:291–298.

Butler, T. (1972). A clinical study of bubonic plague: observations on the 1970 Vietnam epidemic with emphasis on coagulation studies, skin histology and electrocardiograms. *Am. J. Med.* 53:268–276.

Butler, T. (1983). *Plague and Other Yersinia Infections.* Plenum Press, New York.

Butler, T. (1994). *Yersinia* infections: centennial of the discovery of the plague bacillus. *Clin. Infect. Dis.* 19:655–663.

Butler, T., and Dennis, D.T. (2004). *Yersinia* infections, including plague. In: Mandell, G.L., Bennett, J.E., and Dolin, R. (eds.), *Principles and Practice of Infectious Diseases,* 6th ed. Churchill Livingstone, New York.

Butler, T., and Hudson, B.W. (1977). The serological response to *Yersinia pestis* infection. *Bull. World Health Org.* 55:39–42.

Butler, T., Bell, W.R., Nguyen, N.L., Nguyen, D.T., and Arnold, K. (1974). *Yersinia pestis* infection in Vietnam. I. Clinical and hematological aspects. *J. Infect. Dis.* 129(Suppl):S78–S84.

Butler, T., Levin, J., Nguyen, N.L., Duong, M.C., Adickman, M., and Arnold, K. (1976). *Yersinia pestis* infection in Vietnam. II. Quantitative blood cultures and detection of endotoxin in the cerebrospinal fluid of patients with meningitis. *J. Infect. Dis.* 133:493–499.

Byrne, W.R., Welkos, S.L., Pitt, M.L., Davis, K.J., Brueckner, R.P., Ezell, J.W., Nelson, G.O., Vaccaro, J.R., Battersby, L.C., and Friedlander, A.M. (1998). Antibiotic treatment of experimental pneumonic plague in mice. *Antimicrob. Agents Chemother.* 42:675–681.

Campbell, G.L., and Hughes, J.M. (1995). Plague in India: a new warning from an old disease. *Ann. Intern. Med.* 122:151–153.

Carniel, E. (2003). Evolution of pathogenic *Yersinia*, some lights in the dark. *Adv. Exp. Med. Biol.* 529:3–12.

Casadevall, A. (2002). Passive antibody administration (immediate immunity) as a specific defense against biological weapons. *Emerg. Infect. Dis.* 8:833–841.

Centers for Disease Control. (1984). Plague pneumonia—California. *MMWR Morb. Mortal. Wkly. Rep.* 33:481–483.

Centers for Disease Control and Prevention. (1996). Prevention of plague. Recommendations of the Advisory Committee on Immunization Practices (ACIP). *MMWR Morb. Mortal. Wkly. Rep.* 45(RR-14):1–15.

Centers for Disease Control and Prevention. (1997). Fatal human plague—Arizona and Colorado, 1996. *MMWR Morb. Mortal. Wkly. Rep.* 46:617–620.

Centers for Disease Control and Prevention. (2000). Biological and chemical terrorism: strategic plan for preparedness and response. Recommendations of the CDC Strategic Planning Workgroup. *MMWR Morb. Mortal. Wkly. Rep.* 49 (RR-4):1–14.

Centers for Disease Control and Prevention. (2002). Laboratory security and emergency response guidance for laboratories working with select agents. *MMWR Morb. Mortal. Wkly. Rep.* 51 (RR-19): 1–6.

Centers for Disease Control and Prevention, Bioterrorism Program. (2003). http://www.bt.cdc.gov.

Centers for Disease Control and Prevention, Bioweapons Laboratory Issues. (2001). http://www.bt.cdc.gov/Agent/Plague/ype_la_cp/123010.pdf.

Centers for Disease Control and Prevention, Hospital Infections Program. (1997). http://www.cdc.gov/ncidod/hip/ISOLAT/isopart2.htm.

Centers for Disease Control and Prevention, Hospital Infections Program. (2003). http://www.cdc.gov/ncidod/hip/enviro/guide.htm.

Chanteau, S., Rahalison, L., Ralafiarisoa, L., Foulon, J., Ratsitorahina, M., Ratsifasoamanana, L., Carniel, E., and Nato, F. (2003). Development and testing of a rapid diagnostic test for bubonic and pneumonic plague. *Lancet* 361:211–216.

Chanteau, S., Rahalison, L., Ratsitorahina, M., Mahafaly, Rasolomahoro, M., Boisier, P., O'Brien, T., Aldrich, J., Keleher, A., Morgan, C., and Burans, J. (2000). Early diagnosis of bubonic plague using F1 antigen capture ELISA assay and rapid immunogold dipstick. *Int. J. Med. Microbiol.* 290:279–283.

Chanteau, S., Ratsifasoamanana, L., Rasoamanana, B., Rahalison, L., Randriambelosoa, J., Roux, J., and Rabeson, D. (1998). Plague, a reemerging disease in Madagascar. *Emerg. Infect. Dis.* 4:101–104.

Christie, A.B., Chen, T.C., and Elberg, S.S. (1980). Plague in camels and goats: their role in human epidemics. *J. Infect. Dis.* 141:724–726.

Craven, R.B., Maupin, G.O., Beard, M.L., Quan, T.J., and Barnes, A.M. (1993). Reported cases of human plague infections in the United States, 1970–1991. *J. Med. Entomol.* 30:758–761.

Crook, L.D., and Tempest, B. (1992). Plague—a clinical review of 27 cases. *Arch. Intern. Med.* 152:1253–1256.

Cunha, B.A. (2003). Doxycyline for community-acquired pneumonia. *Clin. Infect. Dis.* 37:870.

Dellinger, R.P. (2003). Inflammation and coagulation: implications for the septic patient. *Clin. Infect. Rev.* 36:1259–1265.

Dennis, D.T. and Chow, C.C. (2004). Plague. *Pediatr. Infect. Dis.* 23:69–71.

Dennis, D., and Meier, F. (1997). Plague. In: Horsburgh, C.R., and Nelson, A.M. (eds.), *Pathology of Emerging Infections*. ASM Press, Washington, D.C., pp. 21–47.

Dennis, D.T. (1994). Plague in India. *Br. Med. J.* 309:893–894.

Dennis, D.T. (1998). Plague as an emerging disease. In: Scheld, W.M., Craig, W.A., and Hughes, J.M. (eds.), *Emerging Infections,* vol. 2. ASM Press, Washington, D.C., pp. 169–183.

Dennis, D.T. (2001). Plague. In: Rakel, R.E. and Bope, E.T. (eds.), *Conn's Current Therapy, 2001*. W.B. Saunders, Philadelphia, pp. 115–117.

Dennis, D.T., and Hughes, J.M. (1997). Multidrug resistance in plague. *N. Engl. J. Med.* 10:702–704.

Dennis, D.T., and Campbell, G.L. (2004). Plague and other Yersinia infections. In: Kasper D.L. (ed.), *Harrison's Principles of Internal Medicine*, 16th ed. McGraw-Hill, New York, pp. 921–929.

Dennis, D.T., and Gage, K.L. (2004). Plague. In: Cohen, J., and Powderly, W.G. (eds.), *Infectious Diseases,* 2nd ed. Mosby, London, pp. 1641–1648.

Derbes, V.J. (1996). De Mussis and the great plague of 1348: a forgotten episode of bacteriological war. *J.A.M.A.* 196:59–62.

Doll, J.M., Zeitz, P.S., Ettestad, P., Bucholz, A.L., Davis, T, and Gage, K. 1994. Cat-transmitted fatal pneumonic plague in a person who traveled from Colorado to Arizona. *Am. J. Trop. Med. Hyg.* 51:109–114.

Eyles, J.E., Williamson, E.D., Spiers, I.D., Stagg, A.J., Jones, S.M., and Alpar, H.O. (2000). Generation of protective immune responses to plague by mucosal administration of microspore co-encapsulated recombinant sub-units. *J. Contr. Rel.* 63:191–200.

Ferguson, J.R. (1997). Biological weapons and the law. *J.A.M.A.* 278:357–360.

Fidler, D.P. (2001). The malevolent use of microbes and the rule of the law: legal challenges presented by bioterrorism. *Clin. Infect. Dis.* 33:686–689.

Finegold, M.J. (1968). Pathogenesis of plague deaths in the United States during the last decade. *Am. J. Med.* 45:549–553.

Frean, J., Klugman, K.P., Arntzen, L., and Bukofzer, S. (2003). Susceptibility of *Yersinia pestis* to novel and conventional antimicrobial agents. *J. Antimicrob. Chemother.* 52:294–296.

Frean, J.A., Arntzen, L., Capper, T., Brysier, A., and Klugman, K.P. (1996). In vitro activities of 14 antibiotics against 100 human isolates of *Yersinia pestis* from a southern African plague focus. *Antimicrob. Agents Chemother.* 40:2646–2647.

Fritz, C.L., Dennis, D.T., Tipple, M.A., Campbell, G.L., McCance, C.R., and Gubler, D.J. (1996). Surveillance for pneumonic plague in the United States during an international emergency: a model for control of imported emerging diseases. *Emerg. Infect. Dis.* 2:30–36.

Gabastou, J.M., Proano, J., Vimos, A., Jaramillo, G., Hayes, E., Gage, K., Chu, M., Guarner, J., Zaki, S, Bowers, J., Guillemard, C., Tamayo, H., and Ruiz, A. (2000). An outbreak of plague including cases with probable pneumonic infection, Ecuador (1998). *Trans. Roy. Soc. Trop. Med. Hyg.* 94:387–391.

Gage, K.L. (1998). Plague. In: Collier, L., Balows, A., Sussman, M., and Hausler, W.J. (eds.), *Topley and Wilson's Microbiology and Microbial Infections*, vol. 3, 9th ed. Arnold Publications, London, pp. 885–903.

Gage, K.L. (1999). National health services in prevention and control. In: *Plague Manual: Epidemiology, Distribution, Surveillance and Control.* World Health Organization, Geneva, pp. 167–171.

Gage, K.L., Dennis, D.T., Orloski, K.A., Ettestad, P., Brown, T.L., Reynolds, P.J., Pape, W.J, Fritz, C.L., Carter, L.G., and Stein, J.D. (2000). Cases of cat-associated plague in the western U.S., 1977–1998. *Clin. Infect. Dis.* 30:893–900.

Galimand, M., Guiyoule, A., Gerbaud, G., Rasoamanana, B., Chanteau, S., Carniel, E., and Courvalin, P. (1997). Multidrug resistance in *Yersinia pestis* mediated by a transferable plasmid. *N. Engl. J. Med.* 10:677–680.

Garner, J.S. (1996). Guidelines for isolation precautions in hospitals: Hospital Infection Control Practices Advisory Committee. *Infect. Control Hosp. Epidemiol.* 17:53–80.

Gasquet, F.A. (1908). *The Black Death.* George Bell and Sons, London, pp. 1–17.

Glass, T.A., and Schoch-Spana, M. (2002). Bioterrorism and the people: how to vaccinate a city against panic. *Clin. Infect. Dis.* 34:217–223.

Gratz, N.G. (1999a). Control of plague transmission. In: *Plague Manual: Epidemiology, Distribution, Surveillance and Control.* World Health Organization, Geneva, pp. 97–134.

Gratz, N.G. (1999b). Rodent reservoirs and flea vectors of natural foci of plague. In: *Plague Manual: Epidemiology, Distribution, Surveillance and Control.* World Health Organization, Geneva, pp. 63–96.

Grow, R.W., and Rubinson, L. (2003). The challenge of hospital infection control during a response to bioterrorist attacks. *Biosecurity and Bioterrorism: Bodefense Strategy, Practice, and Science* 1:215–220.

Guarner, J., Shieh, W.J., Greer, P.W., Gabastau, J.M., Chu, M.C., Hayes, E., Nolte, K.B., and Zaki, S.R. (2002). Immunohistochemical detection of *Yersinia pestis* in formalin-fixed paraffin-embedded tissue. *Am. J. Clin. Pathol.* 177:20–209.

Guiyoule, A., Grimont, F., Iteman, I., Grimont, P.A., Lefevre, M., and Carniel, E. (1994). Plague pandemics investigated by ribotyping of *Yersinia pestis* strains. *J. Clin. Microbiol.* 32:634–641.

Guiyoule, A., Rasoamanana, B., Buchrieser, C., Miche, P., Chanteau, S., and Carniel, E. (1997). Recent emergence of new variants of *Yersinia pestis* in Madagascar. *J. Clin. Microbiol.* 35:2826–2833.

Harris, S. (1992). Japanese biological research on humans: a case study of microbiology and ethics. *Ann. N. Y. Acad. Sci.* 555:21–52.

Henchal, E.A., Teska, J.D., Ludwig, G.V., Shoemaker, D.R., and Ezzell, J.W. (2001). Current laboratory methods for biological threat agent identification. *Clin. Lab. Med.* 21:661–678.

Hinnebusch, B.J. (1997). Bubonic plague: a molecular genetic case history of the emergence of an infectious disease. *J. Mol. Med.* 75:645–652.

Hinnebusch, B.J., Rudolph, A.E., Cherepenov, P., Dixon, J.E., Schwan, T.G., and Forsberg, A. (2002). Role of murine toxin in survival of *Yersinia pestis* in the midgut of the vector flea. *Science* 296:733–735.

Hoffman, R.E. (2003). Preparing for a bioterrorist attack: legal and administrative strategies. *Emerg. Infect. Dis.* 9:241–245.

Hoffman, R.E., and Norton, J.E. (2000). Lessons learned from a full-scale bioterrorism exercise. *Emerg. Infect. Dis.* 6:652–653.
Hull, H.F., Montes, J.M., and Mann, J.M. (1986). Plague masquerading as gastrointestinal illness. *West J. Med.* 145:485–487.
Hull, H.F., Montes, J.M., and Mann, J.M. (1987). Septicemic plague in New Mexico. *J. Infect. Dis.* 155:113–118.
Inglesby, T.V., Dennis, D.T., Henderson, D.A., Bartlett, J.G., Ascher, M.S., Eitzen, E. Fine, A.D., Friedlander, A.M., Hauer, J., Koerner, J.F., Layton, M., McDade, J., Osterholm, M.T., O'Toole, T., Parker, G., Perl, T. M., Russell, P.K., Schoch-Spana, M., and Tonat, K. (2000). Plague as a biological weapon: medical and public health management. *J.A.M.A.* 283:2281–2290.
Inglesby, T.V., Grossman, R., and O'Toole, T. (2001). A plague on your city: observations from TOPOFF. *Clin. Infect. Dis.* 32:436–445.
Johns Hopkins Center for Civilian Biodefense Strategies. (2003). http://www.hopkins-biodefense.org.htm.
Kahn, J. (2002) Shouting the pain from Japan's germ attacks. *New York Times*, November 23, 2002, p. A3.
Khan, A.S., and Ashford, D.A. (2001). Ready or not—preparedness for bioterrorism. *N. Engl. J. Med.* 345:287–289.
Khan, A.S., Morse, S., and Lillibridge, S. (2000). Public health preparedness for biological terrorism in the USA. *Lancet* 356:1179–1182.
Koornhof, H.J., Smego, R.A., Jr., and Nicol, M. (1999). Yersiniosis. II. The pathogenesis of *Yersinia* infections. *Eur. J. Clin. Microbiol. Infect. Dis.* 18:87–112.
Kuberski, T., Robinson, L., and Schurgin, A. (2003). A case of plague successfully treated with ciprofloxacin and sympathetic blockade for treatment of gangrene. *Clin. Infect. Dis.* 36:521–523.
Last, J.M. (1983). *A Dictionary of Epidemiology*. Oxford University Press, Oxford.
Laforce, F.M., Acharya, I.L., Stott, G., Brachman, P.S., Kaufman, A.F., Clapp, R.F., & Shah, N.K. (1971). Clinical and epidemiological observations on an outbreak of plague in Nepal. *Bull. Wld. Hlth. Org.* 45:693–706.
Legters, L.J., Cottingham, A.J., Jr., and Hunter, D.H. (1970). Clinical and epidemiological notes on a defined outbreak of plague in Vietnam. *Am. J. Trop. Med. Hyg.* 19:639–52.
Link, V.B. (1955). *A history of plague in the United States of America*. Public Health Monograph No. 26, Government Printing Office, Washington, D.C., 1955.
Loiez, C., Herwegh, S., Wallet, F., Armand, S., Guinet, F, and Courcol, R.J. (2003). Detection of *Yersinia pestis* in sputum by real-time PCR. *J. Clin. Microbiol.* 41:4873–4875.
Mair, J.S., Sapsin, J., and Teret, S. (2002). The Model State Emergency Health Powers Act and beyond. *Biodefense Q.* 3:1–2, 11.
Mann, J.M., Hull, H.F., Schmid, G.P., and Droke, W.E. (1984). Plague and the peripheral blood smear. *J.A.M.A.* 251:953.
Mann, J.M., and Moskowitz, R. (1977). Plague and pregnancy: a case report. *J.A.M.A.* 237:1854–1855.
Mann, J.M., Shandler, L., and Cushing, A.H. (1982). Pediatric plague. *Pediatrics* 69:762–767.
Marshall, J.D., Jr., Quy, D.V., and Gibson, F.L. (1967). Asymptomatic pharyngeal plague infection in Vietnam. *Am. J. Trop. Med. Hyg.* 16:175–177.
Martin, G.J., and Marty, A.M. (2001). Clinicopathologic aspects of bacterial agents. *Clin. Lab. Med.* 21:513–548.
Marty, A.M. (2001). History of the development and use of biological weapons. *Clin. Lab. Med.* 21:421–434.
Marty, A.M., Conran, R.M., and Kortepeter, M.G. (2001). Recent challenges in infectious diseases: biological pathogens as weapons and emerging endemic disease threats. *Clin. Lab. Med.* 21:411–420.

McDonnell, W.M., and Askari, F.K. (1996). Molecular medicine: DNA vaccines. *N. Engl. J. Med.* 334:42–45.

Meyer, K.F. (1961). Pneumonic plague. *Bacteriol. Rev.* 25:249–261.

Miller, J.M. (2001). Agents of bioterrorism: preparing for bioterrorism at the community health care level. *Infect. Dis. Clin. N. Am.* 15:1127–1155.

Morse, S.A., Kellogg, R.B., Perry, S., Meyer, R.F., Bray, D., Nichelson, D., and Miller, J.M. (2003). Detecting biothreat agents: the laboratory response network. *ASM News* 69:433–437.

O'Toole, T., and Inglesby, T.V. (2001). Epidemic response scenario: decision making in a time of plague. *Pub. Health. Rep.* 116(Suppl 2):92–103.

Osterholm, M.T., and Schwartz, J. (2000). *Living Terrors: What America Needs to Know to Survive the Coming Bioterrorist Catastrophe.* Delacorte Press, New York.

Parkhill, J., Wren, B.W., Thompson, N.R., Titball, R.W., Holden, M.T.G., Prentice, M.B., Sebhalia, M., James, K.D., Churcher, C., Mungall, K.L., Baker, S., Basham, D., Bentley, S.D., Brooks, K., Cerdeno-Tarraga, A.M., Chillingworth, T., Cronin, A., Davies, R.M., Davis, P., Dougan, G., Feltwell, T., Hamlin, N., Holroyd, S., Jagels, K., Karlyeshev, A.V., Leather, S., Moule, S., Oyston, P.C.F., Quail, M., Rutherford, K., Simmonds, M., Skelton, J., Stevens, K., Whitehead, S., and Barrell, B.G. (2001). Genome sequence of *Yersinia pestis*, the causative agent of plague. *Nature* 413:523–527.

Perry, R.D. (2003). A plague of fleas—survival and transmission of *Yersinia pestis*. *ASM News* 69:336–340.

Perry, R.D., and Fetherston, J.D. (1997). *Yersinia pestis*—etiologic agent of plague. *Clin. Microbiol. Rev.* 10:35–66.

Poland, J.D. (1989). Plague. In: Hoeprich, P.D., and Jordan, M.C. (eds.), *Infectious Diseases: A Modern Treatise of Infectious Processes.* J.B. Lippincott Co., Philadelphia, pp. 1296–1306.

Poland, J.D., and Dennis, D.T. (1999). Treatment of plague. In: *Plague Manual: Epidemiology, Distribution, Surveillance and Control.* World Health Organization, Geneva, pp. 55–62.

Pollitzer, R. (1954). *Plague.* World Health Organization, Geneva.

Radnedge, L., Agron, P.G., Worsham, P.L., and Andersen, G.L. (2002) Genome plasticity in *Yersinia pestis. Microbiology* 148:1687–1698.

Radnedge, L., Gamez-Chin, S., McCready, P.M., Worsham, P.L., and Andersen, G.L. (2001). Identification of nucleotide sequences for the specific and rapid detection of *Yersinia pestis. Appl. Environ. Microbiol.* 67:3759–3762.

Rahalison, L., Vololonirina, E., Ratsitorahina, M., and Chanteau, S. (2000). Diagnosis of bubonic plague by PCR in Madagascar under field conditions. *J. Clin. Microbiol.* 38:260–263.

Ramalingaswami, V. (2001). Psychosocial effects of the 1994 plague outbreak in Surat, India. *Military Med.* 166(Suppl 2):29–30.

Rasoamanana, B., Coulanges, P., Michel, P., and Raolofonirina, N. (1989). Sensitivity of *Yersinia pestis* to antibiotics: 277 strains isolated in Madagascar between 1926 and 1989. *Arch. Inst. Pasteur Madagascar* 56:37–53.

Ratsitorahina, M., Chanteau, S., Rahalison, L., Ratsifasoamana, L., and Boisier, P. (2000a). Epidemiological and diagnostic aspects of the outbreak of pneumonic plague in Madagascar. *Lancet* 355:111–113.

Ratsitorahina, M., Rabarijaona, L., Chanteau, S., and Boisier, P. (2000b). Seroepidemiology of human plague in the Madagascar highlands. *Trop. Med. Int. Health* 5:94–98.

Rider, T.H., Petrovick, M.S., Nargi, F.E., Harper, J.D., Schwoebel, E.D., Mathews, R.H. Blanchard, D.J., Bortolin, L.T., Young, A.M., Chen, J., and Hollis, M.A. (2003). A B cell-based sensor for rapid identification of pathogens. *Science* 301:213–215.

Romanov, V.E., Evstigneev, V.I., Vasil'ev, N.T., Shabalin B.A., and Paramanov, V.E. (2001a). Evaluation of the effectiveness of antibacterial substances in treating an experimental form of bubonic plague in monkeys. *Antibiot. Khimoter.* 466–468 [abstract].

Romanov, V.E. Vasil'ev, N.T., Shabalin, B.A., and Mironin, A.V. (2001b). Effect of antibacterial therapy on the epidemic threat of experimental pneumonic plague in monkeys. *Antibiot. Khimoter.* 46:16–18 [abstract].

Rotz, L.D., Khan, A.S., Lillibridge, S.R., Ostroff, S.M., and Hughes, J.M. (2002). Public health assessment of potential biological terrorism agents. *Emerg. Infect. Dis.* 8:225–230.

Russell, P., Eley, S.M., Bell, D.L. Manchee, R.J., and Titball, R.W. (1996). Doxycycline or ciprofloxacin prophylaxis and therapy against experimental *Y. pestis* infection in mice. *Antimicrob. Agents Chemother.* 37:769–774.

Smego, R.A., Frean, J., and Koornhof, H.J. (1999). Yersiniosis. I. Microbiological and clinicoepidemiological aspects of plague and non-plague *Yersinia* infections. *Eur. J. Clin. Microbiol. Infect. Dis.* 18:1–15.

Smith, M.D., Vinh, D.X., Nguyen, T.T., Wain, J., Thung, D., and White, N.J. (1995). In vitro antimicrobial susceptibilities of strains of *Yersinia pestis*. *Antimicrob. Agents Chemother.* 39:2153–2154.

Tieh, T.H., Lindauer, E., Miyagawa, F., Kobayashi, G., and Okayasu, G. (1948). Primary pneumonic plague in Mukden, 1946, and report of 39 cases with 3 recoveries. *J. Infect. Dis.* 82:52–58.

Titball, R.W., and Williamson, E.D. (2003). Second and third generation plague vaccines. *Adv. Exp. Med. Biol.* 529:397–406.

Titball, R.W., Williamson, E.D., and Dennis, D.T. (2004) Plague. In: Plotkin, S.A., and Orenstein, W.A. (eds.), *Vaccines,* 4th ed. Saunders, Philadelphia, pp. 99–1010.

U.S. Department of Health and Human Services, Centers for Disease Control and Prevention. (1996). Additional requirements for facilities transferring or receiving select agents: final rule. *Fed. Reg.* 61:55190.

Welty, T.K., Grabman, J., Kompare, E., Wood, G., Welty, E., Van Duzen, J., Rudd, P., and Poland, J. (1985). Nineteen cases of plague in Arizona: a spectrum including ecthyma gangrenosum due to plague and plague in pregnancy. *West. J. Med.* 142:641–646.

Wheeler, A.P., and Gordon, R.B. (1999). Treating patients with severe sepsis. *N. Engl. J. Med.* 340:207–214.

Williamson, E.D. (2001). Plague vaccine research and development. *J. Appl. Microbiol.* 91:606–608.

Wilmoth, B.A., Chu, M.C., and Quan, T.C. (1996). Identification of *Yersinia pestis* by BBL crystal enteric/nonfermenter identification system. *J. Clin. Microbiol.* 34:2829–2830.

Wong, J.D., Barash, J.R., Sandfort, R.F., and Janda, J.M. (2000). Susceptibilities of *Yersinia pestis* strains to 12 antimicrobial agents. *Antimicrob. Agents Chemother.* 44:1995–1996.

World Health Organization. (1970). *Health Aspects of Chemical and Biological Weapons.* World Health Organization, Geneva, pp. 98–109.

World Health Organization. (1980). Plague surveillance and control. *WHO Chronicle* 34:139–143.

World Health Organization. (1983). *International Health Regulations (1969).* World Health Organization, Geneva.

World Health Organization. (2003). Human plague in 2000 and 2001. *Wkly. Epidemiol. Rec.* 16:130–135.

Wu, L.-T., Chun, J.W.H., and Pollitzer, R. (1922). Clinical observations upon the second Manchurian plague epidemic, 1920–1921. *Natl. Med. J. China.* 8:225–249.

Wu, L.-T. (1926). *A Treatise on Pneumonic Plague.* League of Nations Health Organization, Geneva.

Wynne-Griffith, G. (1948). Pneumonic plague in Rangoon. *Lancet* 1:625–627.

3

Tularemia and Bioterrorism

Lisa Hodges and Robert L. Penn

1. INTRODUCTION

Francisella tularensis is a Gram-negative, facultative aerobic coccobacillus and the etiologic agent of tularemia. Historically, it has been referred to as "rabbit fever," "deer-fly fever," and "market men's disease" (Penn, 2005). Tularemia in the United States is a seldom-diagnosed zoonosis that causes disease mostly in animals, occasionally infecting people who come into contact with the organism through a natural reservoir, a sick or dead animal, or are bitten by an arthropod vector. Previously occurring as sporadic epidemics, its incidence has decreased substantially since the 1950s (Penn, 2005). Although *F. tularensis* is one of the most infectious pathogens known, natural human cases of tularemia are uncommon and do not appear to confer a selective advantage in the larger context of the pathogen's evolutionary history.

The Japanese began germ warfare research using *F. tularensis* as early as 1932 and conducted biological warfare research with it on prisoners of war (Dennis *et al.*, 2001; Harris, 1992). Both the former Soviet Union and the United States had weaponized *F. tularensis* by the mid-20th, and both countries experimented with streptomycin-resistant strains (Dennis *et al.*, 2001). In fact, a former Soviet Union biological weapons scientist has indicated that tularemia outbreaks on the Eastern European front during World War II were the result of intentional use (Dennis *et al.*, 2001). He also disclosed that the Soviets continued their biological weapons research into the 1990s, engineering strains of *F. tularensis* to be resistant to vaccines as well as antibiotics. A large number of animal and human studies on tularemia were performed at the U.S. Army Research Institute of Infectious Diseases at Fort Detrick, Maryland. The site contained a 1-million-liter aerosolization chamber where volunteers were exposed to *F. tularensis* and other agents to study the efficacy of prophylactic therapy and vaccines (Christopher *et al.*, 1997). After the 1972 Convention on the Prohibition of the Development, Production, and Stockpiling of Bacteriological and Toxin Weapons, all United

Bioterrorism and Infectious Agents
Edited by Fong and Alibek, Springer Science+Business Media, Inc., New York, 2005

States stockpiles were destroyed (Christopher *et al.*, 1997). Since that time, the United States has continued defensive research in the areas of immunology, genomics, and vaccine research (Franz *et al.*, 1997).

The U.S. Centers for Disease Control and Prevention (CDC) has classified *F. tularensis* as a Category A bioterrorism agent, one most likely to be used as a weapon due to its ease of dissemination, potential to cause disease, probable mortality, impact on public health, and likelihood of creating social disruption (Sewell, 2003). However, *F. tularensis* is less desirable than other organisms as a weapon because it does not have a stable spore phase and is difficult to handle without infecting those processing and dispersing the pathogen (Cunha, 2002). Nonetheless, *F. tularensis* has become a public health issue because of its potential use in biological warfare by organized military or independent terrorist groups. Familiarity with the organism's epidemiology, pathophysiology, and clinical disease will enable health care professionals to better identify unusual patterns of tularemia that may result from its use as a bioweapon.

In the event of a bioterrorist attack, tularemia would likely be released in an aerosolized form in a densely populated urban setting (Dennis *et al.*, 2001). In 1970, the World Health Organization (WHO) estimated that the release of 50 kg of *F. tularensis* by aircraft over a population of 500,000 people would result in 30,000 deaths and 125,000 incapacitated (World Health Organization, 1970). They also noted that vaccinated individuals would only be partially protected from an aerosol exposure (World Health Organization, 1970). In 1997, the CDC estimated that exposing 100,000 persons to an aerosol of *F. tularensis* would result in 82,500 cases of tularemia (82.5% attack rate) and 6,188 deaths (6.2% death rate), and cost between $456 million and $561.8 million (Kaufmann *et al.*, 1997).

2. MICROBIOLOGY

2.1. Taxonomy

F. tularensis are small, aerobic, nutritionally fastidious, pleomorphic, Gram-negative coccobacilli, ranging in size from 0.2×0.2 μm to 0.2×0.7 μm. Pleomorphism is increased after active growth (Eigelsbach and McGann, 1984). A distinguishing feature is its cell wall, of which more than 50% (w/v) is comprised of lipids. *Francisella* is the only genus of the family *Francisellaceae*, belonging to the γ-subclass of proteobacteria (Penn, 2005). It was formerly classified in the genus *Pasteurella* along with *Yersinia pestis*, the etiologic agent of plague. Both are zoonotic infections that cause fever and lymphadenopathy in humans. Man is an accidental host for *F. tularensis*.

The genus *Francisella* contains two species: *F. tularensis*, the animal and human pathogen; and *F. philomiragia*, an opportunistic pathogen associated with water that rarely causes disease in humans (Penn, 2005). *F. tularensis* has four subspecies based on epidemiology, biochemical reactions, and virulence testing; these are *tularensis, holarctica, novicida,* and *mediaasiatica*. However, the only subspecies of *F. tularensis* that are common human pathogens are *tularensis* and *holarctica* (Penn, 2005). Microarray studies on genomic DNA have distinguished five distinct hybridization patterns with strains clustering according to their subspecies (Broekhuijsen *et al.*, 2003). There is a high degree of genetic conservation among the *Francisella*, possibly due to the intracellular nature of the organism (Broekhuijsen *et al.*, 2003). The most virulent subspecies in humans is *F. tularensis* subsp. *tularensis*, also known as type A, and this is the predominant cause of tularemia in North America. *F. tularensis* subspecies *holarctica* (type B) causes a milder form of illness, compared with subspecies *tularensis*. It accounts for almost all of the naturally occurring disease in Europe, Asia, and the former Soviet Union, although it also causes disease in North America. Exposure to as few as 10–50 *F. tularensis* subspecies *tularensis* organisms by either the subcutaneous or aerosol route can cause severe disease in humans, whereas exposure to 12,000 subspecies *holarctica* organisms produces only a mild, self-limited infection (Hornick, 2001).

2.2. Virulence

Relatively little is known about the virulence determinants of *F. tularensis*. Its high degree of infectivity has been demonstrated by the high rates of infection in laboratory personnel who work with the organism (Dennis *et al.*, 2001). Early studies showed that inoculation with as few as 10 organisms either subcutaneously or via aerosol is sufficient to cause tularemia in humans (Saslaw *et al.*, 1961a, 1961b). The bacteria do not possess any toxins or secretion systems typical of *Yersinia, Shigella, Salmonella, Legionella,* or *Brucella* species. Instead, virulence appears to be dependent on its ability to escape host defenses.

The lipid-rich capsule of *F. tularensis* protects it against complement-mediated lysis in the extracellular environment (Sandstrom *et al.*, 1988). It is unknown whether or not the capsule affords any advantage once inside the macrophage. The capsule is neither toxic nor immunogenic (Penn, 2005).

The lipopolysacharide (LPS) has an unusual O-side chain for a Gram-negative organism, and it does not possess the potent endotoxin activity characteristic of other Gram-negative species (Penn, 2005). The length of the LPS O-antigen chain contributes to serum resistance and intracellular replication capacity (Penn, 2005). The live vaccine strain (LVS) of *F. tularensis* has the ability to alter its expression of LPS from a stimulatory chemotype that induces macrophage nitric oxide production that limits its intracellular growth, to the wild type that restores intracellular growth capacity (Cowley *et al.*, 1996). Interleukin-1 (IL-1) and tumor necrosis factor-α (TNF-α) responses to *F. tularensis* LPS are significantly less than those generated in response to *Escherichia coli* LPS.

Methods for gene replacement in *F. tularensis* have been developed and used to attenuate the LVS strain for mice (Golovliov *et al.*, 2003). Toll-like receptors on the surface of mononuclear phagocytes respond to bacterial LPS, initiating a cascade of signaling events that leads to inflammatory cytokine production and antimicrobial activity. A 23-kD protein of *F. tularensis* LVS is required to effectively block toll-like receptor signaling through NF-κB and mitogen-activated protein kinase pathways in murine macrophages, resulting in suppression of IL-1β and TNF-α production (Telepnev *et al.*, 2003). The 23-kD protein, a newly discovered virulence factor, shows no homology with proteins of other intracellular or extracellular pathogens. The ability of *F. tularensis* to inhibit cytokine release from macrophages may contribute to its success as an intracellular pathogen and may explain why few stress proteins are synthesized during macrophage infection (Telepnev *et al.*,

2003). Analyzing specific gene contributions to virulence will be facilitated by the development of additional isogenic mutants of *F. tularensis*.

3. PATHOGENESIS

3.1. Pathophysiology

F. tularensis is capable of surviving and multiplying within macrophages and other cells. It enters cells through a cytochalasin B-insensitive pathway and does not trigger the respiratory burst. *F. tularensis* is well adapted to the intracellular environment, residing in the phagosome where it is capable of acidifying vacuoles, thus creating an environment necessary for iron acquisition (Fortier *et al.*, 1995). After sufficient multiplication has taken place, viable *F. tularensis* can induce cell death by apoptosis (Lai *et al.*, 2001), releasing the bacteria to infect new cells.

The infectious dose in humans depends on the portal of entry: 10–50 bacteria when injected intradermally or inhaled, and 10^8 bacteria when ingested (Penn, 2005). Although most commonly acquired as the result of a biting insect or traumatic inoculation through the skin, *F. tularensis* also may enter through unapparent microabrasions in the skin or intact mucous membranes. After an incubation period of 2–5 days, a small red papule appears at the inoculation site. In approximately 10% of cases, primary lesions do not occur at the portal of entry. The papule enlarges and ulcerates, discharging a necrotic core, leaving behind an indolent "punched-out" ulcer with raised edges and a ragged base. Most ulcers occur on the upper extremities, face, and eye, but can also be found in the nares and pharyngeal cavity. In the preantibiotic era, primary lesions took longer than a month to heal (Foshay, 1940).

F. tularensis invades lymphatic vessels and may cause a regional lymphadenitis and possible bubo formation. If the bacteria invade superficial lymphatics, nodular lymphangitis similar in appearance to sporotrichosis may develop. Affected lymph nodes demonstrate follicular hyperplasia with conglomerates of macrophages, epitheliod-cell granulomas, and caseous necrosis, findings compatible with tuberculosis and sarcoidosis (Tarnvik, 1989). The specific microbial properties responsible for inducing this inflammatory response remain largely unknown.

A transient initial bacteremia seeds the reticuloendothelial system and other organs, commonly resulting in the formation of scattered foci of necrosis in the liver, spleen, lungs, and distant lymph nodes. Within days of dissemination, Th1-related cytokines are produced in infected tissues (Ellis *et al.*, 2002). In the lung, areas of focal necrosis can progress to bronchopneumonia or become discrete nodules that may caseate. Lesions in the liver and spleen evolve in a similar fashion. Healing of the granulomatous lesions occurs with scar formation. In some instances, heavy pulmonary dispersion creates diffuse miliary foci mimicking those of miliary tuberculosis (Foshay, 1940).

In the preantibiotic era, patients that did not survive usually had a protracted illness that terminated in a fatal septicemia. This bacteremia is initiated from one of the established primary foci of necrosis that ulcerates into the wall of a blood vessel or by lymphatic access to the bloodstream. Autopsies of fatal human cases reveal multiple necrotic foci and granuloma formation in the bone marrow, serosae, kidneys, adrenal glands, intestines, meninges, and brain in addition to the liver, spleen, lung, and lymph nodes (Foshay, 1940; Tarnvik, 1989).

Findings in the lung vary over time. Monkeys who inhale a virulent strain of *F. tularensis* initially develop an acute bronchiolitis. Within 72 hours, inflammation is present in the peribronchial tissues and alveolar septa. Bronchopneumonia becomes most pronounced in those exposed to the smallest particles. Areas of inflammation are spread throughout the lungs and are accompanied by tracheobronchial lymphadenopathy; this progresses to consolidation, granuloma formation, and eventually chronic interstitial fibrosis (White *et al.*, 1964). In human lungs, biopsy and autopsy specimens reveal an intense inflammatory edema that accompanies regions of necrotizing bronchiolitis and bronchopneumonia. Necrotic exudates are often found along the bronchial tree and along interlobular septa. The involvement of vessels causes narrowing of their lumina, often resulting in complete occlusion with thrombosis (Permar and Maclahlan, 1931; Shapiro and Mark, 2000).

Mice exposed to a high dose aerosol of *F. tularensis* developed pulmonary infection, followed by dissemination to the liver and spleen, similar to that seen in humans with primary inhalational tularemia (Conlan *et al.*, 2003). However, mice infected with a low dose aerosol of either a type A or B strain showed limited areas of pulmonary tissue damage despite dissemination to other organs (Conlan *et al.*, 2003).

Thus, the degree of pulmonary involvement is determined by the dose and route of inoculation, virulence of the infecting strain, disease duration, and damage to small and medium sized blood vessels. Persons exposed to low-dose aerosols of virulent strains potentially may become ill from dissemination to other organs without significant pulmonary disease. However, it also is possible that inhalation of a large number of *F. tularensis*, as could occur in a bioterrorism event, would cause a rapidly fatal pneumonia with or without extensive involvement of other organs (Conlan *et al.*, 2003).

3.2. Host Immunity

Naturally occurring infection confers long-lasting immunity to the host, and reinfection has been well-documented in only a few persons (Penn, 2005). Although no immunodominant antigens have been found, host immune responses to *F. tularensis* are directed against numerous cell wall antigens, including membrane proteins, LPS, and carbohydrates (Penn, 2005).

3.2.1. Humoral Immunity

The role of the humoral immune response to *F. tularensis* appears to be supportive, at best. Immunoglobulins are synthesized in response to carbohydrate determinants of *F. tularensis*. Opsonizing antibodies may appear early in the disease course; they can be demonstrated in monkeys 4 days after inhalation of LVS organisms (Tarnvik, 1989). Opsonized bacteria may be more easily phagocytosed by activated mononuclear cells and neutrophils (Sandstrom *et al.*, 1988). Agglutinating antibodies appear in the third week of illness, and, interestingly, IgM is produced simultaneously with IgG. Antibody titers peak

2 months after onset of infection and decline thereafter, but have been detected decades after initial infection. The stimulus for continued immunoglobulin synthesis is currently unknown.

A killed vaccine was effective in inducing an agglutinating antibody response, but this was not sufficient to protect laboratory personnel against tularemia (Tarnvik, 1989). However, immune serum from vaccinated persons and a monoclonal antibody to the LPS of *F. tularensis* have been effective therapies in the murine model of infection (Drabick *et al.*, 1994; Narayanan *et al.*, 1993). This suggests that future immunotherapeutic agents may prove useful against human tularemia infection.

3.2.2. Cellular Immunity

Initial cellular defenses are dependent on neutrophils and local cytokine production. Neutrophils successfully phagocytose the attenuated *F. tularensis* LVS strain in the presence of normal serum, and neutrophil killing of this strain is mediated by hypochlorous acid. Wild-type strains of *F. tularensis*, however, are resistant to neutrophil oxidative killing mechanisms (Tarnvik, 1989).

A protective immune response appears about 2 weeks after disease onset. Complete eradication of the pathogen requires intact cell-mediated immunity. Activated macrophages are a crucial part of host defense through bacterial killing and control of the host inflammatory response.

3.2.2.a. Early responses. The innate immune responses that protect the host before the development of conventional cell-mediated immunity are dependent on neutrophils, TNF-α, and interferon-γ (IFN-γ). In mice, these early responses are in part genetically determined and involve the *Bcg* locus (Hernychova *et al.*, 1997; Kovarova *et al.*, 2000). Other cellular mechanisms are also operative in the early postinfection period. Mice inoculated intradermally with the vaccine strain develop specific immunity within 2 days that is independent of α/β T cells, and that requires B cells, TNF-α, and IFN-γ (Culkin *et al.*, 1997). B cells and IFN-γ are also required for the early immunity that develops in mice injected with small amounts of LPS from the vaccine strain, although antibodies to LPS are not involved (Dreisbach *et al.*, 2000). CpG motifs derived from bacterial DNA or synthetic oligonucleotides also induce early protection against *F. tularensis* LVS in mice; this involves IFN-γ and interleukin-12 (IL-12), and is dependent on B and T lymphocytes (Elkins *et al.*, 1999).

3.2.2.b. Cell-mediated immunity. As with other intracellular bacteria, expression of full immunity to tularemia is dependent on classical cell-mediated immunity involving specific T-lymphocytes, cytokines, and activated macrophages. Activated macrophages kill *F. tularensis* through production of reactive nitrogen products, including nitric oxide (Green *et al.*, 1993). Activation of murine macrophages to inhibit the organism involves IFN-γ and TNF-α, although IFN-γ–independent mechanisms also exist (Cowley and Elkins, 2003). IL-12 also is required to fully resolve murine *F. tularensis* LVS infection (Elkins *et al.*, 2002).

Cutaneous delayed-type hypersensitivity (DTH), an in vivo measure of cell-mediated immunity, historically has been used as a means of diagnosing tularemia, as well as demonstrating evidence of successful vaccination. DTH has been detected as early as 4 days after the onset of clinical disease (Foshay, 1932), and as long as 40 years later (Buchanan *et al.*, 1971; Koskela and Herva, 1980). Cell-mediated immunity can also be detected by in vitro T-lymphocyte stimulation in response to *F. tularensis* antigens (Koskela and Herva, 1980, 1982). T lymphocytes from 70% of volunteers receiving the LVS vaccine showed in vitro stimulation in response to heat-killed organisms, compared with 100% of patients with naturally occurring disease (Koskela and Herva, 1980; Tarnvik and Lofgren, 1975). It therefore appears that natural infection induces a more robust cellular immune response than does vaccination, and this may account for the low rate of reinfection after natural disease.

Protective cell-mediated immunity to tularemia is dependent on MHC-restricted α/β T-cell responses to bacterial proteins (Tarnvik, 1989). Using a capsule-deficient mutant of *F. tularensis* LVS, 4 major polypeptides were isolated which stimulated T lymphocytes from both recipients of LVS vaccines and those with a history of naturally occurring infection (Tarnvik, 1989). Although lymphocyte-stimulating membrane proteins are hidden under the capsular surface, the bacterial capsule is only loosely attached to the outer membrane (Tarnvik, 1989).

An increase in the concentration of γ/δ T cells has been found in patients with tularemia. These cells are not MHC-restricted and do not require antigen processing for stimulation. In a study of 108 patients with ulceroglandular tularemia, the percentage of circulating γ/δ T cells began to rise 8 days after the onset of illness (Poquet *et al.*, 1998). The Vγ9Vδ2 subset accounted for the majority of the increase, and they remained elevated for 18 months after successful resolution of illness (Poquet *et al.*, 1998). The stimuli for Vγ9Vδ2 T-cell expansion are phosphoantigens found in *F. tularensis*. In contrast with patients with naturally acquired infection, γ/δ T-cell expansion did not occur in vaccine recipients despite the demonstration of similar phosphoantigens in extracts of the vaccine strain (Poquet *et al.*, 1998). The delayed expansion of Vγ9Vδ2 T cells after the acute phase of illness and their persistence after recovery suggests a potential regulatory function or a contributory role in T-cell memory. However, the T lymphocytes responding to *F. tularensis* heat shock proteins 10–30 years after natural infection are α/β T cells and not γ/δ T cells (Ericsson *et al.*, 2001).

3.2.3. Immune Responses in the Lungs

Immune mechanisms responsible for controlling systemic infection in a murine model should not be assumed for primary pneumonic tularemia. Experimental murine inhalational tularemia is similar to that seen in primates; the first few days after inhalation exposure neutrophils migrate to foci of infection in the alveoli. However, depletion of either neutrophils or IFN-γ does not exacerbate inhalational tularemia, suggesting that other host defenses also help control primary pulmonary infection (Conlan *et al.*, 2002a).

Mice immunized with the vaccine strain LPS O-side chain conjugated to bovine albumin are protected against intradermal challenge with type B organisms, only partially protected against inhalational challenge with type B organisms, and not protected against inhalational challenge with type A organisms (Conlan *et al.*, 2002b). BALB/c and C57BL/6 mice vaccinated intradermally with live *F. tularensis* LVS are protected against intradermal and aerosol challenge with LVS (Chen *et al.*, 2003). However, neither mouse strain is protected against a low-dose type A strain, and only BALB/c mice showed some protection against aerosol challenge with a type B strain (Chen *et al.*, 2003). This supports the notion that different immune responses may be necessary for protection against inhalation-acquired tularemia, or that the immune responses in the lungs may be attenuated by comparison to other organs. Furthermore, the LVS strain (type B) may lack specific antigens needed to control disease due to a type A strain; resistance to an aerosol challenge with a type A strain was increased in LVS vaccinated mice surviving an intradermal challenge with the type A strain, compared with mice only immunized with LVS (Chen *et al.*, 2003).

4. EPIDEMIOLOGY

F. tularensis is found in many parts of the world, and is most common in the Northern Hemisphere between 30° and 71° north latitude (Penn, 2005). It is endemic in northern and central Europe, particularly in Scandinavian countries and the former Soviet Union. Tularemia has been reported in every state in the United States except Hawaii, with the majority of cases localized to the south-central and western states (Penn, 2005).

Between 1927 and 1948, there were 22,812 human cases of tularemia in the United States with a case fatality of 7.7% (Larson, 1970). The incidence peaked in 1939 when about 2,300 cases were reported (Hornick, 2001) and has declined steadily since the 1950's. From 1990 to 2000, the average annual number of reported cases was 124; Arkansas, Missouri, Oklahoma, and South Dakota accounted for greater then half of all cases (Centers for Disease Control, 2002b). The declining incidence in the United States is due in part to eradication of the wild rabbit market and a heightened awareness of the risks imposed by sick rabbits (Hornick, 2001). However, tularemia is still believed to be under-recognized and underreported.

F. tularensis is perpetuated in nature through an enzootic cycle, particularly involving rodents, lagamorphs, and blood-sucking arthropods. The organism has been identified in more than 100 species of wild animals, nine species of birds, and several species of fish and amphibians (Hopla, 1974). Reservoirs for subspecies *holarctica* include rodents, voles, muskrats, and beavers, as well as mud and water contaminated by these animals (Hornick, 2001). Animals harboring *F. tularensis* subspecies *tularensis* infections in the United States are primarily rabbits and hares, infected through the bites of flies, ticks, and mosquitoes or from environmental exposures (Dennis *et al.*, 2001). Other animal hosts include tree squirrels, opossums, coyote, deer, red fox, sage hen, and the bull snake. In the majority of cases, the animals become sick and die, or their immune response effectively eradicates the organism (Tarnvik and Berglund, 2003).

The bacteria can survive for weeks in soil, carcasses, and hides; for months in water and mud; and for years in frozen rabbit meat (Franz *et al.*, 1997). There is evidence that the bacteria can persist in watercourses, leading to waterborne outbreaks of the disease. Contamination of natural water sources is maintained by beavers and muskrats in North America, by lemmings and beavers in Scandinavia, and by rats in parts of the former Soviet Union. *F. tularensis* may persist in a viable but nonculturable form in water (Ellis *et al.*, 2002) and has been shown to successfully replicate in the free-living ameba *Acanthamoeba castellanii* (Abd *et al.*, 2003). It has been hypothesized that protozoa are a potential environmental reservoir for *F. tularensis*, and may play a role in the natural transmission of the pathogen (Abd *et al.*, 2003).

Several arthropod vectors are responsible for transmission between mammalian hosts. In the United States, ticks are important vectors east of the Rocky Mountains, while biting flies are common vectors in Utah, Nevada, and California (Penn, 2005). *Dermacentor reticulates* and *Ixodes ricinus* are important vectors in Europe, ticks and mosquitoes carry the bacteria in Scandinavia, and *Ixodes* species and mosquitoes are recognized vectors in the former Soviet Union. Although ticks may serve both as vectors and reservoirs of the organism, transovarial transmission is debated and infected ticks show a high mortality rate (Abd *et al.*, 2003).

Humans come into contact with the organism most commonly by biting insect vectors, or via skin contact with infected animals, inhalation of aerosolized organisms, or by pharyngeal or intestinal exposure to contaminated food or water. There also are reports of human transmission of *F. tularensis* from cat bites, the animal's oropharynx presumably being colonized from infected prey (Penn, 2005). The only report of human-to-human transmission occurred in 1926 in a woman who acquired tularemia after pricking her thumb while dressing her son's infected wound (Harris, 1926).

Human outbreaks have been preceded by epizootics that sicken or kill large numbers of animal hosts. Thus, an increase in animal tularemia may herald exposure from a bioterrorist event. There is a seasonal incidence of tularemia in the United States, with most cases in the past occurring from June through August, and in December (Penn, 2005). The summer peak corresponds to the prevalence of tick vectors and the winter cases parallel rabbit hunting season. However, the winter peak has not been prominent in more recent years (Centers for Disease Control, 2002b; Chang *et al.*, 2003).

The largest recorded outbreak of airborne tularemia occurred in 1966–1967 in a rural farming area in Sweden. More than 600 people were infected with a type B strain, previously known as *F. tularensis* subspecies *palearctica*, which was aerosolized during the sorting and transporting of rodent-infested hay (Syrjala *et al.*, 1985). There have been two reported outbreaks of pneumonic tularemia in the United States, both occurring on Martha's Vineyard (Feldman *et al.*, 2001; Teutsch *et al.*, 1979). Two of the eight cases in the 1978 outbreak occurred in gardeners (Teutsch *et al.*, 1979). A case-control study of the 2000 outbreak revealed an association with mowing and brush-cutting activities (Feldman *et al.*, 2001). A serosurvey of 132 landscapers in the area conducted during the summer of 2001 demonstrated a tularemia seroprevalence of 9.1%, compared with 0.3% of resident controls (Feldman *et al.*, 2003). Positive tularemia serologies were more common in persons who used a power blower and in those who worked more hours mowing and weed-whacking, and thus were at greater risk for airborne exposures (Feldman *et al.*, 2003).

Investigators at the German Armed Medical Forces Academy in Munich developed a model to assess the probability that an outbreak from a biological warfare agent was due to bioterrorism, rather than naturally acquired infection (Grunow and Finke, 2002; Reintjes *et al.*, 2002). The procedure considers local geographic factors, the existence of a potential biological risk, the seasonal occurrence of cases and their distribution, the population groups affected, and clinical manifestations. The model also considers current political, military, and social factors. The 1999–2000 tularemia outbreak in Kosovo occurred in an area where this disease had been rare. The WHO and local public health officials evaluated more than 900 suspected cases in patients with fever, cervical lymphadenopathy, and pharyngitis with or without skin ulceration (Reintjes *et al.*, 2002). Analysis of the outbreak using the model was done to determine if there had been an intentional release, and it was concluded that it resulted from infected field mice and rat populations near human dwellings that contaminated local drinking water, grain, and food supplies (Grunow and Finke, 2002; Reintjes *et al.*, 2002). This or similar analyses may prove useful in cases where naturally occurring disease is more prevalent, but an intentional event cannot initially be excluded.

The dispersal of anthrax through the postal system in 2001 illustrates the importance of timely and complete physician and health care reporting of syndromes associated with potential bioterrorist agents, and immediate notification of public health authorities. Formulas have been derived from historical trends of the epidemiology of tularemia that can be used to estimate the probability that a given series of cases would have the typical distribution of age, gender, ethnicity, and seasonal occurrence (Chang *et al.*, 2003). Some states are currently evaluating the application of statistical aberration detection algorithms to aid the rapid identification of unusual disease incidence patterns (Hutwagner *et al.*, 2000). Of selected endemic infections caused by potential bioterrorism-related agents, tularemia and brucellosis were the most frequently reported to the National Notifiable Diseases Surveillance System between 1992 and 1999 (Chang *et al.*, 2003). Such disease-specific epidemiologic trends for tularemia and other potential biowarfare-related illnesses establish a baseline against which future disease incidence may be compared.

5. CLINICAL MANIFESTATIONS

Tularemia is classified into six arbitrary clinical forms based on their predominant clinical manifestations (Penn, 2005). Cutaneous exposure results in an ulcerative lesion with regional lymphadenopathy, termed ulceroglandular tularemia. In glandular tularemia, there is no visible primary lesion, but it resembles the ulceroglandular form in all other respects. Oropharyngeal tularemia follows ingestion of contaminated food or water and leads to pharyngeal ulceration and gastrointestinal symptoms. Inoculation of the eye results in conjunctivitis with local lymphadenopathy, the oculoglandular form. Pneumonic tularemia may result from inhalation of the bacteria or from hematogenous seeding of the lung. Typhoidal tularemia is the most difficult to diagnose and refers to systemic tularemia without an obvious external lesion or regional lymph node swelling. A bioterrorist release of *F. tularensis* may result in any of the forms of tularemia, although inhalational tularemia is expected to predominate (Dennis *et al.*, 2001).

5.1. Nonpneumonic Tularemia

The ulceroglandular form is the most common presentation of naturally acquired tularemia (Penn, 2005). It results from infection acquired through the skin during contact with an infected animal or from the bite of an arthropod vector. Dermal abrasions from knives or bone fragments may inoculate the hands of hunters while skinning or dressing rabbits. The primary lesion occurs from 2 to 5 days after inoculation and begins as a papule that rapidly becomes edematous and tender, suppurating in the center to release a necrotic core, and resulting in an ulcer with a necrotic base and a sharply punched-out border. Tender regional lymphadenopathy develops within a few days of onset. Patients may experience a prodrome of pain at the site of a future papule or bubo prior to disease onset (Foshay, 1940), and glandular pain may precede findings at the inoculation site by about a day. Fever is typically high (102°–104°F); after an initial rise lasting from 1 to 4 days, it may remit for 1–5 days, and be followed by a secondary spike lasting 1–2 weeks in untreated cases. The average duration of fever was 4 weeks before the availability of antibiotic treatment. The skin ulcer heals slowly, and this is delayed further by attempts to incise a lesion when there is little purulent material to drain. Untreated, lymphadenopathy will either spontaneously drain or remain firm and tender, resolving over 2–5 months. Suppuration of lymph nodes often leads to protracted morbidity (Penn, 2005). Secondary hematogenous pneumonia developed in 12% to 15% of these patients before antibiotics were available.

Patients with glandular tularemia present with tender regional lymphadenopathy, but without an evident cutaneous lesion. In most cases, a skin lesion either healed prior to presentation or was minimal and overlooked, but this is otherwise similar to ulceroglandular disease. Glandular tularemia accounts for up to 20% of cases in the United States (Penn, 2005).

Oropharyngeal tularemia is ulceroglandular disease that inoculates the pharynx instead of the skin. It occurs as a result of ingesting undercooked infected meat, contaminated water, or via inhalation of the organism; thus, it may predominate in outbreaks spread by contaminated food and water (Reintjes *et al.*, 2002). Patients develop fever, throat pain, and cervical lymphadenopathy with infraauricular nodal involvement; pharyngeal ulcers are common; and an exudative or membranous pharyngotonsillitis similar to that seen in diphtheria may be present (Penn, 2005; Stupak *et al.*, 2003). Organisms can also invade intestinal mucosa, producing ulceration and mesenteric adenopathy. These patients often develop abdominal pain, nausea, vomiting, diarrhea, and possible sepsis (Penn, 2005).

The oculoglandular type is pathologically similar to the ulceroglandular, with the primary site of inoculation being the conjunctival sac. It is less common, comprising up to 5% of all cases (Penn, 2005). A papule forms, usually on the inferior palpebral conjunctiva, with edema, excessive lacrimation, and pain. Regional lymphadenopathy can occur in the preauricular, parotid, postauricular, submaxillary, or anterior cervical nodes.

Typhoidal tularemia indicates any case of systemic tularemia that does not fit another form, with or without pneumonia. The typhoidal form accounts for up to 30% of naturally occurring cases (Penn, 2005). Most laboratory-acquired cases are typhoidal. Fever is the primary symptom, and patients are systemically ill with chills, prostration, and headache. Occasionally, cough, pleuritic chest pain, and dyspnea are prominent, evidence that a coexisting pneumonia is present (Franz *et al.*, 1997). Patients with typhoidal tularemia also may experience abdominal pain, vomiting, diarrhea, meningismus, and delirium. A relative bradycardia has been described in both tularemia and Legionnaire's disease; however, in contrast to Legionnaire's disease patients with tularemia usually have a clear sensorium (Cunha, 2002; Dennis *et al.*, 2001). Typhoidal tularemia is often accompanied by pneumonia (Giddens *et al.*, 1957). The typhoidal form has been complicated by rhabdomyolysis, renal dysfunction, osteomyelitis, endocarditis, meningitis, and hepatitis (Penn and Kinasewitz, 1987; Rodgers *et al.*, 1998). Involvement of the mouth or pharynx has been reported in up to 35% of cases of typhoidal disease in one study, including pharyngitis, ulcerative tonsillitis, and pharyngeal abscess (Dienst, 1963). Renal dysfunction is common, and in severe cases may progress to oliguria and acute renal failure. Acute renal failure is frequently an important contributing factor in fatal cases of typhoidal tularemia. Gastrointestinal symptoms include anorexia, nausea, vomiting, watery diarrhea, and abdominal pain with tenderness (Dienst, 1963; Zaidi and Singer, 2002). Diarrhea may be severe with reports of intestinal necrosis in some cases (Zaidi and Singer, 2002). Hepatic involvement may occur with increasing frequency as the disease progresses. In an early review of fatal cases, liver abnormalities occurred in up to 75% of patients and included elevated transaminases, marked cholestasis with jaundice, and occasionally granulomatous hepatitis (Gundry and Warner, 1933). The illness may persist for weeks or months, and there is an increased risk of relapse (Penn and Kinasewitz, 1987). Untreated, the case fatality rate for typhoidal tularemia is approximately 35% (Franz *et al.*, 1997).

Dermatologic manifestations of tularemia are probably underrecognized and may occur in as many as 35% of cases (Syrjala *et al.*, 1984). These include diffuse maculopapular and vesiculopapular eruptions, erythema multiforme, and urticaria. Erythema nodosum is the skin finding that most often has been associated with pneumonic tularemia (Feldman *et al.*, 2001; Syrjala *et al.*, 1984, 1985).

5.2. Pneumonic Tularemia

Primary tularemia pneumonia results when lung involvement follows inhalation of *F. tularensis* as the mode of acquisition of the organism. Secondary tularemia pneumonia can occur with any other form of tularemia as a result of hematogenous spread to the lung from another site. The first reported case of pneumonic tularemia occurred in 1931 in a man who had skinned a rabbit, but never developed any skin lesions or lymphadenopathy (Permar and Maclahlan, 1931). He developed rapid onset of fever, hemoptysis, relative bradycardia, and progressive lobar consolidation that ended fatally 17 days into his illness.

Fever, headache, dry cough, dyspnea, and pleuritic chest pain are common symptoms of tularemia pneumonia. Hemoptysis is not common although it may occur (Penn, 2005). Physical findings may include rales, signs of consolidation, effusion, and respiratory failure. The degree of lung involvement seen on the chest radiograph is often much more impressive than the clinical findings. Pneumonia caused by *F. tularensis* subspecies *tularensis* is more severe clinically than that caused by *F. tularensis* subspecies *holarctica* (Tarnvik and Berglund, 2003). Disease from subspecies *tularensis* may begin abruptly, and patients can be profoundly ill with either primary or secondary pneumonia. In contrast, disease from subspecies *holarctica* is less likely to be severe, pulmonary infiltrates are seen less often after inhalational exposure, and secondary pneumonias seldom develop (Tarnvik and Berglund, 2003). Some cases have been misdiagnosed as Legionnaire's disease, and false-positive *Legionella* direct fluorescent antibody (DFA) results have been reported in tularemia pneumonia (Penn, 2005).

In a review of 600 patients with tularemia in the preantibiotic era, some degree of pulmonary changes were noted in almost 90% of cases (Foshay, 1940). However, in only 18% of cases were lung lesions prominent enough to cause clinical symptoms. Consolidation was more common in the hilar regions with extension into the lower lobes, and pleurisy occurred even in the absence of detectable pneumonia (Foshay, 1940). In an early autopsy series of tularemia patients, 73% had pulmonary involvement (Stuart and Pullen, 1945). Pneumonia has been found in up to 25% of more recent cases of tularemia in the United States (Scofield *et al.*, 1992). Risk factors for pneumonia in a retrospective review were older age, no exposure history, typhoidal illness, and positive cultures (Scofield *et al.*, 1992). Typhoidal illness is often accompanied by lung infiltrates whether they are primary from inhalation of the organism or secondary from hematogenous spread. As many as 83% of patients with typhoidal tularemia have some pleuropulmonary involvement (Evans *et al.*, 1985). Although debated, it has been proposed that most cases of typhoidal and pneumonic tularemia are acquired through the respiratory tract, but with differing predominant manifestations of disease (Avery and Barnett, 1967; McCrumb, 1961; Overholt *et al.*, 1961). Prior to antibiotic therapy, 30–60% of patients with pleuropulmonary tularemia died (Stuart and Pullen, 1945).

Bronchoscopic findings in acute illness include local or diffuse hemorrhagic inflammation, edema, and increased mucous secretion that may persist for several weeks (Syrjala *et al.*, 1986). Subepithelial hemorrhages may account for the occasional hemoptysis reported in inhalational disease. These early changes may progress to granulomatous inflammation, indistinguishable from that seen in tuberculosis or sarcoidosis (Syrjala *et al.*, 1985, 1986).

Radiographic findings may be bilateral and include bronchial cuffing, bronchopneumonia, lobar consolidation, hilar adenopathy, and pleural effusion (Ketai *et al.*, 2003). Mediastinal lymphadenopathy can cause airway obstruction, as well as extension into the pericardium. Ovoid-shaped densities previously described as characteristic for pneumonic tularemia, miliary lesions, apical infiltrates, cavities, and bronchopleural fistulas are less common. Thus, the radiographic features vary and may be confused with tuberculosis, fungal disease, endemic community-acquired pneumonias, other bioterrorism-related illnesses, sarcoidosis, and lymphoma or cancer. Correct diagnosis requires a high index of suspicion, even in areas endemic for tularemia.

Pleural effusions accompanying pneumonic tularemia are common, reported in as many as 64–82% of cases (Boudreau and Dennis, 1957). Hematogenous spread can lead to multiple nodular lesions in the lung and pleura, and are common in fatal cases (Dienst, 1963). Effusions may be detected as early as 3 days after the onset of symptoms (Miller and Bates, 1969; Rubin, 1978). Isolated effusions without noticeable parenchymal infiltrates also may occur. Pleural effusions caused by *F. tularensis* are characteristically exudative, appear turbid or serosanguinous, have an elevated protein concentration, and a lymphocytic leukocytosis. Other possible findings that also are similar to tuberculosis include a high adenosine deaminase concentration in the pleural fluid and granulomas seen on pleural biopsy (Penn, 2005).

5.3. Spectrum of Disease following Intentional Release of *F. tularensis*

It is anticipated that an intentional release of *F. tularensis* would be by aerosolization of the organism to cause the most profound disease in the largest number of people (Dennis *et al.*, 2001). Following such an inhalational exposure, the most common initial presentations will be either primary tularemia pneumonia or typhoidal tularemia. Importantly, children are theorized to be more susceptible than adults to an aerosolized agent due to their more rapid respiratory rate, permeable skin, and higher skin-to-mass ratio (Patt and Feigin, 2002). Symptoms will be similar to those described previously for cases of naturally acquired pneumonic and typhoidal tularemia. It is expected that many patients will rapidly progress to pneumonia, pleuritis, and possibly respiratory failure with the systemic inflammatory response syndrome (SIRS).

Although inhalation exposure may cause primarily pneumonic disease, the aerosol can also inoculate the eye, penetrate broken skin, or cause oropharyngeal disease (Dennis *et al.*, 2001). Thus, patients intentionally exposed to aerosolized *F. tularensis* could present with either pneumonic, typhoidal, ulceroglandular, oculoglandular, glandular, or pharyngeal tularemia. During the Swedish airborne outbreak of type B tularemia, most patients presented with headache, fever, and malaise, and only 10% had symptoms consistent with pneumonia (Dahlstrand *et al.*, 1971). However, 26% had conjunctivitis, 31% reported pharyngitis, and some had oral ulcers. Also of note, 32% developed an exanthem, of which erythema multiforme and erythema nodosum were the most common (Dahlstrand *et al.*, 1971).

5.4. Complications

The most frequent complication of tularemia is suppuration of involved lymph nodes. It occurred in almost half of untreated cases (Foshay, 1940), and may occur despite appropriate antibiotic therapy (Jacobs *et al.*, 1985). Suppuration may result in chronic draining lymph nodes and cause protracted debility, and is therefore an indication for surgical drainage. Nonsuppurative adenopathy should not be surgically incised as this will prolong healing. Chest tube drainage of an empyema is uncommonly required. Severe disease can be complicated by renal failure, respiratory failure, SIRS, disseminated intravascular coagulation, rhabdomyolysis, and hepatitis. Meningitis is an uncommon complication, characterized by a mononuclear cell pleocytosis, elevated protein, and hypoglycorrhachia (Rodgers *et al.*, 1998). Involvement of the heart, bones, peritoneum, and prosthetic materials are only rarely encountered (Penn, 2005).

The advent of effective antimicrobial therapy has been shown to decrease overall mortality from 33% to less than 4% (Evans *et al.*, 1985; Giddens *et al.*, 1957). Pneumonic tularemia is associated with greater morbidity and a higher mortality risk than other clinical forms of tularemia. Defervescence appears to be a rapid and reliable clinical indicator of response to therapy (Giddens *et al.*, 1957). The average duration of fever in the preantibiotic era was 31 days (Foshay, 1940), but fever and malaise may persist for weeks in severely ill patients despite antibiotic therapy (Dahlstrand *et al.*, 1971). Relapses usually occur after treatment with a bacteriostatic agent, such as tetracycline or chloramphenicol, but occasionally may follow other recommended antibiotics.

6. DIAGNOSIS

The diagnosis of tularemia is frequently delayed due to its often nonspecific clinical manifestations, the unique growth requirements of the organism, a lack of supporting epidemiological history, and misdiagnosis of a more commonly encountered illness. In fact, one of the sentinel cases of pneumonic tularemia during the 2000 outbreak in Martha's

Vineyard was initially diagnosed with community-acquired pneumonia and treated with clarithromycin (Dembek *et al.*, 2003). The patient was a 57-year-old man who presented with eye irritation, rhinorhea, fever, fatigue, diarrhea, anorexia, and a 20-pound weight loss over the previous week. Although he had recently used a side discharge mower to cut his lawn, had tick exposures, and had a right middle lobe infiltrate, the correct diagnosis was only retrospectively obtained by serologic testing after public health authorities investigated the outbreak (Dembek *et al.*, 2003).

Endemic tularemia is diagnosed based on clinical suspicion, and confirmed by a 4-fold or greater rise in serologies between acute and convalescent sera. Cultures are usually not performed to diagnose suspected endemic cases because of their low yield and the potential dangers posed to laboratory personnel. Antibodies to *F. tularensis* do not appear until after the first 7–10 days of illness and peak by 4–5 weeks (Dennis *et al.*, 2001). Agglutination and ELISA techniques can reliably detect serum antibodies, but cannot distinguish infections caused by subspecies *tularensis* from subspecies *holarctica*. A single-tube agglutination titer of ≥1:160 or a microagglutination titer of ≥1:128 can be considered positive in a patient with a compatible illness (Chu and Weyant, 2003). Both IgM and IgG may persist for decades after primary infection; therefore, the presence of IgM in a single serum sample does not indicate acute disease and must be confirmed with convalescent sera. *F. tularensis, Brucella, Yersinia,* and *Proteus* share common antigens, explaining the occurrence of potential serological cross-reactions with these organisms (Penn, 2005).

F. tularensis is isolated from routine cultures in a minority of cases due in part to its unusual growth requirements and overgrowth by commensal bacteria. The organism is rarely seen on gram-stained smears or in tissue biopsies, but may be recovered from respiratory secretions, pleural fluid, ulcerative lesions, lymph nodes, blood, tissue biopsies, and gastric aspirates if cultured on supportive media. It is crucial that the laboratory is notified about the possibility of tularemia prior to receiving clinical specimens so that supportive media are used and appropriate precautions are taken. Routine processing of specimens may be performed using Biosafety Level 2 (BSL-2) procedures (Chu and Weyant, 2003). However, further work-up of suspicious isolates should be performed following Biosafety Level 3 (BSL-3) protocols until they are shown not to be a *Francisella* species (Chu and Weyant, 2003; Robinson-Dunn, 2002; Sewell, 2003).

The bacteria require cysteine for growth, and thus cannot be isolated on typical Gram-negative selective media such as MacConkey or eosin methylene blue agars (Penn, 2005). Solid media available in most laboratories that will support growth of the organism include chocolate agar, modified Thayer-Martin medium, and buffered charcoal-yeast extract agar. Standard blood culture media and thioglycolate broth also will support growth of the organism. Specimens should be incubated at 35°C and do not require supplemental CO_2. Smooth mucoid colonies will form on solid media within 2–5 days. Adding penicillin, cyclohexamide, or polymyxin B to the media can deter commensal overgrowth and enhance the recovery of *F. tularensis* from contaminated specimens (Chu and Weyant, 2003). Almost all strains are β-lactamase positive. Variant clinical strains not requiring a cysteine-enriched media for growth have been mistakenly identified as *Haemophilus* or *Actinobacillus* species (Bernard *et al.*, 1994). If tularemia is suspected after incubation on primary culture media, the plates should be sealed and opened only in a Biosafety cabinet using BSL-3 precautions. Standard identification kits and automated systems should not be used, as they may generate aerosols and incorrectly identify the bacteria as *Actinobacillus actinomycetemcomitans* or *Haemophilus influenzae* (Centers for Disease Control, 2001).

DNA-based methods allow detection with inactivated samples, decreasing the risk imposed on laboratory personnel. Polymerase chain reaction (PCR) assays of clinical samples have greater sensitivity, compared with cultures (Johansson *et al.*, 2000a; Sjostedt *et al.*, 1997). When wound swabs from 40 patients with ulceroglandular tularemia were used for diagnosis in a recent study, a PCR assay was 75% sensitive and culture was 62% sensitive (Johansson *et al.*, 2000a). In the event of a suspected outbreak or intentional bioterrorist attack, PCR would be the most readily available means for rapid diagnosis. PCR has detected the organism in human tissue even after initiation of antimicrobial therapy (Dolan *et al.*, 1998). PCR and other molecular methods can distinguish among the clinically relevant species and subspecies (Penn, 2005). These methods are being applied to develop rapid identification tests for field use (Grunow *et al.*, 2000); however, there are no standardized PCR tests readily available at present. DFA and immunohistochemical staining can detect *F. tularensis* in clinical specimens, and are available in the United States through specialized laboratories in the National Public Health Response Network (Dennis *et al.*, 2001).

Timely diagnosis requires a high index of suspicion and a focused history to elicit possible infected animal exposure, vector exposure (biting flies, mosquitoes, ticks) in a region known to contain the pathogen, and other high-risk activities. If tularemia is considered in the absence of the expected epidemiologic features, then a bioterrorist event should be suspected. The physician should quickly alert local laboratory and infection control personnel, and state and local public health authorities should be consulted.

The differential diagnosis of most bioterrorism-related cases will include other infectious causes of community-acquired pneumonia and typhoidal syndromes, as well as other possible bioterrorism-related diseases, such as anthrax and plague. Tularemia progresses more slowly than pneumonic plague, and *F. tularensis* does not exhibit the characteristic bipolar appearance on Gram stain that is typical of *Y. pestis* (Dennis *et al.*, 2001). Inhalational anthrax can be differentiated by the characteristic widened mediastinum and absence of bronchopneumonia seen with this disease on imaging studies of the chest, and the characteristic Gram stain appearance of *B. anthracis* (Dennis *et al.*, 2001).

7. TREATMENT

7.1. Treatment of Endemic Tularemia

Since it was first reported to be effective in 1946, streptomycin has become the drug of first choice for all forms of tularemia except meningitis (Penn, 2005). The minimum effective dose is 7.5–10 mg/kg of streptomycin intramuscularly every 12 hours for 7–14 days. An alternative dose is 15 mg/kg intramuscularly every 12 hours for the first 3 days, followed by 7.5 mg/kg intramuscularly every 12 hours to complete the same duration of therapy. In severely ill patients, 15 mg/kg intramuscularly every 12 hours may be given for a 7- to 10-day total course. The efficacy of streptomycin therapy is not increased by daily doses of more than 2 g (Penn, 2005). The pediatric dose is 30–40 mg/kg/day intramuscularly divided into 2 doses for a total of 7 days, or 40 mg/kg/day divided into 2 doses intramuscularly for the first 3 days followed by 20 mg/kg/day in 2 divided doses for the next 4 days. Gentamicin is an acceptable alternative for both adults and children. In adults, gentamicin may have a higher risk of failure and relapse than streptomycin, but in children it appears to be equally as effective (Cross *et al.*, 1995; Enderlin *et al.*, 1994). Tularemia in adults has been successfully treated with 3–5 mg/kg/day of gentamicin given intravenously in three divided doses (Evans *et al.*, 1985; Sanford, 1983). The in vitro inhibitory concentration of gentamicin is usually less than 2 μg/mL (Baker *et al.*, 1985; Ikaheimo *et al.*, 2000), a level that easily can be achieved in the serum with these standard dosing schedules. However, the concentration needed for bactericidal activity can be higher (Mason *et al.*, 1980; Maurin *et al.*, 2000); thus, we recommend dosing in adults to achieve peak serum levels of at least 5 μg/mL. Successfully treated cases in children received an average gentamicin dose of 6 mg/kg/day and had peak serum concentrations of at least 7 μg/mL (Cross *et al.*, 1995). The efficacy of single-daily dosing of the aminoglycosides for tularemia has not been reported. Dosing of streptomycin and gentamicin should be adjusted for renal insufficiency. Streptomycin and gentamicin do not penetrate the cerebrospinal fluid effectively, and they should not be used as single agents to treat suspected tularemic meningitis. Regimens reported to be successful as therapy for meningitis have included chloramphenicol in combination with streptomycin, and doxycycline in combination with gentamicin (Penn, 2005).

Tetracycline and chloramphenicol are bacteriostatic for *F. tularensis*. This helps explain why they are associated with higher rates of failure and relapse than streptomycin or gentamicin. Therapy with these agents should be given for at least 14 days to minimize treatment failure and relapse, but they have the advantage of being available for oral dosing.

Fluoroquinolones are promising alternatives that achieve adequate serum and intracellular concentrations when given intravenously or orally. Naturally occurring North American strains of *F. tularensis* subspecies *tularensis* and *holarctica* have comparably low minimum inhibitory concentration values for six common fluoroquinolones (Johansson *et al.*, 2002), and aerosol delivery of liposome-encapsulated ciprofloxacin is effective in eradication of respiratory *F. tularensis* LVS infection in mice (Conley *et al.*, 1997). Nonetheless, reported clinical experience in treating *F. tularensis* subspecies *tularensis* infections with the fluoroquinolones is limited. Both ciprofloxacin and norfloxacin have been used successfully to treat ulceroglandular and pulmonary disease (Syrjala *et al.*, 1991). Ciprofloxacin use during the initial tularemia epidemic in Spain was associated with the least number of overall therapeutic failures and the fewest side effects (Perez-Castrillon *et al.*, 2001). However, relapses occurred in 7 of 14 patients who received ciprofloxacin for either primary or secondary therapy at one of the Spanish hospitals (Chocarro *et al.*, 2000). Ciprofloxacin also has proven to be effective in the treatment of children. In one study of 12 children ages 1–10 years, a dose of 15–20 mg/kg/day divided twice daily was successful when patients completed 10–14 days of uninterrupted therapy (Johansson *et al.*, 2000b).

7.2. Treatment of Tularemia Resulting from Bioterrorism

A bioterrorism event may result either in a relatively low numbers of patients easily managed by existing health care systems and referred to as contained casualties, or in mass casualties. The Working Group on Civilian Biodefense consensus recommendations for therapy of tularemia resulting from a bioterrorism event are summarized in Tables 3.1 and 3.2 (Dennis *et al.*, 2001). The recommended regimens differ depending on whether there are contained or mass casualties. Note that a minimum of 14 days is recommended for all regimens used for mass casualties; this is due to the fact that, in this circumstance, oral agents are exclusively used beginning from the initiation of treatment (Table 3.1).

When gentamicin is used for contained casualties with tularemia resulting from bioterrorism, the Working Group recommends once-daily dosing despite the lack of supportive clinical data (Tables 3.1 and 3.2). When the bacteriostatic drugs are used, they should be continued for a longer duration than the bactericidal drugs (Table 3.1). When ciprofloxacin, doxycycline, or chloramphenicol are begun intravenously, they can be changed to oral dosing after substantial clinical improvement is noted (Dennis *et al.*, 2001). Bacteriostatic agents are not recommended for immunosuppressed patients because of higher rates of treatment failure with these agents; giving the entire therapy parenterally should be considered in such patients (Dennis *et al.*, 2001).

For children, the Working Group recommends streptomycin or gentamicin as first-line therapy in a contained casualty situation (Dennis *et al.*, 2001). In the event of a mass casualty situation, oral doxycycline or ciprofloxacin is recommended for both adults and

Table 3.1
Recommendations for Therapy of Casualties and for Postexposure Prophylaxis after an Intentional Release of *F. tularensis*[a]

Patient population	Contained casualties[b]	Mass casualties and postexposure prophylaxis[c]
Adults		
Preferred agents	Streptomycin, gentamicin	Doxycycline, ciprofloxacin,
Alternatives	Doxycycline, ciprofloxacin, chloramphenicol	N/A
Children		
Preferred agents	Streptomycin, gentamicin	Doxycycline, ciprofloxacin
Alternatives	Doxycycline, ciprofloxacin, chloramphenicol	N/A
Pregnancy		
Preferred agents	Gentamicin, streptomycin	Ciprofloxacin, doxycycline
Alternatives	Ciprofloxacin, doxycycline	N/A

[a]Choices are listed in our order of preference for use. See Table 3.2 for dosing recommendations, and see the text for recommendations for immunosuppressed patients. Based on recommendations from the Working Group on Civilian Biodefense (Dennis *et al.*, 2001).
[b]Treatment for contained casualties should be initiated parenterally (see text). Therapy with streptomycin, gentamicin, or ciprofloxacin should be given for 10 days. Therapy with doxycycline or chloramphenicol should be given for 14–21 days.
[c]Treatment for mass casualties and as postexposure prophylaxis should be given orally for 14 days. N/A, not applicable.

Table 3.2
Recommended Dosing of Antibiotics Used for Therapy and Postexposure Prophylaxis after an Intentional Release of *F. tularensis*[a]

Antibiotic	Adults	Children	Pregnancy
Streptomycin[b]	1 g IM every 12 hr	15 mg/kg IM every 12 hr, up to a maximum of 2 g/day	1 g IM every 12 hr
Gentamicin[b]	5 mg/kg IM or IV once daily	5 mg/kg IM or IV once daily	5 mg/kg IM or IV once daily
Ciprofloxacin[b]	400 mg IV or 500 mg PO every 12 hr	15 mg/kg IV or PO every 12 hr, up to a maximum of 1 g/day	400 mg IV or 500 mg PO every 12 hr
Doxycycline	100 mg IV or PO every 12 hr	If weight ≥45 kg, then 100 mg IV or PO every 12 hr If weight <45 kg, then 2.2 mg/kg IV or PO every 12 hr	100 mg IV or PO every 12 hr
Chloramphenicol	15 mg/kg IV or PO every 6 hr	15 mg/kg IV or PO every 6 hr	Not recommended

IM, intramuscularly; IV, intravenously; PO, orally.
[a]Indicated doses assume normal renal function. See Table 3.1 for selecting among the choices and the duration of therapy. Based on recommendations from the Working Group on Civilian Biodefense (Dennis *et al.*, 2001).
[b]Dosing of these agents should be adjusted for renal insufficiency.

children (Table 3.1). Short courses of fluoroquinolones and tetracyclines have been used safely in children, although these agents have possible adverse effects in children and immature laboratory animals. Their potential risks should be weighed in proportion to their potential benefits in these settings.

Pregnant women should receive gentamicin in a contained casualty event (Table 3.1). Damage to fetal nerves and kidneys has not been reported with gentamicin. Oral ciprofloxacin is the most reasonable alternative in pregnancy when there are mass casualties (Table 3.1).

Susceptibility testing of any isolates should be done to detect unusual resistance patterns that may indicate a strain engineered for antibiotic resistance.

8. INFECTION CONTROL

Isolation of patients with tularemia is not indicated, as human-to-human transmission does not occur. Standard precautions are sufficient for hospital care of these patients, and potentially contaminated linens may be handled according to standard protocols (Dennis *et al.*, 2001).

Hospital and other institutions need to familiarize their staff with their comprehensive plans for handling bioterrorism before an event takes place, including how to manage casualties with tularemia. It is crucial that microbiology and pathology personnel be notified about potentially infected specimens before they receive them. Even in the best circumstances, notification is going to be inconsistent (Shapiro and Schwartz, 2002); therefore, it is imperative that hospitals have procedures in place to handle unexpected infectious agents. Clinical specimens may be processed using BSL-2 procedures. However, manipulation of suspected isolates or any procedure with a potential for aerosol or droplet production (e.g., grinding, centrifugation, vigorous shaking, growing cultures in volume, and animal studies) must be performed using BSL-3 procedures (Chu and Weyant, 2003; Robinson-Dunn, 2002; Sewell, 2003). If the organism is identified in a laboratory without BSL-3 capabilities, it should be forwarded to a BSL-3 laboratory for confirmation and susceptibility testing (Centers for Disease Control, 2000). If a patient dies, the body should be handled with standard precautions. If an autopsy is performed, procedures that are likely to cause aerosols like bone sawing should be avoided when possible (Centers for Disease Control, 2000).

The infectivity of aerosolized *F. tularensis* declines by 10% over 3 hours (Hornick, 2001). Thus, the Working Group anticipates that intentionally aerosolized *F. tularensis* would have a short half-life due to dessication, solar radiation, and oxidation; this should limit the occurrence of secondary dispersal (Dennis *et al.*, 2001). Wet inanimate surfaces may be sprayed with a 10% bleach solution, and followed in 10 minutes with a 70% alcohol solution to further clean the area and limit any harmful effects of bleach (Dennis *et al.*, 2001). Persons with direct exposure to contaminated powders, liquids, or aerosols should wash themselves and clothing with soap and water. Standard levels of chlorine in municipal water sources should protect against waterborne infection (Dennis *et al.*, 2001). Contaminated natural water sources will need to be quickly identified and contained to prevent human and animal exposures. In an event where animals become sick or die, the public needs to be educated not to handle them and to take protective measures against potential insect vectors.

9. PREVENTION

9.1. Antibiotic Prophylaxis

Volunteers exposed to an aerosol of *F. tularensis* were protected when taking tetracycline 1 g daily for 14 or 28 days, whereas 20% of those receiving only 5 days of tetracycline became symptomatic after therapy was stopped (Sawyer *et al.*, 1966). Tetracycline, doxycycline, and ciprofloxacin also can be effective as preemptive therapy in experimental models of inhalational tularemia (Russell *et al.*, 1998). In part based on these findings, the Working Group (Dennis *et al.*, 2001) has recommended prophylactic therapy for persons with documented exposures identified during the incubation period using 14 days of either oral ciprofloxacin or doxycycline (Tables 3.1 and 3.2). However, it is unlikely that a bioterrorist attack with *F. tularensis* via aerosol or through contamination of water sources would be known early enough to provide exposed individuals with antibiotic prophylaxis. If an attack is discovered after persons become ill, all those with potential exposures should begin a fever watch. Anyone who develops a fever or flu-like illness within 14 days of the exposure should begin therapy as listed in Tables 3.1 and 3.2.

Antibiotic prophylaxis after tick bites or other exposures with an unknown risk is not recommended. In the past, intramuscular streptomycin was found to be effective in preventing illness in laboratory personnel accidentally exposed to *F. tularensis* (Evans *et al.*, 1985; Sawyer *et al.*, 1966); the efficacy of gentamicin for this purpose is unknown. More recently, 14 days of either oral doxycycline or ciprofloxacin is currently recommended for adults with suspected or proven high-risk natural exposures (Centers for Disease Control, 2002a). Vaccinated individuals with exposures may be observed without antibiotic prophylaxis. There is no need to provide prophylaxis to exposed individuals who have had tularemia in the past, or to those whose only exposure is close contact with an infected patient.

9.2. Vaccination

An early phenol-killed vaccine developed in the United States was eventually found to be less effective than the live attenuated vaccine in preventing tularemia (Burke, 1977). This is explained in part by the fact that the killed vaccine does not induce a cell-mediated response and humoral immunity alone is inadequate to resolve this intracellular infection (Tarnvik, 1989).

Multiple tularemia epidemics in the former Soviet Union stimulated development of live attenuated vaccines, and mass vaccination programs were begun in 1946 (Tigertt, 1962). By the 1950s, tens of millions of people living in endemic areas within the former Soviet Union had been vaccinated (Dennis *et al.*, 2001); 4.3% of the unvaccinated population in endemic regions developed tularemia, but only 0.36% of those who were vaccinated became ill (Tigertt, 1962).

In 1956, two of these live vaccine strains were brought to the United States (Ellis *et al.*, 2002). Two colony types were observed under oblique light, one blue and one gray in appearance. The blue colony variant was chosen because it had greater immunogenicity in mice and guinea pigs. An LVS was derived from passage of the blue colony variant through mice, and a newer lot has been developed (Waag *et al.*, 1996). The LVS offers some protection against typhoidal and ulceroglandular tularemia (Burke, 1977). The incidence of typhoidal tularemia decreased from 5.70 to 0.27 cases per 1,000 at-risk employee-years following the introduction of the LVS vaccine at the U.S. Army Medical Research Institute of Infectious Diseases at Fort Detrick, Maryland (Burke, 1977). The incidence of ulceroglandular tularemia remained unchanged, but the clinical symptoms were modified in those vaccinated. Determining the efficacy of LVS against aerosol challenge has been more difficult. Immunized volunteers were incompletely protected against an aerosol challenge with a virulent type A strain of *F. tularensis* (Hornick and Eigelsbach, 1966; Saslaw *et al.*, 1961a). The LVS vaccine does not adequately protect guinea pigs and monkeys against an aerosol challenge with larger numbers of bacteria (Eigelsbach and Downs, 1961). Aerosolized exposure to LVS induces an equivalent immune response, compared with percutaneous vaccination without causing clinical disease (Hornick and Eigelsbach, 1966; Tigertt, 1962). However, it is not known if this route of immunization would improve or prolong protection against inhalational tularemia (Eigelsbach *et al.*, 1967).

It is difficult to make firm conclusions on the reliability of the LVS vaccine based on previous studies for several reasons. Manufacturing a consistent vaccine strain and standardization of strain pathogenicity is difficult. The critical antigenic determinants for inducing essential immune responses have not been characterized. A question that remains unanswered is to what degree the efficacy of the LVS vaccine depends on its ability to multiply in the host. In addition, the molecular basis for the LVS strain attenuation for humans remains unknown (Tarnvik, 1989). The duration of protective immunity following vaccination has not been determined. However, persons challenged with an aerosol containing up to 2,000 organisms of the type A Schu S4 strain had some protection 1 year after vaccination (Eigelsbach *et al.*, 1967).

Despite its limitations, the LVS vaccine is still the most efficacious vaccine developed and remains the standard against which any replacement vaccines must be compared. However, the live tularemia vaccine is no longer available in the United States. A new Investigational New Drug protocol pending approval would permit the LVS vaccine to be used at Fort Detrick to vaccinate at-risk personnel through a cooperative research agreement with the U.S. Army Research Institute of Infectious Diseases (USAMRIID). Inquiries should be directed to the Chief, Medical Division, USAMRIID, Fort Detrick, Frederick, Maryland 21702, USA.

Vaccination is not recommended for postexposure prophylaxis due to the short incubation period of tularemia and incomplete protection of the current vaccine against inhalational tularemia. The Working Group recommends use of the LVS only for at-risk laboratory personnel (Dennis *et al.*, 2001). Persons who have recovered from previous tularemia are not vaccine candidates.

10. FUTURE DIRECTIONS

Completing the full genome sequence of *F. tularensis* will aid efforts to construct a defined, attenuated, safe, and more effective live vaccine. This information – together with techniques for gene modification – will stimulate ongoing efforts to identify key virulence factors and other antigens that could be used in subunit vaccines (Ellis *et al.*, 2002). There is an urgent need for rapid but accurate tools for diagnosis, and also to detect the organism in environmental samples. Determining the molecular characteristics of the organism will be crucial to rapidly detect strains engineered for antibiotic resistance or enhanced virulence. Immunotherapeutic agents for tularemia also may be possible in the future.

References

Abd, H., Johansson, T., Golovliov, I., Sandstrom, G., and Forsman, M. (2003). Survival and growth of *Francisella tularensis* in *Acanthamoeba castellanii*. *Appl. Environ. Microbiol.* 69:600–606.

Avery, F.W., and Barnett, T.B. (1967). Pulmonary tularemia. A report of five cases and consideration of pathogenesis and terminology. *Am. Rev. Respir. Dis.* 95:584–591.

Baker, C.N., Hollis, D.G., and Thornsberry, C. (1985). Antimicrobial susceptibility testing of *Francisella tularensis* with a modified Mueller-Hinton broth. *J. Clin. Microbiol.* 22:212–215.

Bernard, K., Tessier, S., Winstanley, J., Chang, D., and Borczyk, A. (1994). Early recognition of atypical *Francisella tularensis* strains lacking a cysteine requirement. *J. Clin. Microbiol.* 32:551–553.

Boudreau, R.P., and Dennis, J.M. (1957). Pleuropulmonary tularemia: its roentgen manifestations. *Radiology* 68:25–30.

Broekhuijsen, M., Larsson, P., Johansson, A., Bystrom, M., Eriksson, U., Larsson, E., Prior, R.G., Sjostedt, A., Titball, R.W., and Forsman, M. (2003). Genome-wide DNA microarray analysis of *Francisella tularensis* strains demonstrates extensive genetic conservation within the species but identifies regions that are unique to the highly virulent *F. tularensis* subsp. *tularensis*. *J. Clin. Microbiol.* 41:2924–2931.

Buchanan, T.M., Brooks, G.F., and Brachman, P.S. (1971). The tularemia skin test. 325 skin tests in 210 persons: serologic correlation and review of the literature. *Ann. Intern. Med.* 74:336–343.

Burke, D.S. (1977). Immunization against tularemia: analysis of the effectiveness of live *Francisella tularensis* vaccine in prevention of laboratory-acquired tularemia. *J. Infect. Dis.* 135:55–60.

Centers for Disease Control. (2000). Biological and chemical terrorism: strategic plan for preparedness and response. Recommendations of the CDC Strategic Planning Workgroup. *MMWR Recomm. Rep.* 49:1–14.

Centers for Disease Control. (2001). *Laboratory Response Network (LRN): Level A laboratory procedures for identification of Francisella tularensis.* Accessed June 1, 2003, Atlanta, (updated December 13, 2001): www.bt.cdc.gov/Agent/Tularemia/ftu_la_cp_121301.pdf.

Centers for Disease Control. (2002a). Outbreak of tularemia among commercially distributed prairie dogs, 2002. *MMWR Morb. Mortal. Wkly. Rep.* 51:688–699.

Centers for Disease Control. (2002b). Tularemia—United States, 1990–2000. *MMWR Morb. Mortal. Wkly. Rep.* 51:181–184.

Chang, M.H., Glynn, M.K., and Groseclose, S.L. (2003). Endemic, notifiable bioterrorism-related diseases, United States, 1992–1999. *Emerg. Infect. Dis.* 9:556–564.

Chen, W., Shen, H., Webb, A., KuoLee, R., and Conlan, J.W. (2003). Tularemia in BALB/c and C57BL/6 mice vaccinated with *Francisella tularensis* LVS and challenged intradermally, or by aerosol with virulent isolates of the pathogen: protection varies depending on pathogen virulence, route of exposure, and host genetic background. *Vaccine* 21:3690–3700.

Chocarro, A., Gonzalez, A., and Garcia, I. (2000). Treatment of tularemia with ciprofloxacin. *Clin. Infect. Dis.* 31:623.

Christopher, G.W., Cieslak, T.J., Pavlin, J.A., and Eitzen, E.M., Jr. (1997). Biological warfare. A historical perspective. *J.A.M.A.* 278:412–417.

Chu, M.C., and Weyant, R.S. (2003). *Francisella* and *Brucella*. In: Murray, P.R., Baron, E.J., Jorgensen, J.H., Pfaller, M.A., and Yolken, R.H. (eds.), *Manual of Clinical Microbiology*, vol. 1, 8th ed. ASM Press, Washington, D.C., pp. 789–808.

Conlan, J.W., Chen, W., Shen, H., Webb, A., and KuoLee, R. (2003). Experimental tularemia in mice challenged by aerosol or intradermally with virulent strains of *Francisella tularensis*: bacteriologic and histopathologic studies. *Microb. Pathog.* 34:239–248.

Conlan, J.W., KuoLee, R., Shen, H., and Webb, A. (2002a). Different host defences are required to protect mice from primary systemic vs pulmonary infection with the facultative intracellular bacterial pathogen, *Francisella tularensis* LVS. *Microb. Pathog.* 32:127–134.

Conlan, J.W., Shen, H., Webb, A., and Perry, M.B. (2002b). Mice vaccinated with the O-antigen of *Francisella tularensis* LVS lipopolysaccharide conjugated to bovine serum albumin develop varying degrees of protective immunity against systemic or aerosol challenge with virulent type A and type B strains of the pathogen. *Vaccine* 20:3465–3471.

Conley, J., Yang, H., Wilson, T., Blasetti, K., Di Ninno, V., Schnell, G., and Wong, J.P. (1997). Aerosol delivery of liposome-encapsulated ciprofloxacin: aerosol characterization and efficacy against *Francisella tularensis* infection in mice. *Antimicrob. Agents Chemother.* 41:1288–1292.

Cowley, S.C., and Elkins, K.L. (2003). Multiple T cell subsets control *Francisella tularensis* LVS intracellular growth without stimulation through macrophage interferon gamma receptors. *J. Exp. Med.* 198:379–389.

Cowley, S.C., Myltseva, S.V., and Nano, F.E. (1996). Phase variation in *Francisella tularensis* affecting intracellular growth, lipopolysaccharide antigenicity and nitric oxide production. *Mol. Microbiol.* 20:867–874.

Cross, J.T., Jr., Schutze, G.E., and Jacobs, R.F. (1995). Treatment of tularemia with gentamicin in pediatric patients. *Pediatr. Infect. Dis. J.* 14:151–152.

Culkin, S.J., Rhinehart-Jones, T., and Elkins, K.L. (1997). A novel role for B cells in early protective immunity to an intracellular pathogen, *Francisella tularensis* strain LVS. *J. Immunol.* 158:3277–3284.

Cunha, B.A. (2002). Anthrax, tularemia, plague, ebola or smallpox as agents of bioterrorism: recognition in the emergency room. *Clin. Microbiol. Infect.* 8:489–503.

Dahlstrand, S., Ringertz, O., and Zetterberg, B. (1971). Airborne tularemia in Sweden. *Scand. J. Infect. Dis.* 3:7–16.

Dembek, Z.F., Buckman, R.L., Fowler, S.K., and Hadler, J.L. (2003). Missed sentinel case of naturally occurring pneumonic tularemia outbreak: lessons for detection of bioterrorism. *J. Am. Board. Fam. Pract.* 16:339–342.

Dennis, D.T., Inglesby, T.V., Henderson, D.A., Bartlett, J.G., Ascher, M.S., Eitzen, E., Fine, A.D., Friedlander, A.M., Hauer, J., Layton, M., Lillibridge, S.R., McDade, J.E., Osterholm, M.T., O'Toole, T., Parker, G., Perl, T.M., Russell, P.K., and Tonat, K. (2001). Tularemia as a biological weapon: medical and public health management. *J.A.M.A.* 285:2763–2773.

Dienst, F.T., Jr. (1963). Tularemia: a perusal of three hundred thirty-nine cases. *J. Louisiana State Med. Soc.* 115:114–124.

Dolan, S.A., Dommaraju, C.B., and DeGuzman, G.B. (1998). Detection of *Francisella tularensis* in clinical specimens by use of polymerase chain reaction. *Clin. Infect. Dis.* 26:764–765.

Drabick, J.J., Narayanan, R.B., Williams, J.C., Leduc, J.W., and Nacy, C.A. (1994). Passive protection of mice against lethal *Francisella tularensis* (live tularemia vaccine strain) infection by the sera of human recipients of the live tularemia vaccine. *Am. J. Med. Sci.* 308:83–87.

Dreisbach, V.C., Cowley, S., and Elkins, K.L. (2000). Purified lipopolysaccharide from *Francisella tularensis* live vaccine strain (LVS) induces protective immunity against LVS infection that requires B cells and gamma interferon. *Infect. Immun.* 68:1988–1996.

Eigelsbach, H.T., and Downs, C.M. (1961). Prophylactic effectiveness of live and killed tularemia vaccines. I. Production of vaccine and evaluation in the white mouse and guinea pig. *J. Immunol.* 87:415–425.

Eigelsbach, H.T., Hornick, R.B., and Tulis, J.J. (1967). Recent studies on live tularemia vaccine. *Med. Ann. Dist. Columbia* 36:282–286.

Eigelsbach, H.T., and McGann, V.G. (1984). Genus *Francisella* Doroféev 1947, 176^{AL}. In: Krieg, N.R., and Holt, J.G. (eds.), *Bergey's Manual of Systematic Bacteriology.* Williams & Wilkins, Baltimore, pp. 394–399.

Elkins, K.L., Bosio, C.M., and Rhinehart-Jones, T.R. (1999). Importance of B cells, but not specific antibodies, in primary and secondary protective immunity to the intracellular bacterium *Francisella tularensis* live vaccine strain. *Infect. Immun.* 67:6002–6007.

Elkins, K.L., Cooper, A., Colombini, S.M., Cowley, S.C., and Kieffer, T.L. (2002). In vivo clearance of an intracellular bacterium, *Francisella tularensis* LVS, is dependent on the p40 subunit of interleukin-12 (IL-12) but not on IL-12 p70. *Infect. Immun.* 70:1936–1948.

Ellis, J., Oyston, P.C., Green, M., and Titball, R.W. (2002). Tularemia. *Clin. Microbiol. Rev.* 15: 631–646.

Enderlin, G., Morales, L., Jacobs, R.F., and Cross, J.T. (1994). Streptomycin and alternative agents for the treatment of tularemia: review of the literature. *Clin. Infect. Dis.* 19:42–47.

Ericsson, M., Kroca, M., Johansson, T., Sjostedt, A., and Tarnvik, A. (2001). Long-lasting recall response of CD4+ and CD8+ alphabeta T cells, but not gammadelta T cells, to heat shock proteins of *Francisella tularensis*. *Scand. J. Infect. Dis.* 33:145–152.

Evans, M.E., Gregory, D.W., Schaffner, W., and McGee, Z.A. (1985). Tularemia: a 30-year experience with 88 cases. *Medicine* (Baltimore) 64:251–269.

Feldman, K.A., Enscore, R.E., Lathrop, S.L., Matyas, B.T., McGuill, M., Schriefer, M.E., Stiles-Enos, D., Dennis, D.T., Petersen, L.R., and Hayes, E.B. (2001). An outbreak of primary pneumonic tularemia on Martha's Vineyard. *N. Engl. J. Med.* 345:1601–1606.

Feldman, K.A., Stiles-Enos, D., Julian, K., Matyas, B.T., Telford, S.R., 3rd, Chu, M.C., Petersen, L.R., and Hayes, E.B. (2003). Tularemia on Martha's Vineyard: seroprevalence and occupational risk. *Emerg. Infect. Dis.* 9:350–354.

Fortier, A.H., Leiby, D.A., Narayanan, R.B., Asafoadjei, E., Crawford, R.M., Nacy, C.A., and Meltzer, M.S. (1995). Growth of *Francisella tularensis* LVS in macrophages: the acidic intracellular compartment provides essential iron required for growth. *Infect. Immun.* 63:1478–1483.

Foshay, L. (1932). Tularemia. Accurate and earlier diagnosis by means of the intradermal reaction. *J. Infect. Dis.* 51:286–291.

Foshay, L. (1940). Tularemia: a summary of certain aspects of the disease including methods for early diagnosis and the results of serum treatment in 600 patients. *Medicine* (Baltimore) 19:1–83.

Franz, D.R., Jahrling, P.B., Friedlander, A.M., McClain, D.J., Hoover, D.L., Bryne, W.R., Pavlin, J.A., Christopher, G.W., and Eitzen, E.M., Jr. (1997). Clinical recognition and management of patients exposed to biological warfare agents. *J.A.M.A.* 278:399–411.

Giddens, W.R., Wilson, J.W., Dienst, F.T., Jr., and Hargrove, M.D. (1957). Tularemia: an analysis of one hundred forty-seven cases. *J. Louisiana State Med. Soc.* 109:93–98.

Golovliov, I., Sjostedt, A., Mokrievich, A., and Pavlov, V. (2003). A method for allelic replacement in *Francisella tularensis*. *FEMS Microbiol. Lett.* 222:273–280.

Green, S.J., Nacy, C.A., Schreiber, R.D., Granger, D.L., Crawford, R.M., Meltzer, M.S., and Fortier, A.H. (1993). Neutralization of gamma interferon and tumor necrosis factor alpha blocks in vivo synthesis of nitrogen oxides from L-arginine and protection against *Francisella tularensis* infection in *Mycobacterium bovis* BCG-treated mice. *Infect. Immun.* 61:689–698.

Grunow, R., and Finke, E.J. (2002). A procedure for differentiating between the intentional release of biological warfare agents and natural outbreaks of disease: its use in analyzing the tularemia outbreak in Kosovo in 1999 and 2000. *Clin. Microbiol. Infect.* 8:510–521.

Grunow, R., Splettstoesser, W., McDonald, S., Otterbein, C., O'Brien, T., Morgan, C., Aldrich, J., Hofer, E., Finke, E.J., and Meyer, H. (2000). Detection of *Francisella tularensis* in biological specimens using a capture enzyme-linked immunosorbent assay, an immunochromatographic handheld assay, and a PCR. *Clin. Diagn. Lab. Immunol.* 7:86–90.

Gundry, L.P., and Warner, C.G. (1933). Fatal tularemia: review of autopsied cases with report of a fatal case. *Ann. Intern. Med.* 7:837–852.

Harris, C.E. (1926). Tularemia. *Colo. Med.* 23:328–334.

Harris, S. (1992). Japanese biological warfare research on humans: a case study of microbiology and ethics. *Ann. N. Y. Acad. Sci.* 666:21–52.

Hernychova, L., Kovarova, H., Macela, A., Kroca, M., Krocova, Z., and Stulik, J. (1997). Early consequences of macrophage—*Francisella tularensis* interaction under the influence of different genetic background in mice. *Immunol. Lett.* 57:75–81.

Hopla, C.E. (1974). The ecology of tularemia. *Adv. Vet. Sci. Comp. Med.* 18:25–53.

Hornick, R. (2001). Tularemia revisited. *N. Engl. J. Med.* 345:1637–1639.

Hornick, R.B., and Eigelsbach, H.T. (1966). Aerogenic immunization of man with live tularemia vaccine. *Bacteriol. Rev.* 30:532–538.

Hutwagner, L.C., Thompson, W., Groseclose, S.L., and Willamson, G.D. (2000). An evaluation of alternative methods for detection of aberrations in public health surveillance data. *Proceedings of the Biometrics Sections*, American Statistical Association, Baltimore, pp. 82–85.

Ikaheimo, I., Syrjala, H., Karhukorpi, J., Schildt, R., and Koskela, M. (2000). In vitro antibiotic susceptibility of Francisella tularensis isolated from humans and animals. *J. Antimicrob. Chemother.* 46:287–290.

Jacobs, R.F., Condrey, Y.M., and Yamauchi, T. (1985). Tularemia in adults and children: a changing presentation. *Pediatrics* 76:818–822.

Johansson, A., Berglund, L., Eriksson, U., Goransson, I., Wollin, R., Forsman, M., Tarnvik, A., and Sjostedt, A. (2000a). Comparative analysis of PCR versus culture for diagnosis of ulceroglandular tularemia. *J. Clin. Microbiol.* 38:22–26.

Johansson, A., Berglund, L., Gothefors, L., Sjostedt, A., and Tarnvik, A. (2000b). Ciprofloxacin for treatment of tularemia in children. *Pediatr. Infect. Dis. J.* 19:449–453.

Johansson, A., Urich, S.K., Chu, M.C., Sjostedt, A., and Tarnvik, A. (2002). In vitro susceptibility to quinolones of *Francisella tularensis* subspecies *tularensis*. *Scand. J. Infect. Dis.* 34:327–330.

Kaufmann, A.F., Meltzer, M.I., and Schmid, G.P. (1997). The economic impact of a bioterrorist attack: are prevention and postattack intervention programs justifiable? *Emerg. Infect. Dis.* 3:83–94.

Ketai, L., Alrahji, A.A., Hart, B., Enria, D., and Mettler, F., Jr. (2003). Radiologic manifestations of potential bioterrorist agents of infection. *AJR Am. J. Roentgenol.* 180:565–575.

Koskela, P., and Herva, E. (1980). Cell-mediated immunity against *Francisella tularensis* after natural infection. *Scand. J. Infect. Dis.* 12:281–287.

Koskela, P., and Herva, E. (1982). Cell-mediated and humoral immunity induced by a live *Francisella tularensis* vaccine. *Infect. Immun.* 36:983–989.

Kovarova, H., Hernychova, L., Hajduch, M., Sirova, M., and Macela, A. (2000). Influence of the *bcg* locus on natural resistance to primary infection with the facultative intracellular bacterium *Francisella tularensis* in mice. *Infect. Immun.* 68:1480–1484.

Lai, X.H., Golovliov, I., and Sjostedt, A. (2001). *Francisella tularensis* induces cytopathogenicity and apoptosis in murine macrophages via a mechanism that requires intracellular bacterial multiplication. *Infect. Immun.* 69:4691–4694.

Larson, C.L. (1970). Tularemia. In: Tice, F., and Sanford, J.P. (eds.), *Tice's Practice of Medicine*, vol. 3. Harper & Row Publishers, Inc., Hagerstown, MD, pp. 663–676.

Mason, W.L., Eigelsbach, H.T., Little, S.F., and Bates, J.H. (1980). Treatment of tularemia, including pulmonary tularemia, with gentamicin. *Am. Rev. Respir. Dis.* 121:39–45.

Maurin, M., Mersali, N.F., and Raoult, D. (2000). Bactericidal activities of antibiotics against intracellular *Francisella tularensis*. *Antimicrob. Agents Chemother.* 44:3428–3431.

McCrumb, F.R.J. (1961). Aerosol infection of man with *Pasteurella tularensis*. *Bacteriol. Rev.* 25: 262–267.

Miller, R.P., and Bates, J.H. (1969). Pleuropulmonary tularemia. A review of 29 patients. *Am. Rev. Respir. Dis.* 99:31–41.

Narayanan, R.B., Drabick, J.J., Williams, J.C., Fortier, A.H., Meltzer, M.S., Sadoff, J.C., Bolt, C.R., and Nacy, C.A. (1993). Immunotherapy of tularemia: characterization of a monoclonal antibody reactive with *Francisella tularensis*. *J. Leukoc. Biol.* 53:112–116.

Overholt, E.L., Tigertt, W.D., Kadull, P.J., Ward, M.K., Charkes, N.D., Rene, R.M., Salzman, T.E., and Stephens, M. (1961). An analysis of forty-two cases of laboratory-acquired tularemia. Treatment with broad spectrum antibiotics. *Am. J. Med.* 30:785–806.

Patt, H.A., and Feigin, R.D. (2002). Diagnosis and management of suspected cases of bioterrorism: a pediatric perspective. *Pediatrics* 109:685–692.

Penn, R.L. (2005). *Francisella tularensis* (tularemia). In: Mandell, G., Bennett, J., and Dolan, R. (eds.), *Mandell, Douglas, and Bennett's Principles and Practice of Infectious Dieseases*, 6th ed. Elsevier Churchill Livingston, Philadelphia, pp. 2674–2685.

Penn, R.L., and Kinasewitz, G.T. (1987). Factors associated with a poor outcome in tularemia. *Arch. Intern. Med.* 147:265–268.

Perez-Castrillon, J.L., Bachiller-Luque, P., Martin-Luquero, M., Mena-Martin, F.J., and Herreros, V. (2001). Tularemia epidemic in northwestern Spain: clinical description and therapeutic response. *Clin. Infect. Dis.* 33:573–576.

Permar, H.H., and Maclahlan, W.W.G. (1931). Tularemia pneumonia. *Ann. Intern. Med.* 5:687–698.

Poquet, Y., Kroca, M., Halary, F., Stenmark, S., Peyrat, M.A., Bonneville, M., Fournie, J.J., and Sjostedt, A. (1998). Expansion of Vγ9Vδ2 T cells is triggered by *Francisella tularensis*-derived phosphoantigens in tularemia but not after tularemia vaccination. *Infect. Immun.* 66:2107–2114.

Reintjes, R., Dedushaj, I., Gjini, A., Jorgensen, T.R., Cotter, B., Lieftucht, A., D'Ancona, F., Dennis, D.T., Kosoy, M.A., Mulliqi-Osmani, G., Grunow, R., Kalaveshi, A., Gashi, L., and Humolli, I. (2002). Tularemia outbreak investigation in Kosovo: case control and environmental studies. *Emerg. Infect. Dis.* 8:69–73.

Robinson-Dunn, B. (2002). The microbiology laboratory's role in response to bioterrorism. *Arch. Pathol. Lab. Med.* 126:291–294.

Rodgers, B.L., Duffield, R.P., Taylor, T., Jacobs, R.F., and Schutze, G.E. (1998). Tularemic meningitis. *Pediatr. Infect. Dis. J.* 17:439–441.

Rubin, S.A. (1978). Radiographic spectrum of pleuropulmonary tularemia. *AJR Am. J. Roentgenol.* 131:277–281.

Russell, P., Eley, S.M., Fulop, M.J., Bell, D.L., and Titball, R.W. (1998). The efficacy of ciprofloxacin and doxycycline against experimental tularaemia. *J. Antimicrob. Chemother.* 41:461–465.

Sandstrom, G., Lofgren, S., and Tarnvik, A. (1988). A capsule-deficient mutant of *Francisella tularensis* LVS exhibits enhanced sensitivity to killing by serum but diminished sensitivity to killing by polymorphonuclear leukocytes. *Infect. Immun.* 56:1194–1202.

Sanford, J.P. (1983). Landmark perspective: tularemia. *J.A.M.A.* 250:3225–3226.

Saslaw, S., Eigelsbach, H.T., Prior, J.A., Wilson, H.E., and Carhart, S. (1961a). Tularemia vaccine study. II. Respiratory challenge. *Arch. Intern. Med.* 107:702–714.

Saslaw, S., Eigelsbach, H.T., Wilson, H.E., Prior, J.A., and Carhart, S. (1961b). Tularemia vaccine study. I. Intracutaneous challenge. *Arch. Intern. Med.* 107:689–701.

Sawyer, W.D., Dangerfield, H.G., Hogge, A.L., and Crozier, D. (1966). Antibiotic prophylaxis and therapy of airborne tularemia. *Bacteriol. Rev.* 30:542–550.

Scofield, R.H., Lopez, E.J., and McNabb, S.J. (1992). Tularemia pneumonia in Oklahoma, 1982–1987. *J. Oklahoma State Med. Assoc.* 85:165–170.

Sewell, D.L. (2003). Laboratory safety practices associated with potential agents of biocrime or bioterrorism. *J. Clin. Microbiol.* 41:2801–2809.

Shapiro, D.S., and Mark, E.J. (2000). Case records of the Massachusetts General Hospital. Weekly clinicopathological exercises. Case 14-2000. A 60-year-old farm worker with bilateral pneumonia. *N. Engl. J. Med.* 342:1430–1438.

Shapiro, D.S., and Schwartz, D.R. (2002). Exposure of laboratory workers to *Francisella tularensis* despite a bioterrorism procedure. *J. Clin. Microbiol.* 40:2278–2281.

Sjostedt, A., Eriksson, U., Berglund, L., and Tarnvik, A. (1997). Detection of *Francisella tularensis* in ulcers of patients with tularemia by PCR. *J. Clin. Microbiol.* 35:1045–1048.

Stuart, B.M., and Pullen, R.L. (1945). Tularemia pneumonia. Review of American literature and report of 15 additional cases. *Am. J. Med. Sci.* 210:223–236.

Stupak, H.D., Scheuller, M.C., Schindler, D.N., and Ellison, D.E. (2003). Tularemia of the head and neck: a possible sign of bioterrorism. *Ear Nose Throat J.* 82:263–265.

Syrjala, H., Karvonen, J., and Salminen, A. (1984). Skin manifestations of tularemia: a study of 88 cases in northern Finland during 16 years (1967–1983). *Acta Derm. Venereol.* 64: 513–516.

Syrjala, H., Kujala, P., Myllyla, V., and Salminen, A. (1985). Airborne transmission of tularemia in farmers. *Scand. J. Infect. Dis.* 17:371–375.

Syrjala, H., Schildt, R., and Raisainen, S. (1991). In vitro susceptibility of *Francisella tularensis* to fluoroquinolones and treatment of tularemia with norfloxacin and ciprofloxacin. *Eur. J. Clin. Microbiol. Infect. Dis.* 10:68–70.

Syrjala, H., Sutinen, S., Jokinen, K., Nieminen, P., Tuuponen, T., and Salminen, A. (1986). Bronchial changes in airborne tularemia. *J. Laryngol. Otol.* 100:1169–1176.

Tarnvik, A. (1989). Nature of protective immunity to *Francisella tularensis*. *Rev. Infect. Dis.* 11:440–451.

Tarnvik, A., and Berglund, L. (2003). Tularaemia. *Eur. Respir. J.* 21:361–373.

Tarnvik, A., and Lofgren, S. (1975). Stimulation of human lymphocytes by a vaccine strain of Francisella tularensis. *Infect. Immun.* 12:951–957.

Telepnev, M., Golovliov, I., Grundstrom, T., Tarnvik, A., and Sjostedt, A. (2003). *Francisella tularensis* inhibits toll-like receptor-mediated activation of intracellular signalling and secretion of TNF-alpha and IL-1 from murine macrophages. *Cell Microbiol.* 5:41–51.

Teutsch, S.M., Martone, W.J., Brink, E.W., Potter, M.E., Eliot, G., Hoxsie, R., Craven, R.B., and Kaufmann, A.F. (1979). Pneumonic tularemia on Martha's Vineyard. *N. Engl. J. Med.* 301: 826–828.

Tigertt, W.D. (1962). Soviet viable *Pasteurella tularensis* vaccines. A review of selected articles. *Bacteriol. Rev.* 26:354–373.

Waag, D.M., Sandstrom, G., England, M.J., and Williams, J.C. (1996). Immunogenicity of a new lot of *Francisella tularensis* live vaccine strain in human volunteers. *FEMS Immunol. Med. Microbiol.* 13:205–209.

White, J.D., Rooney, J.R., Prickett, P.A., Derrenbacher, E.B., Beard, C.W., and Griffith, W.R. (1964). Pathogenesis of experimental respiratory tularemia in monkeys. *J. Infect. Dis.* 114:277–283.

World Health Organization. (1970). *Health Aspects of Chemical and Biological Weapons: Report of a WHO Group of Consultants*, World Health Organization, Geneva.

Zaidi, S.A., and Singer, C. (2002). Gastrointestinal and hepatic manifestations of tickborne diseases in the United States. *Clin. Infect. Dis.* 34:1206–1212.

4

Melioidosis and Glanders as Possible Biological Weapons

David Allan Brett Dance

1. INTRODUCTION

Glanders and melioidosis are caused by two closely related Gram-negative bacilli, known as *Burkholderia mallei* and *Burkholderia pseudomallei*, respectively. Both are considered to have potential as bioweapons, and glanders has actually been used deliberately against animals and humans in the past. Of the two, melioidosis has been far more studied in recent years, as it has been recognized increasingly as an important disease in some parts of the tropics (Dance, 1991, 2001), whereas naturally occurring glanders is extremely rare nowadays. Over the past few years, there has been huge progress in our understanding of the basic biology and pathogenesis of these species, and the sequencing of both their genomes has now been completed. This contrasts starkly with our lack of understanding of many aspects of their ecology, epidemiology, and the clinical diseases that they cause. Although glanders has the longer pedigree as a biothreat agent, this review will particularly focus on melioidosis, about which more is known, as the agents are so closely related and their clinical manifestations are very similar.

2. HISTORY, DISTRIBUTION, AND EPIDEMIOLOGY

2.1. Melioidosis

Melioidosis was first recognized by Whitmore in 1911 as a glanders-like disease among autopsies he conducted in Rangoon, Burma (Whitmore, 1913). The disease was subsequently identified in humans and other animals throughout south and southeast Asia and, since 1949,

Bioterrorism and Infectious Agents
Edited by Fong and Alibek, Springer Science+Business Media, Inc., New York, 2005

northern Australia (Cottew, 1950). Cases have also been described in Africa, central and South America, and the Caribbean (Dance, 1991, 2001). It is likely that the true incidence of the disease is considerably underestimated in many tropical countries, due to a lack of familiarity and the absence of laboratory facilities capable of making the diagnosis. The greatest number of cases is reported from Thailand, where it has been estimated that some 2,000–5,000 cases occur each year (Dharakul and Songsivilai, 1996). An annual incidence of 4.4/100,000 has been estimated for Ubon Ratchatani province in rural northeast Thailand (Suputtamongkol *et al.*, 1994a), although this is almost certainly an underestimate. A not dissimilar incidence has been reported in the contrastingly urban environment of Singapore, where surveillance is likely to be more complete (Heng *et al.*, 1998). The highest reported annual incidences come from northern Australia and the Torres Straits Islands (up to 42.7/100,000), which again probably reflect better case ascertainment (Currie *et al.*, 2000b, 2000c; Faa and Holt, 2002). Increases in incidence are often associated with particularly heavy monsoons (Merianos *et al.*, 1993). Sporadic cases occur in western and temperate countries as a result of importation from endemic areas (Dance *et al.*, 1999). In this regard, the increasing number of independently confirmed reports of melioidosis in the Americas is important and is a reminder that a sporadic case of melioidosis is far more likely to result from natural importation than deliberate release. Cases have been acquired in Ecuador, Panama, Mexico, El Salvador, Puerto Rico, Brazil, Aruba, Martinique, Guadeloupe, Honduras, Venezuela, and Colombia; and closely related organisms have caused infections that apparently originated within the United States (Dance, 1991; Dorman *et al.*, 1998; McCormick *et al.*, 1977; Miralles *et al.*, 2004; S. Champagne, personal communication). Only once is the disease known to have become established in the environment and transmitted within a temperate country, during the extraordinary outbreak that took place in France during the mid-1970s that became known as "L'affaire du Jardin des Plantes," although this appears eventually to have been self-limiting (Mollaret, 1988).

Burkholderia pseudomallei is an environmental saprophyte, frequently found in mud and surface water in endemic areas, from which man and other animals acquire infection (Inglis *et al.*, 2001). The disease attracted considerable attention as a cause of infection among French and American soldiers serving in southeast Asia, and its propensity for long periods of latency led to the nickname 'Vietnamese Time Bomb' (Goshorn, 1987). Most infections nowadays, however, are seen in indigenous peoples in regular contact with soil and water (rice farmers in northeast Thailand, aboriginals in northern Australia), with the vast majority presenting during the rainy season (Currie *et al.*, 2000b, 2000c; Supputamongkol *et al.*, 1994a). The natural route of infection is thought usually to be inoculation of the organism from soil or surface water, although a history supporting this can only be elicited in 5–25% of cases (Currie *et al.*, 2000b; Suputtamonkol *et al.*, 1994a). A small proportion (<1%) undoubtedly arises from aspiration of surface water following near-drowning incidents (Achana *et al.*, 1985; Suputtamonkol *et al.*, 1994a). The evidence for inhalation as a route of infection is discussed later. Two recent clusters of melioidosis in northern Australia have been associated with contaminated potable water supplies (Inglis *et al.*, 2000a; Currie *et al.*, 2001), raising the possibility of ingestion as a route of infection, although inhalation from a colonized aerator was felt to be more likely in one of these (Inglis *et al.*, 2000a). Person-to-person and animal-to-human spread, although reported, are vanishingly rare (Abbink *et al.*, 2001; Kunakorn *et al.*, 1991a; McCormick *et al.*, 1975). Early French work suggested that transmission by biting insects might be possible (Blanc and Baltazard, 1941), although there is no evidence of this occurring in nature, but *B. pseudomallei* has recently been isolated from flies in Malaysia (Sulaiman *et al.*, 2000).

The incubation period following relatively well-defined exposures is from 1 day to 21 days (Achana *et al.*, 1985; Currie *et al.*, 2000b), although infection may remain latent for many years (see below). It is likely that the majority of cases exposed to a dense aerosol (such as might occur with a deliberate release) would present relatively acutely (within 1 week of exposure) although cases in victims of the recent tsunami in Southeast Asia have presented up to 38 days after exposure (S. J. Peacock, personal communication).

2.2. Glanders

B. mallei was originally isolated by Loeffler and Schütz in 1882 (Boerner, 1882) and glanders is primarily a disease of equines that is occasionally transmitted to man. Since reliance on horses has decreased, it is now very rare: between 1996 and 2001, animal cases were reported from Latvia, Belarus, Mongolia, Turkey, Brazil, Bolivia, and Iran, and human cases from Turkey, New Caledonia, and Sri Lanka (Arun *et al.*, 1999; Office International des Épizooties, 2003). Like melioidosis, glanders is probably underrecognized and underreported. Even when it was common among horses, cases in man were rare, and occurred predominantly among those in prolonged close contact with animals such as grooms. Until the last century, it was a disease of considerable military importance, which caused major outbreaks resulting in the death or necessitating the slaughter of thousands of horses (Howe, 1950). Control measures were largely based on the identification and destruction of sick or infected horses, with heavy reliance on skin testing with "mallein" (Blancou, 1994; Derbyshire, 2002).

3. MICROBIOLOGY AND PATHOGENESIS

3.1. Taxonomy

Originally named "*Bacillus mallei*" and "*Bacillus pseudomallei*," these bacteria have been assigned to a variety of genera over the years (e.g., *Malleomyces*, *Pfeifferella*, *Loefflerella*, and *Actinobacillus*), and until the early 1990s resided in rRNA homology group II of the genus *Pseudomonas*. They now appear to have found a permanent resting place in the genus *Burkholderia* proposed by Yabuuchi and colleagues (1992). Currently approximately 30 species have been assigned to the genus. Many are plant-associated inhabitants of the rhizosphere where they play a variety of roles, including nitrogen fixation, provision of plant nutrients, and inhibition of pathogens, but others are obligate or opportunistic plant and animal pathogens. Of great interest in recent years has been the recognition of an organism in soil in southeast Asia, which was initially thought to be an arabinose-assimilating *B. pseudomallei* (Wuthiekanun *et al.*, 1996), but is now recognized as a distinct species, *Burkholderia thailandensis* (Brett *et al.*, 1998). This is of particular importance because, despite its strong phenotypic similarity with *B. pseudomallei* (they have very similar metabolic profiles and antibiotic susceptibility, and cross-react serologically), it is completely avirulent (Smith *et al.*, 1997), thus making comparative studies of great interest in terms of identifying virulence determinants.

The similarity between *B. mallei* and *B. pseudomallei* that was originally recognized by Whitmore has been confirmed by modern genetic technology. For some time, it has been known that *B. mallei* and *B. pseudomallei* are very closely related genetically by DNA-DNA hybridization (Rogul *et al.*, 1970), and the differences between the two genomes are now proving to be relatively minor. In a recent study of a large collection of human, animal, and environmental isolates of both species, Godoy *et al.* (2003) used multilocus sequence typing (MLST) to compare the strains. They identified 71 sequence types among 128 *B. pseudomallei* strains and found that five isolates of *B. mallei* clustered within the *B. pseudomallei* isolates, whereas both were clearly distinct from *B. thailandensis*, suggesting that *B. mallei* is actually a clone of *B. pseudomallei* and should not be classified as a separate species. In practice, it is convenient to regard it as such, given its distinct clinical and epidemiological behavior, analogous to the situation with *Yersinia pseudotuberculosis* and *Yersinia pestis*.

3.2. Characteristics

3.2.1. General

B. pseudomallei are Gram-negative rods measuring approximately 0.3–0.8 μm by 1–5 μm and have 2–4 polar flagella per cell. The optimal temperature for growth is 37°C; many strains grow poorly below 25°C, but all will grow at 41°C. Colonial morphology varies between strains and between media, and several different morphotypes are often seen in a single isolate, giving the false impression of a mixed culture. Different colonial types exhibit different cellular morphology, cells from the more common opaque, rugose colonies exhibiting irregular and sometimes bipolar staining, probably reflecting accumulation of poly-β-hydroxybutyrate, whereas cells from "smooth" colonies are more evenly staining and parallel-sided. Cells of *B. mallei* are straight or slightly curved and of similar dimensions, but lack flagella and are thus nonmotile, although flagellin genes are present but not expressed (Sprague *et al.*, 2002). They may also stain irregularly. Both are catalase and oxidase positive (although the latter may be unreliable in *B. mallei*) and aerobic, but can grow anaerobically in the presence of nitrate as an alternative electron acceptor. A detailed discussion of the biochemical characteristics of the two species is outside the scope of this chapter. Culture and identification of the organisms are discussed further below.

3.2.2. Antigenic Structure

B. pseudomallei has the typical structure of a Gram-negative bacterium, and the main antigenic components include lipopolysaccharide (LPS), an extracellular polysaccharide capsule, flagella, and a variety of surface proteins. *B. mallei* is similar but lacks flagella.

Lipopolysaccharide. LPS is one of the immunodominant components of *B. pseudomallei* and is remarkably homogeneous within the species (Pitt *et al.*, 1992). More recently, three different LPS banding patterns on SDS-PAGE have been described (Anuntagool *et al.*, 2000b). Ninety six percent of a collection of 739 clinical isolates had a "typical" ladder pattern, while the remainder either had an "atypical" ladder pattern or no ladder pattern, and failed to cross-react with serum from patients infected with "typical" strains. There is serological cross-reactivity between *B. pseudomallei* and *B. mallei*, *B. thailandensis* and, less so, *B. cepacia* LPS, although monoclonal antibodies specific for the LPS of individual species can be produced (Anuntagool and Sirisinha, 2002; Burtnick *et al.*, 2002). Initial studies suggested that *B. pseudomallei* produced two distinct LPS molecules (Knirel *et al.*, 1992, Perry *et al.*, 1995). However, it now appears that what was originally named "Type I O-PS" actually appears to be an extracellular capsule, with many features similar to group 3 capsules in other bacteria (Isshiki *et al.*, 2001; Reckseidler *et al.*, 2001). A cluster of 15 genes appears to be linked to Type II O-PS production and the *wbiA* locus is required for the *O*-acetylation of the L-6dTal*p* residues that constitutes an antibody recognition site (Brett *et al.*, 2003; DeShazer *et al.*, 1998).

B. mallei strains are said to be more antigenically heterogeneous and form two or three groups, only one of which is related to *B. pseudomallei* (Dodin and Fournier, 1970; Stanton *et al.*, 1927). However, the number of strains of *B. mallei* available for current study is relatively limited. Although *B. mallei* LPS is very similar to *B. pseudomallei* LPS, and cross-reacts serologically with the latter, it differs in lacking an *O*-acetyl group at the 4′ position of the L-talose residue, and some strains also lack *O*-repeating units (Burtnick *et al.*, 2002).

Capsule. Vorachit *et al.* (1995) and Puthucheary *et al.* (1996) both described apparent extracellular capsular material visualized by electron microscopy of *B. pseudomallei* cells in vitro and in vivo. Characterization of a species-specific exopolysaccharide (EPS) of *B. pseudomallei* was started by Steinmetz *et al.* (1995), who raised a mouse monoclonal antibody to a carbohydrate antigen that had an estimated molecular weight >150 kDa and had a capsule-like appearance on electron microscopy. All 12 strains of the species tested reacted with the antibody, and only *B. mallei* was found to cross-react. Later work by Steinmetz's group (Nimtz *et al.*, 1997) characterized the polysaccharide as a linear tetrasaccharide consisting of three galactose residues and one Kdo residue, which was confirmed with a different strain of *B. pseudomallei* by Masoud *et al.* (1997). Kawahara *et al.* (1998), however, reported that two additional capsular polysaccharides were produced by *B. pseudomallei*, depending on the culture conditions, including one that was mainly 1,4-linked glucan, and the other a novel polysaccharide containing galactose, rhamnose, glucose, and a uronic acid – produced only when the organism was grown in glycerol. The tetrasaccharide EPS is not present in *B. thailandensis*, but cross-reactivity occurs with some, but not all, *B. mallei* (Anuntagool and Sirisinha, 2002). A fourth capsular polysaccharide, a 6-deoxy-heptane, has been identified by Isshiki *et al.* (2001).

Outer membrane proteins. Gotoh *et al.* (1994) identified five major outer membrane proteins, present in all strains of *B. pseudomallei* studied. These had molecular weights of 70, 38, 31, 24, and 17 kDa. The 38 kDa protein was peptidoglycan-associated and formed aggregates of 110 kDa – characteristics suggesting it was probably a porin.

Flagella. The flagellar antigens of *B. pseudomallei* are also serologically homogeneous; a polyclonal antiflagellar antiserum reacted with all but one of 65 strains tested by Brett *et al.* (1994). The flagellin protein had a molecular weight of 43.4 kDa and showed homology with flagellin from other Gram-negative species, including *Proteus mirabilis*, *Bordetella bronchiseptica*, and *Pseudomonas aeruginosa*. The genetic control of flagellar synthesis also has similarities to that in other Gram-negative bacteria (DeShazer *et al.*, 1997).

3.3. Ecology and Environmental Survival

B. pseudomallei is well equipped to survive in the environment, and has repeatedly been isolated from soil and surface water in endemic areas, particularly moist sites such as a rice paddy (Nachiangmai *et al.*, 1985; Strauss *et al.*, 1969). It is unevenly distributed within the environment, although the reasons for this are unknown. For example, within Thailand, the high incidence of melioidosis in the northeast of the country correlates with greater environmental contamination (Leelarasamee *et al.*, 1997; Vuddhakul *et al.*, 1999). Laboratory studies have shown that it can survive for up to 30 months in moist clay soil (Thomas and Forbes-Faulkner, 1981) or for more than 3 years in distilled water (Wuthiekanun *et al.*, 1995b), but for a much shorter time in dry sandy soil. In one temperate area of Western Australia, the apparent clonality of isolates of *B. pseudomallei* collected over 25 years suggested environmental persistence throughout this period (Currie *et al.*, 1994). Moisture seems to be particularly critical; Tong *et al.* (1996) found that the organism died within 70 days in soil with a water content of <10%, but was able to survive for up to 726 days if this was >40%. In these experiments, death was also found to occur within 18 days on nutrient medium at 0°C and at pH values below 5 and above 8, and the organisms were found to be more susceptible to ultraviolet rays than permanent soil bacteria. Other workers have reported greater acid-tolerance, with growth occurring readily in media with an initial pH of 4.5 (Dejsirilert *et al.*, 1991). Somewhat counterintuitively, some studies have reported a higher yield of *B. pseudomallei* from soil during the dry season than the wet season (Brook *et al.*, 1997; Wuthiekanun *et al.*, 1995a). It is likely, however, that available techniques for isolation of environmental *B. pseudomallei* are suboptimal, particularly for recovering stressed bacteria, which may even be "viable but noncultivable," and further work in this area needs to be done. Saline environments (sea or estuarine water) appear to be inimical to survival (Chen *et al.*, 2003).

B. mallei appears to be less robust than *B. pseudomallei*, being unable to survive in dried pus for longer than a few days, or for 24 hours when exposed to sunlight, although it can survive in tap water for at least 4 weeks (Howe, 1950; Miller *et al.*, 1948a).

3.4. Antibiotic Susceptibility

Both species have considerable intrinsic resistance to antibiotics, *B. pseudomallei* being the more resistant of the two (Dance *et al.*, 1989b). Resistance to aminoglycosides, polymyxins, and early beta-lactams is the norm in this species, and fluoroquinolones have only moderate activity, with the exception of clinafloxacin (Ho *et al.*, 2002a). Aminoglycoside and macrolide resistance appears to be due to a multidrug efflux system (Moore *et al.*, 1999), and resistance to polymyxins and other cationic peptides to their inability to bind to lipid A due to the protective barrier of *O*-antigen and outer core LPS components (Burtnick and Woods, 1999). Conversely, most strains are susceptible to the carbapenems (imipenem, meropenem, and biapenem), ceftazidime, amoxycillin-clavulanic acid, piperacillin, doxycycline, and chloramphenicol (Dance *et al.*, 1989b; Jenney *et al.*, 2001; Smith *et al.*, 1996), although they may be more resistant when growing as biofilms (Vorachit *et al.*, 1993). Susceptibility to trimethoprim and sulfonamides, both alone and in combination, is very difficult to test reproducibly (Pilourias *et al.*, 2002). The carbapenems have the greatest bactericidal activity in vitro against *B. pseudomallei* (Smith *et al.*, 1994) with a significant postantibiotic effect (Walsh *et al.*, 1995b) and retain activity against ceftazidime-resistant isolates (Smith *et al.*, 1996). Resistance to chloramphenicol can emerge during treatment, often accompanied by reduced susceptibility to tetracyclines and co-trimoxazole, and also, less frequently, to any of the beta-lactams (Dance *et al.*, 1989b). In vitro antagonism between co-trimoxazole and chloramphenicol, tetracyclines, and beta-lactams has been reported, but the clinical significance of this is unknown (Dance *et al.*, 1989c), as is that of small colony variants with reduced susceptibility to various unrelated classes of antimicrobials (Häußler *et al.*, 1999).

Our knowledge of the *B. pseudomallei* genome is contributing to our understanding of beta-lactam resistance in the species (Holden *et al.*, 2004). The intrinsic presence of a clavulanate-sensitive beta-lactamase has been known for some time (Livermore *et al.*, 1987), and three different patterns of acquired beta-lactam resistance have been described (Godfrey *et al.*, 1991). In fact, the sequencing strain, which is assumed to be representative of the species, possesses three beta-lactamase genes (classes A, C, and D). All three enzymes may be involved in the acquisition of beta-lactam resistance, either through overproduction or by mutation (Ho *et al.*, 2002b; Niumsup and Wuthiekanun, 2002; Tribuddharat *et al.*, 2003).

Data on the antibiotic susceptibility of *B. mallei* are scanty, as few strains are available for study. The main difference between *B. mallei* and *B. pseudomallei* susceptibility profiles is that the former retains susceptibility to the aminoglycosides and azithromycin, but otherwise the patterns are similar (Al-Izzi and Al-Bassam, 1989; Heine *et al.*, 2001; Kenny *et al.*, 1999).

3.5. Genomics

The genomes of both species have been sequenced, *B. pseudomallei* by the Sanger Institute in the UK and *B. mallei* by TIGR in the United States. The size of the *B. pseudomallei* genome is 7.24 Mb, comprising two chromosomes of 4.07 Mb and 3.17 Mb, with a G+C content of 68% (Holden *et al.*, 2004). It is rich in insertion sequences and transposable elements (Brown and Beacham, 2000; Brown *et al.*, 2000; Mack and Titball, 1998). The *B. mallei* genome is smaller, with a chromosome of 3.5 Mb and a 2.3 Mb "megaplasmid," perhaps reflecting its greater adaptation to a parasitic lifestyle (Nierman *et al.*, 2004). Insertion sequences and phase-variable genes are also abundant. This rich source of information is already enhancing our understanding of the species, and readers are best advised to consult the respective websites for the most up-to-date information. Plasmids and phages have also been described in these species (Koh *et al.*, 1989; Woods *et al.*, 2002). Techniques for genetic manipulation of *B. pseudomallei* have been developed. Woods' group in Calgary has pioneered the use of a transposon, Tn5-OT182, for insertional mutagenesis studies. This contains ampicillin and tetracycline resistance markers allowing selection of potentially mutagenized clones (DeShazer *et al.*, 1997; Woods *et al.*, 1999). Transformation by electroporation can also be performed if kanamycin is added to the culture medium, with efficiency of up to 8×10^3 transformants per microgram of plasmid DNA (Mack and Titball, 1996).

3.6. Typing Systems

The apparent serological homogeneity among isolates of *B. pseudomallei* and *B. mallei* has meant that no serotyping system has ever been developed. Differentiation between strains has had to await the development of genotyping techniques, many of which have now been applied to the species. There is, as yet, no internationally accepted and standardized system, and so comparisons between strains can only be made within an individual laboratory. Approaches that have been used include ribotyping (Inglis *et al.*, 2002; Lew and Desmarchelier, 1993; Pitt *et al.*, 2000; Sexton *et al.*, 1993), pulsed-field gel electrophoresis (PFGE) (Pitt *et al.*, 2000), random amplified polymorphic DNA analysis (RAPD) (Haase *et al.*, 1995), multilocus enzyme electrophoresis (Norton *et al.*, 1998), and MLST (Godoy *et al.*, 2003). As one might expect, these vary in their discriminatory powers. Ribotyping, using enzymes such as *Bam*HI or *Eco*RI, tends to group most isolates into two or three clusters, with other ribotypes restricted to only a handful or single isolates (Lew and Desmarchelier, 1993). The more common ribotypes can then be separated by a more discriminatory technique such as PFGE or RAPD (Haase *et al.*, 1995; Pitt *et al.*, 2000). Using this approach, some interesting differences between the geographical distributions of different genotypes (Pitt *et al.*, 2000) or their associations with clinical manifestations (Norton *et al.*, 1998; Pitt *et al.*, 2000) have been observed, but the picture is far from complete. MLST is better suited to interlaboratory comparisons, and more extensive use of this sort of approach should greatly enhance our understanding of the epidemiology and pathogenesis of melioidosis, although the remarkable homogeneity of *B. mallei* isolated from three continents over a 30-year period (Godoy *et al.*, 2003) means that alternative approaches will be needed for glanders.

3.7. Bacterial Virulence

Our understanding of features of *B. pseudomallei* that contribute to virulence has been greatly enhanced by genetic techniques, such as subtractive hybridization, comparing virulent *B. pseudomallei* with the avirulent *B. thailandensis*, and transposon mutagenesis. Using the former technique, Brown and Beacham (2000) found that many of the sequences unique to *B. pseudomallei* were of unknown function and some showed similarity to known virulence genes. The latter approach has been extensively used by Professor Don Woods' group in Canada, who have used the transposon Tn5-OT182, a self-cloning, promoter probe that integrates randomly into the *B. pseudomallei* chromosome (DeShazer *et al.*, 1997; Woods *et al.*, 1999).

3.7.1. Endotoxin and Lipids

Clinically, septicemic melioidosis has many of the features of endotoxemia, and it is likely that LPS contributes to pathogenesis. *B. pseudomallei* LPS, however, appears to possess weaker biological activity than enterobacterial LPS, with less pyrogenic activity in rabbits, less toxicity for galactosamine sensitized mice, and weaker activation of murine macrophages, but stronger mitogenic activity on mouse splenocytes, even of LPS-resistant mice (Matsuura *et al.*, 1996). LPS also appears to be involved in resistance to serum bactericidal activity, as serum-sensitive mutants generated by transposon mutagenesis lacked the O-PSII polysaccharide (DeShazer *et al.*, 1998).

A cytotoxic, hemolytic rhamnolipid produced by *B. pseudomallei* has also been described and may contribute to virulence through impairment of macrophage phagocytosis (Häußler *et al.*, 1998, 2003).

3.7.2. Capsule

The importance of extracellular polysaccharide as a virulence factor for both species has been elegantly demonstrated by the work of Reckseidler *et al.* (2001) and DeShazer *et al.* (2001) using subtractive hybridization and mutation. Acapsular mutants of *B. pseudomallei* and *B. mallei* were greatly attenuated (LD_{50} increased by approximately 10^5–10^6-fold in a hamster inoculation model). Similar findings were reported in a mouse model by Atkins *et al.* (2002a) using signature tagged mutagenesis.

3.7.3. Flagella

The evidence for the involvement of flagella in pathogenesis of melioidosis is conflicting. No significant difference was found between the animal virulence of a flagellar mutant and the wild-type strain, suggesting that flagella and/or motility are probably not virulence determinants in the model used (DeShazer *et al.*, 1997). On the other hand, an isogenic mutant with a deletion in *fliC* (aflagellate and non motile) was avirulent for intranasal infection of BALB/c mice, despite having equivalent ability to invade and replicate in cultured human lung cells compared with the wild type (Chua *et al.*, 2003). During invasion of *Acanthamoeba* spp., *B. pseudomallei* initially attaches to the cell surface via flagella, and a nonflagellate isogenic mutant did not demonstrate flagellum-mediated endocytosis in contrast to the wild-type strain (Inglis *et al.*, 2003).

3.7.4. Exotoxins and Enzymes

Early workers suggested that some of the manifestations of *B. pseudomallei* infection were caused by secreted toxins, although recent corroborative evidence for this is scanty. Liu (1957) showed that *B. pseudomallei* produced a lethal and dermonecrotic toxin, and Heckly and Nigg (1958) separated two heat-labile toxic constituents, both of which were lethal for mice when injected intraperitoneally but only one of which was dermonecrotic. Ismail *et al.* (1987) later reported the production by *B. pseudomallei* of a 31 kDa protein exotoxin that was a potent inhibitor of DNA and protein synthesis in cultured macrophages (Mohamed *et al.*, 1989). A low molecular weight (3 kDa) toxin, which produced a lethal cytopathic effect in McCoy cells, was detected in culture filtrates of isolates of *B. pseudomallei* [Haase *et al.* (1997)]. Isolates from soil were found to be less toxic than those from patients with melioidosis encephalitis. Another cytotoxic product was found to be an acidic rhamnolipid of 762 Da, composed of two molecules of rhamnose and two molecules of β-hydroxytetradecanoic acid (Häußler *et al.*, 1998). It is unclear whether any of these toxins correspond to those described earlier by Heckly and Nigg (1958).

A wide range of potentially tissue-damaging enzymes is produced by *B. pseudomallei*, many of which appear to be secreted using the same pathway (DeShazer *et al.*, 1999). A survey of 100 clinical isolates of *B. pseudomallei* showed that 91% produced lecithinase, lipase, and protease, and 93% haemolysin, but none was positive for elastase (Ashdown and Koehler, 1990). Sexton *et al.* (1994) further characterized a *B. pseudomallei* protease as a metalloenzyme of approximately 36 kDa, which was active against various substrates, including immunoglobulins, but only weakly elastolytic, and which showed serological cross-reactivity with *P. aeruginosa* alkaline protease. Furthermore, a protease-deficient mutant was attenuated in a diabetic rat pneumonia model. Lee and Liu (2000), on the other hand, estimated the molecular mass of the *B. pseudomallei* metalloprotease as 50 kDa. Gauthier *et al.* (2000) found no correlation between the level of activity of a 42 kDa protease and the virulence for mice of *B. pseudomallei* injected intraperitoneally.

The acid phosphatase of *B. pseudomallei* may play a role in intracellular survival (Dejsirilert *et al.*, 1989). Furthermore, cell-associated acid phosphatase appears to serve as a receptor for insulin (Kondo *et al.*, 1996). Woods *et al.* (1993) also reported the ability of *B. pseudomallei* to bind insulin, and proposed the inhibitory effect of insulin on the growth of the organism in vitro as a mechanism to explain the strong association between melioidosis and diabetes mellitus. Subsequent work, however, has suggested that this inhibitory effect was probably artefactual (Simpson and Wuthiekanun, 2000), and the majority of diabetics with melioidosis probably do not have insulin deficiency in any case (Currie, 1995; Simpson *et al.*, 2003).

3.7.5. Secretion Systems

A type II secretion gene cluster – which controls the secretion of protease, lipase and phospholipase C – was identified by DeShazer *et al.* (1999), and two genes were found to be necessary for the maximal secretion of these enzymes. *B. pseudomallei* appears to possess at least three type III secretion systems (TTSS). One of these, which is homologous to a TTSS in *Ralstonia solanacearum* (Winstanley *et al.*, 1999), is present in *B. pseudomallei* but not in *B. thailandensis*, suggesting that it may play a role in virulence (Winstanley and Hart, 2000). Another TTSS is similar to a system found in *Salmonella* and *Shigella*, and may play a role in facilitating invasion and modulating the intracellular behavior of *B. pseudomallei* (Stevens *et al.*, 2002, 2003).

3.7.6. Siderophores

B. pseudomallei obtains iron in serum and cells through a hydroxamate siderophore, malleobactin, which is optimally produced under iron-deficient conditions (Yang *et al.*, 1991). The siderophore is capable of sequestering iron from transferrin and lactoferrin at neutral and acid pH, and appears to acquire iron more effectively from these sources than from the host cell (Yang *et al.*, 1993). The ferric uptake regulator gene in *B. pseudomallei* has also been identified and sequenced (Loprasert *et al.*, 2000).

3.7.7. Adhesion

As previously described, flagella appear to be involved in the attachment of *B. pseudomallei* to *Acanthamoeba* spp. (Inglis *et al.*, 2003), but there is as yet no evidence that they play a similar role with eukaryotic cells. Attachment, albeit weak, of the organism to pharyngeal epithelial cells has also been documented, and electron microscopy suggested the involvement of a thin electron-dense layer, possibly capsular polysaccharide, but not fimbriae (Ahmed *et al.*, 1999). Kanai *et al.* (1997) reported a specific receptor-ligand system, involving the binding of *B. pseudomallei* acid phosphatase to the tissue glycolipids gangliotetraosylceramide (asialo GM1) and gangliotriaosylceramide (asialo GM2). They suggested that the acid phosphatase was most probably a protein tyrosine phosphatase enzyme that might be involved in a two-component signal transfer mechanism. Adherence of *B. pseudomallei* grown at 30°C to human epithelial cell lines was found to be greater than that of bacteria grown at 37°C (Brown *et al.*, 2002.), although the nature of the adhesins in this model was not investigated.

3.7.8. Intracellular Growth

There is considerable evidence that *B. pseudomallei* is able to survive within mammalian cells, and this may be related to the recalcitrant nature of the infection. It is interesting to speculate why such an ability should have evolved in an environmental saprophyte, and it is possible that this relates to growth within environmental amebae and protozoa. Inglis *et al.* (2000b, 2003) found that bacterial adhesion to *Acanthamoeba* species trophozoite surface was followed sequentially by rapid bacillary rotation, endocytosis, the appearance of vacuoles containing single bacilli, then multibacillary vacuoles, and finally formation of external bacillary tufts. Intracellular bacteria remained motile for up to 72 hours, perhaps suggests a symbiotic relationship in the natural environment.

Structurally intact bacteria have been observed within macrophages in experimental melioidosis (Narita *et al.*, 1982). Intracellular survival and growth have also been demonstrated in a range of phagocytic and nonphagocytic cells in vitro (Egan and Gordon, 1996; Harley *et al.*, 1998a; Jones *et al.*, 1996; Pruksachartvuthi *et al.*, 1990). Jones *et al.* (1997) found that invasion of A459 cells could be reduced by transposon mutagensis of two loci, the predicted protein products of which shared considerable homology with a two-component regulatory system involved in heavy metal resistance in other organisms, and these were linked to Cd2+ and Zn2+ resistance in the wild-type strain. However, both the mutant and wild type exhibited similar virulence for animals, suggesting that this property was not linked to virulence. Utaisincharoen *et al.* (2001) found that ingestion of *B. pseudomallei* by the mouse macrophage cell line RAW 264.7 did not induce the production of nitric oxide synthase to the same degree as *Escherichia coli* and *Salmonella typhi*, and suggested that intracellular survival of *B. pseudomallei* might thus reflect a failure to trigger substantial macrophage activation.

There is conflicting evidence regarding the intracellular location of ingested bacteria, some groups suggesting that they remain within membrane-bound vacuoles (Jones *et al.*, 1996). Harley *et al.* (1998a), however, found that *B. pseudomallei* ingested by mouse peritoneal macrophages appeared initially in phagosomes with intact membranes, but fusion with lysosomes was not observed, and loss of the membrane occurred rapidly and the organisms escaped into the cytoplasm. The same group also reported that *B. pseudomallei* induced the formation of multinucleate giant cells in culture (Harley *et al.*, 1998b), as did Utaisincharoen *et al.* (2003) and Kespichayawattana *et al.* (2000), who demonstrated that this resulted from direct cell-to-cell fusion. The pathogenic significance of this phenomenon is unknown, although multinucleated giant cells have also been reported in the tissues of patients with melioidosis (Wong *et al.*, 1995). On the other hand, Hoppe *et al.* (1999) found evidence of intracellular replication occurring within membrane-bound phagosomes. Cell-to-cell spread of *B. pseudomallei* appears to occur in a manner analogous to, but different from, that which occurs with *Listeria monocytogenes*, involving actin rearrangement into a comet tail appearance within membrane protrusions, and leading to cell death through apoptosis (Breitbach *et al.*, 2003; Kespichayawattana *et al.*, 2000). By analogy with *L. monocytogenes*, the 73 kDa phospholipase C of *B. pseudomallei* might also play a role in intracellular survival and spread (Korbrisate *et al.*, 1999).

3.8. Host Defense

Melioidosis is an opportunistic infection, and some 60–80% of cases have recognized underlying diseases that predispose them to infection (Currie *et al.*, 2000b; Suputtamongkol *et al.*, 1994a). The fact that exposure to the organism in the environment is common in all ages, but that disease most frequently presents from the fifth decade of life onwards, suggests that a failure of host defense is the most important determinant of the outcome of exposure. Well-described associations include diabetes mellitus, chronic renal disease, malignancy, steroid therapy, alcoholism and liver disease, chronic lung disease (including cystic fibrosis), pregnancy and, in Australia, kava consumption (Chaowagul *et al.*, 1989; Currie *et al.*, 2000b; O'Carroll *et al.*, 2003; Suputtamongkol *et al.*, 1994a). In a case-control study in northeast Thailand, diabetes mellitus, preexisting renal disease, and thalassemia were all confirmed to be significant independent risk factors for melioidosis (Suputtamongkol *et al.*, 1999). The immune dysfunctions found in these various systemic predisposing conditions are multiple and complex, making it difficult to determine which are most important in determining susceptibility. Interestingly, the immunocompromise associated with human immunodeficiency virus infection does not appear to predispose to melioidosis (Chierakul *et al.*, 2004). The strongest association is with diabetes, which may increase the relative risk of acquiring melioidosis by more than 100-fold in some age groups (Suputtamongkol *et al.*, 1994a), and up to a half of all cases in some series have been people with diabetes. In Thailand, the majority are late-onset diabetics, albeit with low body mass, and it is not uncommon for their diabetes to be first recognized when they present with septicemic melioidosis.

There is also some evidence that genetic make-up may contribute to the susceptibility of an individual to melioidosis. Dharakul *et al.* (1998) reported that there was a significant association between HLA DRB1*1602 and melioidosis, a negative correlation between DQA1*03 and septicemic melioidosis, and trends toward an association between DRB1*0701, DQA1*0201, and DQB1*0201 with relapsing cases. Nuntayanuwat *et al.* (1999) found a significant correlation between the possession of a relatively uncommon allele (TNF2) in the promoter region of the tumor necrosis factor (TNF)-α gene and both the development of, and death from, melioidosis.

An association between glanders and underlying immunocompromise is not apparent from the historical literature, and so it is unclear whether *B. mallei* behaves the same way as *B. pseudomallei* in this respect. The fact that the disease was relatively rare in humans even when glanders was common implies a degree of intrinsic resistance to infection. The American microbiologist who recently contracted laboratory-acquired glanders was diabetic (Srinavasan *et al.*, 2001), so it remains possible that the disease might preferentially affect similar patients to those who are susceptible to melioidosis.

3.8.1. Humoral Immunity

Most patients with melioidosis mount a vigorous antibody response to the organism. Indeed, in endemic areas, the background seropositivity rate is very high (Kanaphun *et al.*, 1993). In most patients, IgG (predominantly IgG1 and IgG2), IgA, and IgM are all produced (Vasu *et al.*, 2003). There is some evidence that humoral immunity may provide some protection against melioidosis, although the occurrence of severe, progressive infections and relapses in patients with high titer antibody suggests that this is only partial, and evidence regarding the protective antigens is conflicting. Early evidence of such a role came from animal experiments demonstrating the passive transfer of protection by serum from animals immunized with LPS, flagellin, and flagellin-LPS conjugate (Brett *et al.*, 1994; Brett and Woods, 1996; Bryan *et al.* 1994). A protective polyclonal antiserum, recognizing O-PSI (i.e. capsule) and O-PSII, and an IgM monoclonal antibody specific for O-PSII, both mediated phagocytic killing by polymorphonuclear leucocytes (Ho *et al.*, 1997). Monoclonal antibodies to the capsular polysaccharide were shown to be significantly more protective in passive immunization studies than antibodies to surface proteins and LPS (Jones *et al.*, 2002). Recently, it has been suggested that some of the differences in susceptibility to melioidosis between BALB/c and C57BL/6 mice may relate to the production of systemic and local antibodies (Liu *et al.*, 2002). In human infections, Charuchaimontri *et al.* (1999) found that the level of antibody to O-PSII, but not O-PSI, was significantly higher in patients who survived melioidosis than in those who died.

3.8.2. Intrinsic and Cellular Immunity

Both intrinsic and acquired cellular immunity appear important in controlling the outcome of infection with *B. pseudomallei* (Ketheesan *et al.*, 2002). The organism is a potent activator of complement, both by the classical and alternative pathways, and complement opsonization enhances phagocytosis (Egan and Gordon, 1996). Macrophages activated by γ-interferon (IFN-γ) exhibit enhanced, dose-dependent inhibition of intracellular growth of *B. pseudomallei*, which appears to be mediated by reactive nitrogen intermediates and less so to reactive oxygen-dependent killing mechanisms (Miyagi *et al.*, 1997).

In an experimental mouse model, IFN-γ appears absolutely crucial to the control of *B. pseudomallei* infection. Depletion of IFN-γ by the administration of antibody reduced the LD_{50} from $>5 \times 10^5$ to approximately 2 colony-forming units (CFUs), and resulted in a marked increase in bacterial counts in the liver and spleen (Santanirand *et al.*, 1999). Neutralization of either TNF-α or interleukin (IL)-12, but not granulocyte-macrophage colony-stimulating factor, increased susceptibility to infection to a lesser extent. Depletion of IFN-γ even precipitated acute, overwhelming sepsis when delayed until 7 days after the bacterial inoculation.

The differential behavior of C57BL/6 and BALB/c mice has proved invaluable in understanding the response to a variety of intracellular pathogens. This model has been studied extensively by groups in Queensland (Leakey *et al.*, 1998) and Hanover (Hoppe *et al.*, 1999), who both found that C57BL/6 mice were considerably more resistant to infection by

B. pseudomallei than BALB/c mice. The latter are exquisitely susceptible to melioidosis, with LD_{50} values as low as only four bacterial cells (Leakey *et al.*, 1998). Following intravenous injection, BALB/c mice developed a progressive bacteremia leading to death within 96 hours, whereas C57BL/6 mice typically remained asymptomatic for up to 6 weeks. However, the course of infection was highly dependent on the infective dose and the bacterial strain used (Hoppe *et al.*, 1999). Even at 12 hours after infection, bacterial loads in the C57BL/6 mice were lower than those in BALB/c mice, suggesting the involvement of an innate mechanism that results in an inability to contain the infection at sites of infection (Barnes *et al.*, 2001). The characteristics of the antibody response (higher IgG2a/IgG1 ratio) initially suggested a predominantly T helper type 1 immune response in the C57BL/6 animals (Hoppe *et al.*, 1999). More recent characterization of cytokine response profiles, however, suggests that the patterns do not follow polarized Th1 or Th2 patterns (Ulett *et al.*, 2000).

3.8.3. Immunopathogenesis

Host responses may be both protective and damaging. Serum levels of TNF-α, IFN-γ, and soluble IL-2 receptors, as well as IL-6, IL-8, IL-10, IL-12p40, IL-15, IL-18, granzymes A and B, procalcitonin, and CXC chemokines have all been found to be elevated in patients with severe melioidosis, and some of these correlate with disease severity and outcome (Friedland *et al.*, 1992; Lauw *et al.*, 1999, 2000a, 2000b; Simpson *et al.*, 2000b; Smith *et al.*, 1995a; Suputtamongkol *et al.*, 1992). Support for a role for hyperproduction of IFN-γ and other proinflammatory cytokines in pathogenesis of septic shock in melioidosis has also been obtained from animal models (Liu *et al.*, 2002; Ulett *et al.*, 2000).

4. CLINICAL SPECTRUM

Although there are some features that appear characteristic of glanders, the two diseases have many similarities, as Whitmore noted long before genomic analysis had quantified the relatedness between the two species. The similarities are likely to be even more marked following aerosol release, the most likely route of deliberate exposure. Melioidosis will be discussed in greater detail as representative of the two diseases, as there has been a considerable amount of recent literature describing its clinical features.

4.1. Melioidosis

Melioidosis has an extremely broad spectrum of clinical manifestations, as a result of which it has been nicknamed "The Remarkable Imitator" (Poe *et al.*, 1971). The majority of exposures appear to result in mild or inapparent infection that never comes to medical attention. When it does, it usually presents as a febrile illness, ranging from an acute overwhelming septicemia accompanied by widespread abscesses, to a chronic localized condition with granulomatous inflammation. Various attempts have been made to group patients into distinct categories, but one form of the disease may lead to another and individual patients are often difficult to categorise. Several reviews have summarized the clinical manifestations of melioidosis (Currie *et al.*, 2000c; Dance, 1990; Howe *et al.*, 1971; Leelarasamee and Bovornkitti, 1989; Putucheary *et al.*, 1992; White, 2003), and readers requiring more detail are referred to these.

4.1.1. Mild and Subclinical Infections

Serological studies suggest that the most common outcome of contact with *B. pseudomallei* in an immunocompetent host is asymptomatic seroconversion. In northeast Thailand, seropositivity rates were found to increase by 20% per year in the first few years of life, so that low-level antibodies to *B. pseudomallei* were found in approximately 80% of hospitalized children by the time they were 4 years old (Kanaphun *et al.* 1993), even though they had had no features of the disease. The assay used in this study was crude and of questionable specificity, however, and cross-reactivity with *B. thailandensis* may have occurred. A flu-like illness associated with seroconversion has also been reported from Australia (Ashdown, 1989).

4.1.2. Latent Infections

One of the more remarkable features of *B. pseudomallei* is its ability to remain latent for many years before causing disease, an unusual property in a bacterial infection. The longest recorded latent period between return from an endemic area to the development of melioidosis is 62 years (Ngauy *et al.*, 2005). Infection usually manifests itself at times of intercurrent stress (e.g., the development of acute viral infections, burns or trauma, malignancies or diabetes mellitus) (Mays and Ricketts, 1975). We have little understanding of where the organism resides in the intervening period and what precise immunological mechanisms underlie this process. Clinically silent, chronic, localized foci of melioidosis in the lung, liver, or spleen have been reported in animals (Dance *et al.*, 1992; Low Choy *et al.*, 2000), and intracellular survival may also play a part. In experimental animals, inapparent melioidosis can be induced to become fulminant by depletion of IFN-γ (Santanirand *et al.*, 1999). The proportion of seropositive individuals who harbor latent infection, and are therefore at risk of relapse, is also uncertain, although the highly seasonal nature of melioidosis suggests that most cases have been recently acquired. Currie *et al.* (2000a) found evidence of reactivation in only 3% of cases seen in northern Australia.

4.1.3. Clinical Disease

Only a small proportion of *B. pseudomallei* infections are sufficiently severe to come to medical attention. The manifestations of those that do are so varied that the literature is full of individual case reports, and pyogenic or granulomatous infection in any organ could potentially be due to melioidosis. Several published case series exemplify the more frequent clinical manifestations (Chaowagul *et al.*, 1989; Currie *et al.*, 2000b; Everett and Nelson 1975; Guard *et al.*, 1984; Kosuwon *et al.*, 1993; Lumbiganon and Viengnondha, 1994; Putucheary *et al.*, 1992; Punyagupta, 1989; Rode and Webling, 1981; Stanton and Fletcher, 1932; Whitmore, 1913; Vatcharapreechasakul *et al.*, 1992). The clinical features of melioidosis seen in Thailand and Australia are broadly similar, with some key differences described in Table 4.1.

Table 4.1
Key Differences Between Melioidosis in Thailand and Australia

	Thailand	Australia
Male:female	3:2	3:1
Mortality	51%	19%
Bacteremia	62%	46%
GU presentation	5%	15%
Encephalomyelitis	0.2%	4%
Parotitis	5.2%	0%

Septicemic melioidosis. Approximately 50–60% of cases of melioidosis have positive blood cultures (Currie *et al.*, 2000b; Suputtamongkol *et al.*, 1994a). The majority of these present with an acute community-acquired septicemia, with characteristic features of "sepsis syndrome." These patients usually have a short history (median 6 days; range 1 day to 2 months) of high fever and rigors (Chaowagul *et al.*, 1989). Approximately half will have a history and signs that point to a primary focus of infection, usually pneumonia or skin or soft tissue infection. Other prominent features frequently include confusion, delirium and altered level of consciousness, jaundice, and diarrhea. Acute respiratory distress syndrome may also be a prominent feature (Puthucheary *et al.*, 2001). Basic laboratory investigations usually reveal anemia, a neutrophil leukocytosis, coagulopathy, and evidence of renal and hepatic impairment. These patients often undergo rapid clinical deterioration, with the development of widespread abscesses, particularly in the lungs, liver, and spleen, and metabolic acidosis with Kussmaul's respiration, and death is common within the first 48 hours of hospital admission irrespective of treatment, particularly if septic shock is present. Absence of fever, leukopenia, azotemia, and abnormal liver function tests are all associated with a poor outcome, as are the levels of a number of proinflammatory cytokines (see above) and high levels of bacteremia (>50 CFUs/mL) (Walsh *et al.*, 1995a).

During bacteremia, the organism is seeded extensively throughout the body and, if the patient survives long enough, the manifestations of these metastatic abscesses become apparent. These secondary abscesses may already be present at presentation. Any site or tissue may be involved, but the most common foci are in the lungs, liver, and spleen, and skin and soft tissues. Prostatic abscesses have also been reported commonly in men in Australia (Currie *et al.*, 2000b). The chest X-ray is abnormal in 60–80% of patients, the most common pattern being widespread, nodular shadowing known as "blood-borne pneumonia" (Chaowagul *et al.*, 1989; Dhiensiri *et al.*, 1988; Putucheary *et al.*, 1992). Multiple liver and splenic abscesses may be detected by ultrasound or CT scan (Vatcharapreechasakul *et al.*, 1992). Pustules and abscesses in the skin and subcutaneous tissues are found in 10–20% of cases (Chaowagul *et al.*, 1989). Other well-described secondary foci include bones and joints, elsewhere in the urinary tract, and the brain and spinal cord. A syndrome comprising peripheral motor weakness, brainstem encephalitis, aseptic meningitis, and respiratory failure, known as "neurological melioidosis," was initially thought to be toxin-mediated (Woods *et al.*, 1992), but now appears more likely to reflect direct invasion of the central nervous system with microabscess formation (Currie *et al.*, 2000b). *B. pseudomallei* bacteremia may also cause a less fulminant disease, with no obvious focal involvement. Such patients usually have a high swinging fever, often associated with profound weight loss. Diagnosis of such cases is difficult until the blood culture becomes positive, which may require several sets as the bacteremia is frequently low level and intermittent, although a complete failure of the fever to lyse when the patient is treated with penicillin and gentamicin is a useful clue. Patients with multiple, noncontiguous foci of infection, which probably reflect bacteremia at some stage, behave similarly.

Localized Melioidosis. The organs most frequently involved in localized melioidosis are the lungs. This may take the form of an acute pneumonia, but most often manifests as a subacute cavitating pneumonia accompanied by profound weight loss, which is often confused with tuberculosis or lung abscess (Everett and Nelson, 1975), but fails to respond to empirical antituberculous treatment. Distinguishing features include relative sparing of the apices and the infrequency of hilar adenopathy (Dhiensiri *et al.*, 1988). The upper lobes are most frequently involved, although any lung zone may be affected, and pneumothorax, empyema, and purulent pericarditis may all occur. As with any form of localized melioidosis, it may ultimately lead to septicemia.

In children in northeast Thailand, acute suppurative parotitis is a characteristic manifestation of melioidosis in children, and accounts for approximately one-third of paediatric cases (Dance *et al.*, 1989a). Surprisingly, this form of the disease has rarely been seen outside Thailand, and has never been reported in Australia, for reasons that are unclear, as is the pathogenesis of this unusual age-site association. Most cases are unilateral and result in parotid abscesses that require surgical drainage, although they may rupture spontaneously into the auditory canal. Facial nerve palsy and septicemia are rare complications.

Other well-described forms of localized melioidosis include skin and soft-tissue abscesses, lymphadenitis, bone and joint infections, liver and/or splenic abscesses, cystitis, pyelonephritis, prostatic abscesses, epidydymoorchitis, keratitis, mycotic aneurysms (which are probably more common than previously recognized), nasopharyngeal infections and, central nervous system abscesses. There is no reason to believe that any organ or system is resistant to *B. pseudomallei* infection, and no doubt additional unusual forms of the disease will continue to be described.

4.2. Glanders

Recent clinical descriptions of human glanders are few and far between, and so we are largely dependent on the historical literature for our understanding of the disease. Traditional descriptions of the disease in horses distinguished between true glanders, predominantly an infection of the nasal passages, and farcy, a cutaneous form that spread along the tracks of lymphatics. However, even these distinctions were often blurred, and mixed pictures were seen, along with nonspecific symptoms and signs of systemic infection, and pulmonary involvement was almost invariably found at autopsy.

Glanders is also a febrile illness in humans that is almost always seen in those in regular close contact with horses (Bernstein and Carling, 1909). Initial symptoms are nonspecific, but cutaneous and subcutaneous nodules and pustules that ulcerated are said to be characteristic (Howe, 1950; Hunting, 1908). The disease can run an acute or chronic course, but death was frequently the outcome in the preantibiotic era, and widespread abscesses were often found at autopsy in muscles, lungs, spleen, and liver. One particularly graphic personal account reported a total of 45 surgical operations to drain abscesses over a 3-year period (Gaiger,

1913). The degree of prostration is often out of proportion to the clinical signs (Bernstein and Carling, 1909). Although involvement of the nasal or oral mucosa has been well described, this is by no means invariable, and certainly not as prominent as it is in horses, but pustular lesions around the face appear to be common. In some cases, there may be a definite history of inoculation of material from a glandered horse, but this is by no means the rule. Laboratory-acquired infections are relatively common in comparison with laboratory-acquired melioidosis (Howe and Miller, 1947), leading to the suspicion that *B. mallei* is more infectious in this setting than *B. pseudomallei*. This has been the only source of human glanders in the United States since 1938 (Howe and Miller, 1947; Srinavasan *et al.*, 2001). Sometimes laboratory-acquired glanders has followed obvious exposure such as a broken culture tube or flask (Bernstein and Carling, 1909: Howe and Miller, 1947), but often there is no discrete exposure. The most recent such case involved a military researcher who had failed to wear gloves when working with *B. mallei*, despite the fact that he was diabetic (Srinavasan *et al.*, 2001). This latter case had an illness that closely resembled melioidosis, with initial axillary lymphadenopathy followed by a systemic illness with liver and splenic abscesses. Despite the history of working with *B. mallei*, both the clinical and laboratory diagnoses of this case were delayed, which highlights the potential difficulties of making a diagnosis of these rare infections.

5. ANIMAL MODELS

5.1. Melioidosis

The list of animals that have developed melioidosis in nature is extremely long, ranging from rodents to nonhuman primates, including farm animals such as sheep, goats and pigs, companion species such as dogs and cats, birds [cockatoos, galahs, pigeons, and even a macaroni penguin! (MacKnight *et al.*, 1990)], crocodiles, and aquatic mammals such as killer whales and dolphins (Hicks *et al.*, 2000; Low Choy *et al.*, 2000). Species undoubtedly differ considerably in their susceptibility to melioidosis, with hamsters being exquisitely susceptible and rats relatively resistant. Bovines, such as water buffalo, that are continually immersed in water containing *B. pseudomallei* in endemic areas and yet rarely contract melioidosis, must also be highly resistant to infection. Different strains within a species also demonstrate marked differences in susceptibility. Furthermore, the clinical manifestations vary from species to species, with sheep and goats often displaying nasal and neurological symptoms, and pigs frequently exhibiting mycotic aneurysm formation (Low Choy *et al.*, 2000). It should not be forgotten that a deliberate release of either of these agents could well cause considerable morbidity and mortality among species other than humans, as is the case with anthrax.

It has thus not proved difficult to identify potential animal models, and many of these have been developed over the years (Woods, 2002). Mice, both well-characterized inbred strains and outbred lines, have been used most frequently. Hamsters, which are exquisitely susceptible to *B. pseudomallei* infection, have been used as a model of acute melioidosis (DeShazer and Woods, 1999). Rats, which are intrinsically relatively resistant to melioidosis, can be rendered susceptible by streptozotocin-induced diabetes (Woods *et al.*, 1993).

Other animal models that have been developed include rabbits (Miller and Clinger, 1961; Miller *et al.*, 1948b), guinea pigs and monkeys (Miller *et al.*, 1948b), goats (Narita *et al.*, 1982; Thomas *et al.*, 1988), and chickens (Vesselinova *et al.*, 1996).

Some of the uses of animal models to elucidate the role of host response in protection against melioidosis have already been previously described, particularly the inbred mouse models. Other uses have included the evaluation of chemotherapy and prophylaxis (Fukuhara *et al.*, 1995; Russell *et al.*, 2000; Ulett *et al.*, 2003) and the assessment of potential vaccine candidates. For example, Atkins *et al.* (2002b) used the highly susceptible BALB/c mice to determine the effect of immunization with a mutant auxotrophic for branched-chain amino acids generated by transposon mutagenesis. Mice inoculated intraperitoneally with the auxotroph were found to be protected against challenge with the wild-type strain. Unfortunately, the results of chemotherapy in animal models cannot necessarily be extrapolated to human use, as evidenced by the contradictory results with fluoroquinolones (Chaowagul *et al.*, 1997; Fukuhara *et al.*, 1995).

In an attempt to overcome the need for mammalian models, a number of groups have explored the use of the nematode, *Caenorhabditis elegans* (Gan *et al.*, 2002; O'Quinn *et al.*, 2001). Unfortunately, since both *B. pseudomallei* and *B. thailandensis* are lethal in this model, whereas *B. mallei* is not, its relevance to human infection is doubtful.

5.2. Glanders

Although the natural hosts of glanders are solipeds, a number of other species can also be infected. There has been relatively little recent work published on the use of animal models of glanders (Fritz *et al.*, 2000). Woods' group in Calgary are currently investigating the use of various horse models to study the disease further, but clearly such work is fraught with problems (Lopez *et al.*, 2003).

6. POTENTIAL AS A BIOLOGICAL WEAPON

The bioweapon potential of *B. mallei* and *B. pseudomallei* has been recognized for many years. It has even been suggested that Conan Doyle (1913) was thinking of melioidosis when he wrote *The Adventure of the Dying Detective*, published the year after Whitmore's original description of the disease, in which an attempt is made on Sherlock Holmes' life using an ingenious spring-loaded device to inoculate him with a culture of a an "... infallibly deadly ... coolie disease from Sumatra ..." (Sodeman, 1994; Vora, 2002).

6.1. Glanders

B. mallei is one of the few agents that has actually been used deliberately to infect man and other animals. During World War I, German agents mounted an extensive campaign of sabotage directed toward animals that were being shipped to the Allies from neutral countries (United States, Romania, Spain, Argentina, and Norway). Both glanders and anthrax were used, being administered by various routes, including percutaneous inoculation, ingestion in sugar lumps containing tiny glass vials, and nasal instillation. The success of the program is unknown, although its participants claimed that several hundred animals were successfully infected (Wheelis, 1999). Glanders was later among the agents used to infect human victims by Japan's notorious Unit 731 in Manchuria under the direction of Ishii Shiro, although it was never successfully incorporated into weapons by his unit (Harris, 1999). Unit 100, under Wakamatsu Yujiro, however, is believed to have produced more than 100 kg of *B. mallei*, which it used for field tests by contamination of water supplies, with apparent success (Harris, 1999). The Russian bioweapons development program also studied *B. mallei*, and even got as far as field tests with glanders (Bojtzov and Geissler, 1999). *B. mallei* (codenamed L5) had been high on the list of potential bioweapons considered by the former USSR since the 1930s, a number of early workers apparently succumbing to the disease themselves during experiments (Alibek and Handelman, 1999). Alibek claims that *B. mallei* was subsequently weaponized successfully, manipulated to enhance its antibiotic resistance, and was used by the Russians to attack the mujaheddin in Afghanistan in the 1980s.

6.2. Melioidosis

Like *B. mallei*, *B. pseudomallei* has been classed as a Category B critical biological agent by Centers for Disease Control and Prevention (CDC) (Rotz *et al.*, 2002). In comparison with *B. mallei*, however, there appears to have been less work done on the development of *B. pseudomallei* as a weapon. Both the United States and the former USSR considered both organisms as potential bioweapons; indeed, the Americans regarded melioidosis as a "... terrifyingly effective disease for poisoning a water supply system" (Moon, 1999). The pathogenesis and infectivity of glanders (LA) and melioidosis (HI) were studied in experimental models at Camp/Fort Detrick, although this work appears never to have progressed as far as field trials (Moon, 1999). Certainly, the Russians believed that the Americans were developing these agents as weapons (Levi, 1960). Less work appears to have been done by the Soviets on *B. pseudomallei* (L6) than *B. mallei*, although the virulence of the organism was studied by scientists in Biopreparat (Alibek and Handelman, 1999).

The key issue regarding the suitability of melioidosis as a bioweapon is its ability to initiate infection in normal individuals via aerosol. Indirect evidence for inhalation as a route of naturally acquired infection comes from the fact that a disproportionate number of cases during the Vietnam War occurred among helicopter crews (Howe *et al.*, 1971). At least two well-described instances of laboratory-acquired melioidosis involved lapses in procedure that are likely to have resulted in the liberation of aerosols (Green and Tuffnell, 1968; Schlech *et al.*, 1981), although inoculation may also have occurred. More recently, Inglis *et al.* (2000a) suggested that an outbreak in an aboriginal community in Western Australia might have been caused by an aerator in a potable water supply system, which was shown to be heavily colonized and liberating aerosols.

One recently reported, unexpected, finding from a 12-year study in Northern Australia was that rainfall in the 14 days before admission was a strong, independent risk factor for

Table 4.2
Suitability of *B. mallei* and *B. pseudomallei* as Bioweapons

Feature	*B. mallei*	*B. pseudomallei*
Availability	+/−	+++
Natural infections		
Humans	+	++[a]
Other animals	++	++
Previous studies		
In vitro assessments	++	++
Field trials	+	−
Weaponization	−	−
Infectious by		
Inhalation	+	+
Inoculation	++	++
Ingestion	+	?
Mortality of septicemic infection	?	++
Person-to-person spread	+/−	+/−
Environmental persistence	+/−	++
Intrinsic antibiotic resistance	+	++

[a]Opportunistic infection – infectivity for immunocompetent individuals likely to be greatly reduced.

a fatal outcome in melioidosis ($p < 0.0001$) (Currie and Jacups, 2003). The authors hypothesized that this may reflect exposure by inhalation to a heavier aerosol generated during heavy monsoon rains. In this context, interesting anecdotal reports from Hong Kong indicate that *B. pseudomallei* may occasionally be grown when selective agar plates are simply exposed to the prevailing typhoon winds (R. Kinoshita, personal communication).

Certainly evidence from animal experiments has for some time suggested that melioidosis may be induced by aerosol exposure (Dannenberg and Scott, 1958). Indeed, there is evidence that both *B. mallei* and *B. pseudomallei* may be more lethal by this route than by intraperitoneal inoculation (Jeddeloh *et al.*, 2003; Lever *et al.*, 2003). As one would expect, higher infecting doses produce more acute disease with a shorter incubation period. What is more difficult to answer is whether it is practically feasible to generate aerosols that will infect significant numbers of immunologically competent humans under field conditions.

The suitability of these two bacteria as biological weapons is considered in Table 4.2. On balance, they are probably appropriately placed in the CDC Category B.

7. DIAGNOSIS AND TREATMENT

7.1. Clinical Diagnosis

Given the wide range of clinical manifestations of melioidosis, it is remarkably difficult to make a specific diagnosis of the disease on clinical grounds alone. Naturally acquired melioidosis should be considered in any patient who has been in an endemic area and presents with a septicemic illness accompanied by multiple abscesses in lung, liver, and spleen, particularly if they have diabetes mellitus or present during the rainy season. The unusual feature of parotid abscesses in children is also suggestive, although for unknown reasons this has rarely been described outside Thailand. Following aerosol exposure, it is likely that patients would present with a rapidly progressive pneumonia, and cavitation would probably be a prominent feature in comparison with other causes. Pharyngocervical infections might also be seen. Specific diagnosis, however, is unlikely to occur until the organism is isolated and identified.

Glanders following a deliberate release would probably present similarly to melioidosis, although involvement of the upper respiratory tract with rhinitis or sinusitis might be more prominent.

7.2. Laboratory Diagnosis

7.2.1. Microscopy and Culture

Definitive diagnosis of melioidosis and glanders requires isolation and identification of the causative organisms from clinical samples. A carrier state for *B. pseudomallei* has never been reported, and so its presence implies infection (Wuthiekanun *et al.*, 2001), and the same is likely to be true for *B. mallei*. Both organisms grow on relatively simple laboratory media such as nutrient agar, but as many microbiologists are unfamiliar with them, there are numerous examples in the literature of them being dismissed as contaminants or labeled as "*Pseudomonas* species," or even aerobic spore-bearers, resulting in delays in diagnosis and treatment, sometimes for many years. Numerous instances of laboratory-acquired infection, particularly with *B. mallei* (Howe *et al.*, 1947) have been recorded, and both species are classed as "Category 3" pathogens in the United Kingdom, requiring specific laboratory containment facilities. In the "post-9/11" era, biosecurity issues are also increasingly important in addition to protection of laboratory workers. It is thus important that clinicians alert laboratories if glanders or melioidosis are suspected, and that microbiologists are aware of these agents and trained to recognize them.

B. pseudomallei grows well on simple media, including nutrient, blood, and MacConkey agar, but it does not grow on deoxycholate citrate or salmonella–shigella agars. After overnight incubation on nutrient agar at 37°C, the colonies are 1–2 mm in diameter and rather nondescript, although a metallic surface sheen and a fruity smell – said to be reminiscent of truffles or of wet earth – may be noticed, although cultures should obviously not be sniffed deliberately if *B. pseudomallei* is suspected! As incubation continues, the distinctive features of the species develop, although these are very variable between strains and on different media formulations. The most common appearance of clinical isolates is an opaque, cream-colored colony, which tends to become wrinkled and rugose with prolonged incubation. Confluent growth is usually surrounded by an area of hemolysis on horse blood agar, although individual colonies may not be. Other strains form predominantly tan, smooth colonies with no hemolysis, still others may exhibit frankly mucoid colonies, and a mixture of two or more colony types is not uncommon. Very rarely, a yellow pigment may be produced. Colonies on MacConkey agar are frequently umbonate. On agar containing glycerol, such as that described by Ashdown (1979a), colony wrinkling is enhanced, and the dye in the medium is taken up, facilitating the presumptive identification of *B. pseudomallei* (Dance *et al.*, 1989d).

The organism can be detected in blood, pus, sputum, urine, or any other appropriate specimen according to the site of infection. Smears that contain bipolar or unevenly staining Gram-negative rods may be suggestive, but the Gram stain alone has poor specificity and sensitivity. One of the most useful developments for rapid diagnosis has been direct immunofluorescent microscopy, which has a sensitivity of 73% and a specificity of 99%, compared with culture (Walsh *et al.*,1994). This is currently only available in a small number of centers, however, although laboratory networks preparing to respond to deliberate release would be well advised to consider establishing this method. An immunoperoxidase method has also been developed for use on tissue sections (Wong *et al.*, 1996).

Isolation of *B. pseudomallei* from normally sterile sites (e.g., blood cultures) is straightforward, but the yield from sites with a normal flora can be increased using a variety of selective media. Wuthiekanun *et al.* (1990) found that the isolation rate from such sites was greatly increased by Ashdown's medium (containing gentamicin), compared with blood and MacConkey agars, and preenrichment using broths containing colistin increases the yield still further (Walsh *et al.*, 1995c, Wuthiekanun *et al.*, 1990). Culture of a throat swab alone using these selective techniques has an overall sensitivity of 36% for the diagnosis of melioidosis (79% in sputum-positive patients), which is particularly useful in children or others who cannot produce sputum (Wuthiekanun *et al.*, 2001). A new medium that was recently developed by Howard and Inglis (2003) is said to be less inhibitory for some mucoid isolates and is undergoing further evaluation. Most strains of *B. pseudomallei* will also grow well on commercially selective media for *B. cepacia* (S. J. Peacock, personal communication), although these have not yet been formally evaluated for their ability to detect the organism in clinical samples.

Identification of *Burkholderia* species once grown may be difficult. Simple guidelines and flowcharts for screening suspect colonies have been developed (e.g., American Society for Microbiology, 2003) and laboratories should use the scheme recommended in their own countries. These days, increasingly few diagnostic laboratories use conventional "homemade" biochemical tests for bacterial identification, but fortunately a number of commercial kits will reliably identify *B. pseudomallei*, such as the API 20E (Ashdown, 1979b), Microbact 24E (Thomas, 1983), API 20NE (Dance *et al.*, 1989d), Minitek discs (Ashdown, 1992a), and Vitek 1 (Lowe *et al.*, 2002), although none is more accurate than a few simple tests (colonial appearance on Ashdown's medium, resistance to gentamicin and colistin) in experienced hands (Dance *et al.*, 1989d). Some have questioned the reliability of the API 20NE (Inglis *et al.*, 1998), and early reports have identified deficiencies with the Vitek 2 and Phoenix BD systems in this respect (Koh *et al.*, 2003; Lowe *et al.*, 2002), and so it is important to ensure that the method used has been validated for this species. Alternative approaches to identification that may yield more rapid results include immunological methods, such as immunofluorescence (Naigowit *et al.* 1993), direct agglutination with monoclonal antibodies (Pongsunk *et al.*, 1996; Rugdech *et al.*, 1995), latex agglutination using polyclonal (Smith *et al.*, 1993) or monoclonal antibodies either directly on colonies (Steinmetz *et al.*, 1999) or on culture broths (Anuntagool *et al.*, 2000a; Dharakul *et al.*, 1999; Pongsunk *et al.*, 1999; Samosornsuk *et al.*, 1999), or a molecular approach (Bauernfeind *et al.*, 1998; Gee *et al.*, 2003; Tungpradabkul *et al.*, 1999; Woo *et al.*, 2002). Overall, latex agglutination is probably the most useful approach, being simple, rapid, and applicable both to isolates and broth cultures. Latex agglutination with a monoclonal antibody specific for the extracellular polysaccharide of

B. pseudomallei has the advantage of avoiding cross-reactivity with *B. thailandensis* (Steinmetz *et al.*, 1999; Wuthiekanun *et al.*, 2002). However, since the latter species is never isolated from clinical cases of melioidosis [the very rare arabinose-assimilating clinical isolates reported not having been fully characterized (Lertpatanasuwan *et al.*, 1999)], this is really only a problem when dealing with environmental isolates. Unfortunately these reagents are not commercially available.

Given the rarity of glanders, it is not surprising that far less is known about optimal diagnostic methods for this than for melioidosis. Few if any western microbiologists will ever have encountered the organism, and no comparative studies of diagnostic methods have been conducted on real clinical samples. *B. mallei* grows less well than *B. pseudomallei* on nutrient agar and forms smooth, gray translucent colonies 0.5–1 mm in diameter in 18 hours at 37°C (Miller *et al.*, 1948a). Unlike *B. pseudomallei*, it is said not to grow on MacConkey agar (Holmes *et al.*, 1986).

7.2.2. *Serological Methods*

Antigen detection. To speed up the laboratory diagnosis of melioidosis, attempts have been made to identify *B. pseudomallei* antigens directly in clinical samples. Most successful of these has been immunofluorescence, which can be used on sputum and pus (see previous data). Detection of antigens in urine has also been attempted, but with less success (Anuntagool *et al.*, 1996; Desakorn *et al.*, 1994; Smith *et al.*, 1995b).

Antibody detection. Unfortunately, there is no international standard method for serodiagnosis of melioidosis. The most widely used method in endemic areas has been an indirect hemagglutination assay that used heated culture filtrates, which contains a mixture of poorly characterized antigens, although it has been suggested that it primarily detects IgM responses to polysaccharides (Ashdown, 1987). Various cut-off titers for regarding the test as "positive" have been suggested (Leelarasamee, 1997; Sirisinha, 1991), but most workers in endemic areas have found that it lacks both sensitivity and, more importantly, specificity (Khupulsup and Petchclai, 1986), probably due to recurrent exposure to the ubiquitous *Burkholderia* spp. in the environment, although cross-reactions with unrelated species may occur (Klein, 1980).

Early attempts to improve the serological diagnosis of melioidosis focused on the differentiation of IgG and IgM responses, using techniques such as immunofluorescence (Ashdown, 1981; Khupulsup and Petchclai, 1986; Vadivelu *et al.*, 1995; Vadivelu and Putucheary, 2000), enzyme-linked immunosorbent assay (ELISA) (Ashdown *et al.*, 1989; Chenthamarakshan *et al.*, 2001; Kunakorn *et al.*, 1990; Vadivelu *et al.*, 1995) or gold blot (Kunakorn *et al.*, 1991b). IgM assays were generally found to correlate better with disease activity, albeit at the expense of a slight loss of sensitivity. Then, during the 1990s, several groups attempted to identify individual antigens that might be purified and used in tests to give greater sensitivity and specificity than the mixtures used in the earlier assays. Potential candidates included a 45 kDa protein (Lertmemongkolchai *et al.*, 1991), LPS (Petkanjanapong *et al.*, 1992), a 40 kDa protein (Wongratanacheewin *et al.*, 1993), a 19.5 kDa antigen (Anuntagool *et al.*, 1993), exotoxin (Embi *et al.*, 1993), a glycolipid antigen (Phung *et al.*, 1995), a monoclonal antibody-affinity purified nonprotein surface antigen (Dharakul *et al.*, 1997), and a recombinant 18.7 kDa antigen (Wongprompitak *et al.*, 2001). Some impressive sensitivities, specificities, and predictive values have been found in studies of relatively small numbers of sera, and there is little doubt that some of these newer assays are an improvement on the IHA test (Sermswan *et al.*, 2000;

Wongratanacheewin *et al.*, 2001), but none of these tests has ever been subjected to a large scale, multicenter analysis. Only one commercially produced kit is available, and this performed encouragingly (IgG – sensitivity 100%; specificity 95%: IgM – sensitivity 93%; specificity 95%) in a small-scale evaluation and warrants further assessment (Cuzzubbo *et al.*, 2000).

7.2.3. Molecular Diagnosis

Due to problems with the speed of conventional culture techniques, and the relative sensitivity and specificity of serology, a number of groups have devised methods for detecting *B. pseudomallei* nucleic acids in clinical samples. Although these have been around for more than 10 years, none has yet found their way into widespread routine use in diagnostic laboratories within endemic areas. Lew and Desmarchelier (1994) constructed an oligonucleotide probe following partial sequencing of the 23S rRNA that they used for the identification and detection of *B. pseudomallei* by hybridization or by direct polymerase chain reaction (PCR). The probe hybridized with DNA from *B. mallei*, but did not react with other *Pseudomonas* spp. A combination of the two methods allowed the detection of approximately 10^4 CFUs/mL in spiked blood samples, and sensitivity could be increased 100-fold by concentration of bacteria from 0.5 mL of blood. Kunakorn and Markham (1995) described a seminested PCR targeting the internal transcribed spacers between 16S and 23S rRNA, with products detected quantitatively by enzyme immunoassay. The primers did not amplify DNA from *B. mallei* and a sensitivity of 75 bacteria per mL was claimed. The 16S rRNA gene was targeted by Dharakul *et al.* (1996) in a nested PCR system for the detection of *B. pseudomallei* in septicemic melioidosis. No cross-reactions were evident with other Gram-negative species tested, although these did not include *B. mallei.* The assay claimed to detect as few as two bacteria, both in buffy coat or pus specimens. Rattanathongkom *et al.* (1997) developed yet another system to amplify a 178 bp product from several clinical isolates. This PCR system detected as little as 0.5 fg of DNA, which equated to a sensitivity in spiked whole blood of 1 bacterial cell per mL. This assay was later compared with three serological methods (ELISA, dot immunoassay, and IHA) for the diagnosis of melioidosis in 130 patients in northeast Thailand, 21 of whom had culture-positive melioidosis and 39 had other bacterial infections (Sermswan *et al.* 2000). Using 1 mL of heparinized blood, they found that the PCR had the best sensitivity (95.2%) and specificity (91.7%) of all the assays, with only two false-positives in patients infected with other organisms (*Acinetobacter anitratus* and *Candida* sp.), and positives in three patients who had blood culture-negative melioidosis. Less sensitive was the system developed by Sura *et al.* (1997), who amplified a specific 124 bp product from clinical strains, which had a lower limit of detection of 35 CFUs/mL in bacterial suspensions and approximately 400 CFUs in liver tissue.

A comparison of several of these published methods was carried out by Haase *et al.* (1998). They were unable to detect *B. pseudomallei* at all using the primer set described by Kunakorn and Markham, and found that the primer set used by Lew and Desmarchelier was less sensitive than culture. On the other hand, they found that the 16S rRNA-derived primer set had a sensitivity approaching 100% for clinical samples and identified three culture-negative cases, but unfortunately lacked specificity, giving positives in a third of patients with other diseases. In contrast, Kunakorn *et al.* (2000) reported the sensitivity and specificity of the 16S rRNA PCR to be only 41% and 47%, respectively, on blood from patients with culture-positive septicemic melioidosis. More recently, Hagen *et al.* (2002) developed a seminested PCR assay targeting a genus-specific sequence of the ribosomal protein subunit 21 and a nested PCR to amplify the filament forming flagellin gene (*fliC*) to detect *B. pseudomallei* in fixed tissues, and found that PCR followed by sequencing of the amplicons resulted in high sensitivity and specificity.

In summary, although numerous groups have developed PCRs that have been reported to be able to detect *B. pseudomallei* in clinical samples, the evidence of their success is conflicting and difficult to reproduce, and there have been no published studies of their use prospectively to diagnose melioidosis in a routine setting.

7.3. Treatment

7.3.1. General

Patients require general supportive treatment, including admission to an Intensive Care Unit as appropriate. Correction of volume depletion, management of metabolic disturbances such as hyperglycemia, acidosis, and uremia are all important. Abscesses should be drained where possible to reduce the infective load, and there is no hard evidence that this carries a risk of precipitating septicemia if patients are on appropriate antibiotics.

7.3.2. Specific Chemotherapy

The incidence of melioidosis is sufficiently high in Thailand to enable comparative trials of antibiotic regimens to be carried out, and numerous such studies have been published over the past 15 years. These have been summarized in two recent reviews (Samuel and Ti, 2003; White, 2003), and readers are referred to these for more detail. Treatment is divided into two phases – an initial parenteral phase, during which the objective is to prevent death, and a subsequent eradicative phase, when the aim is to reduce relapse. The overall mortality rate is usually 40–50% in Thailand (range 14–74%) (White, 2003), but lower in Australia (19%) (Currie *et al.*, 2000b). The relapse rate ranges from 1 to 26% (White, 2003).

Acute phase. B. pseudomallei is intrinsically resistant to many antibiotics, including aminoglycosides and early beta-lactams, and a failure to respond to these agents is characteristic of melioidosis (Chaowagul *et al.*, 1989). "Third-generation" cefalosporins, such as ceftriaxone and cefotaxime, are also associated with a poor outcome (mortality >70%) (Chaowagul *et al.*, 1999b). Several recent comparative trials have, however, shown that the mortality of acute severe melioidosis can be substantially reduced by newer beta-lactam agents, such as ceftazidime, imipenem, amoxicillin–clavulanic acid, and cefoperazone-sulbactam, with or without trimethoprim–sulfamethoxazole (co-trimoxazole), compared with more conventional regimens comprising chloramphenicol, doxycycline, and co-trimoxazole (Chetchotisakd *et al.*, 2001b; Simpson *et al.*, 1999; Sookpranee *et al.*, 1992; Suputtamongkol *et al.*, 1994b; White *et al.*, 1989). Very encouraging results have also been reported from Australia using meropenem plus co-trimoxazole (Cheng *et al.*, 2004; Currie *et al.*, 2000b). The greater release of endotoxin that occurs during treatment with ceftazidime as opposed to the carbapenems does not appear to have any adverse clinical consequences (Simpson *et al.*, 2000a). One of these regimens should be given for 2–4 weeks, and occasionally longer, according to the clinical response. The relevant doses, which of course should be adjusted for renal function if necessary, are shown in

Table 4.3
Regimens for the Treatment of Melioidosis

Acute phase (first 2–4 weeks)		
Drug	Dose (mg/kg/day)	Reference
Ceftazidime	120	White *et al.*, 1989
Ceftazidime plus	100	Sookpranee *et al.*, 1992
Co-trimoxazole	8/40	
Imipenem[a]	50	Simpson *et al.*, 1999
Co-amoxiclav	160	Suputtamongkol *et al.*, 1994b
Cefoperazone-sulbactam plus	25	Chetchotisakd *et al.*, 2001b
Co-trimoxazole	8/40	
Chloramphenicol plus	100	White *et al.*, 1989
Doxycycline plus	4	
Cotrimoxazole	10/50	

[a]Meropenem in equivalent doses, with or without co-trimoxazole, is probably equally effective.

Table 4.3. Continuous infusion of ceftazidime may represent the most efficient and effective mode of administration of ceftazidime (Angus *et al.*, 2000). It remains to be determined whether co-trimoxazole added to the beta-lactam is beneficial, as suggested by animal studies (Ulett *et al.*, 2003), and clinical trials to address this issue are underway in Thailand.

Eradication phase. Following parenteral treatment, prolonged oral antibiotics are needed to prevent relapse, which occurs in up to 25% of patients and is more common in patients who have more severe disease (Chaowagul *et al.*, 1993). Again, several prospective trials have been conducted in Thailand to determine the best regimen (Chaowagul *et al.*, 1997, 1999a; Chetchotisakd *et al.*, 2001a; Rajchanuvong *et al.*, 1995). It is clear that fluoroquinolones and doxycycline alone are inadequate for eradication treatment (relapse rates 18–25%) (Chaowagul *et al.*, 1997, 1999a), as is the combination of ciprofloxacin plus azithromycin (relapse rate 22%) (Chetchotisakd *et al.*, 2001a). The proportion of patients who relapse can be reduced to less than 10%, and probably less than 5%, if appropriate antibiotics are given for 20 weeks. Doses are shown in Table 4.4. The lowest relapse rates have been associated with combinations of the "conventional" agents (chloramphenicol, doxycycline, and co-trimoxazole) (Rajchanuvong *et al.*, 1995), although co-amoxiclav is preferable in

Table 4.4
Regimens for the Treatment of Melioidosis

Eradication phase (to complete 20 weeks)		
Drug	Dose (mg/kg/day)	Reference
Chloramphenicol[a] plus	40	Rajchanuvong *et al.*, 1995
Doxycycline plus	4	Chaowagul *et al.*, 1999a
Co-trimoxazole	8–10/40–50	
Co-amoxiclav	60/15	Rajchanuvong *et al.*, 1995
Co-trimoxazole plus	10/50	Chetchotisakd *et al.*, 2001a
Doxycycline	4	

[a]Duration – first 4–8 weeks only.

children and pregnant or lactating women because of potential toxicity. It is possible that the chloramphenicol may be omitted (Chetchotisakd *et al.*, 2001a), and co-trimoxazole alone has been used in Australia (Cheng *et al.*, 2004; Currie *et al.*, 2000b). Comparative studies to determine whether co-trimoxazole alone is adequate are also underway in Thailand. In patients who have mild localized disease, any of the oral regimens described previously may be used, although the optimum duration is uncertain.

7.3.3. Adjunctive Treatments

Since the mortality of severe melioidosis remains high during the early stages of treatment, and antibiotics alone are unlikely to influence this early mortality, attempts have been made to interrupt proinflammatory pathways or augment host defense to improve outcomes. Dramatic reductions in the mortality of melioidosis associated with septic shock have recently been reported from Darwin in the Northern Territory of Australia (Cheng *et al.*, 2004). Hypothesizing that many of the conditions that predispose to melioidosis are associated with at least functional, if not quantitative, neutrophil deficits, this group has used G-CSF as adjunctive treatment in cases of melioidosis with septic shock admitted to their Intensive Care Unit since 1998. The mortality fell from 20 of 21 (95%) cases between 1989–1998 to 2 of 21 (10%) cases treated with G-CSF between 1998–2002. The use of historical controls is not ideal, and a number of other changes that might have contributed to this reduction took place between these periods, including the appointment of an intensivist and the greater use of carbapenems. Laboratory studies in mice have not supported a beneficial effect from the use of G-CSF immunotherapy (Powell *et al.*, 2003). Nonetheless, this dramatic change warrants further evaluation in properly controlled prospective studies before G-CSF can be recommended routinely for the treatment of severe melioidosis.

7.3.4. Outcome and Follow-up

Even with optimal treatment, the mortality from acute severe melioidosis is high (30–50% in Thailand, 19% in Australia) (Currie *et al.*, 2000c; White, 2003). In patients who survive, there is often chronic morbidity resulting both from the disease itself and from the underlying conditions. Patients require long-term follow-up to detect recurrent infection, which is almost always due to relapse rather than re-infection (Desmarchelier *et al.*, 1993). Susceptibility tests should be carried out on isolates obtained during or after treatment, because resistance may emerge in 5–10% of cases, as discussed previously. Both C-reactive protein and sequential measurement of IgM antibodies have been recommended as ways of detecting disease activity and early relapse (Ashdown, 1981, 1992b).

8. INFECTION CONTROL MEASURES

8.1. Secondary Spread and Isolation

Despite the apparent ease with which it may be acquired from the environment, there is remarkably little evidence of the secondary spread of melioidosis from cases of the disease.

Only two well-documented instances of person-to-person spread are recorded in the literature, one in the wife of a man with *B. pseudomallei* prostatitis (who merely seroconverted and never actually became unwell) (McCormick *et al.*, 1975) and the other in a woman who nursed her brother while he was suffering from septicemic melioidosis (Kunakorn *et al.*, 1991a). A handful of anecdotal instances of animal-to-human spread are also recorded (Low Choy *et al.*, 2000). Glanders likewise appears to require close and prolonged contact with infected animals for transmission, but human-to-human spread has been documented (Howe, 1950). Like anthrax, therefore, the risk of secondary spread of either of these infections appears to be low. It would seem prudent, however, to nurse cases in standard isolation where this is practicable. In view of the greatly increased risk of infection in people with diabetes and people with other forms of underlying immunosuppression, staff with such conditions should not generally be allowed to nurse infectious cases.

8.2. Environmental Contamination

Since the ecology of these two agents is so poorly understood, it is difficult to be precise about the risks of environmental contamination. Methods for the detection of *B. pseudomallei* in soil and water have been evaluated, but their true sensitivity is unknown, even though they have been capable of detecting the organism in up to 68% of sites sampled in north east Thailand (Wuthiekanun *et al.*, 1995a). The limited geographical distribution of melioidosis implies that certain environmental conditions must be fulfilled for the organism to become established. The precise factors that would favor long-term survival of the organism and establishment of new foci are not known, although tropical or subtropical climate and high rainfall are likely to be necessary. Nonetheless, protracted environmental contamination was documented following an outbreak of melioidosis in France in the mid-1970s. This eventually disappeared, although it is not known to what extent the use of disinfectants on soil contributed to this. Nosocomial infection and contamination of the hospital environment have been reported by Ashdown (1979c), and contaminated disinfectant solutions were suggested as a possible source of hospital-acquired melioidosis in infants by Lumbiganon *et al.* (1988). A single documented transmission by endoscopy has been reported (Markovitz, 1979). *B. mallei*, a more highly adapted parasite, is thought not to persist in the environment as well as *B. pseudomallei*.

8.3. Antibiotic Prophylaxis and Vaccines

There is limited evidence for the effectiveness of prophylactic antibiotics in preventing the development of infection in animals exposed to *Burkholderia* spp. Russell *et al.* (2000) found that both ciprofloxacin and doxycycline offered, at best, only partial protection when given before or immediately after, but not 24 hours after, intraperitoneal challenge with *B. mallei* and *B. pseudomallei* in hamsters and mice, respectively. It is possible that co-trimoxazole might be effective prophylactically on the basis of its therapeutic efficacy, although this has never been studied. In the absence of further evidence, any of these three agents might be offered as prophylaxis to those exposed to an aerosol of Burkholderia spp., although their likely efficacy is unknown.

Although a number of groups are studying pathogenesis and immunity of these two species with a view to developing vaccines, at this stage none is currently available for use in either humans or animals. Atkins *et al.* (2002b) have generated an auxotrophic mutant that is protective in mice. Inactivated *B. mallei* preparations have so far shown no protective effect in mice, possibly related to the fact that they generate a mixed Th1- and Th2-like cytokine response and a Th2-like subclass immunoglobulin response (Amemiya *et al.*, 2002). This subject has been reviewed recently by Warawa and Woods (2002).

9. FUTURE DIRECTION

On the balance of current evidence, clinicians and microbiologists are more likely to encounter naturally acquired cases of melioidosis than those due to deliberate release. Glanders is so rare these days that even one case should evoke considerable suspicion. Relatively little work has been done on the methodology necessary to generate infectious aerosols of these agents. Nonetheless, it would be wise not to become complacent. Terrorists will not necessarily be deterred from trying something simply because it has not been done before, and *B. pseudomallei* has the considerable advantage of being widely available within the environment in endemic areas. If resources and efforts within developed countries are devoted to enhancing our understanding of these agents as a result of their potential as bioweapons, it is likely that this will have beneficial spin-offs for the poor inhabitants of those rural tropical endemic areas where they represent current and real threats. This can only be welcomed.

Research aimed at a better understanding of bacterial virulence should give rise to the identification of potential vaccine candidates, although it is unlikely that these would ever be used other than in a deliberate release setting, unless targeted specifically at those at high risk (e.g., people with diabetes). Opportunities for novel therapeutic targets may also come from the unraveling of the secrets of the bacterial genomes. More discriminatory genotyping systems will assist epidemiological investigation of natural infections, and may be the "holy grail" of forensic scientists intent on tracing the source of potential deliberate releases, although the ubiquity of environmental isolates may complicate this endeavor. Better and standardized diagnostic tools, particularly for rapid diagnosis allowing specific directed treatment, are also needed. Ongoing clinical studies in areas of high incidence like Thailand should answer important therapeutic questions, like whether additional co-trimoxazole during the acute phase improves the results of beta-lactam therapy, whether co-trimoxazole alone is adequate eradication therapy, and whether adjunctive G-CSF is really beneficial, although the goal of affordable treatment remains well out of reach for many countries where melioidosis is endemic. Animal studies will be needed to evaluate further the potential for prophylaxis in those exposed to these agents, and again I would put co-trimoxazole at the top of the list of agents needing to be assessed.

Finally, I would like to make a plea for more research on the neglected issue of the ecology of *B. pseudomallei*, including the effects of agricultural chemicals on the ability of soil to support the species. We still have very little understanding of what makes a particular environment conducive to the growth and persistence of *B. pseudomallei*, and how it might be removed from a given setting. Not only would a better grasp of the relevant factors enable us to address the issues of decontamination following a deliberate release that have proved so problematic following the anthrax-contaminated letters in the United States in 2001, but also it would help the search for other unrecognized areas of natural melioidosis endemicity and possibly facilitate environmental remediation of such foci.

References

Abbink, F.C., Orendi, J.M., and de Beaufort, A.J. (2001). Mother-to-child transmission of *Burkholderia pseudomallei*. *N. Engl. J. Med.* 344:1171–1172.

Achana, V., Silpapojakul, K., Thininta, W., and Kalnaowakul, S. (1985). Acute *Pseudomonas pseudomallei* pneumonia and septicemia following aspiration of contaminated water: a case report. *Southeast Asian J. Trop. Med. Public Health* 16:500–504.

Ahmed, K., Encisco, H.D.R., Masaki, H., Tao, M., Omori, A., Tharavichikul, P., and Nagatake, T. (1999). Attachment of *Burkholderia pseudomallei* to pharyngeal epithelial cells: a highly pathogenic bacteria with low attachment ability. *Am. J. Trop. Med. Hyg.* 60:90–93.

Alibek, K., and Handelman, S. (1999). *Biohazard*. Random House, New York.

Al Izzi, S.A., and Al Bassam, L.S. (1989). *In vitro* susceptibility of *Pseudomonas mallei* to antimicrobial agents. *Comp. Immunol. Microbiol. Infect. Dis.* 12:5–8.

Amemiya, K., Bush, G.V., DeShazer, D., and Waag, D.M. (2002). Nonviable *Burkholderia mallei* induces a mixed Th1- and Th2-like cytokine response in BALB/c mice. *Infect. Immun.* 70:2319–2325.

American Society for Microbiology. (2003). Sentinel laboratory guidelines for suspected agents of bioterrorism. *Burkholderia mallei* and *B. pseudomallei*. http://www.asm.org/ASM/files/LEFTMARGINHEADERLIST/DOWNLOADFILENAME/0000001141/B_pseudorevJJ_33_80803.pdf.

Angus, B.J., Smith, M.D., Suputtamongkol, Y., Mattie, H., Walsh, A.L., Wuthiekanun, V., Chaowagul, W., and White, N.J. (2000). Pharmacokinetic-pharmacodynamic evaluation of ceftazidime continuous infusion *vs.* intermittent bolus injection in septicaemic melioidosis. *Br. J. Clin. Pharmacol.* 49:445–452.

Anuntagool, A., Intachote, P., Naigowit, P., and Sirisinha, S. (1996). Rapid antigen detection assay for identification of *Burkholderia (Pseudomonas) pseudomallei* infection. *J. Clin. Microbiol.* 34:975–976.

Anuntagool, N., Aramsri, P., Panichakul, T., Wuthiekanun, V., Kinoshita, R., White, N.J., and Sirisinha, S. (2000b). Antigenic heterogeneity of lipopolysaccharide among *Burkholderia pseudomallei* clinical isolates. *Southeast Asian J. Trop Med. Public Health* 31(Suppl 1):146–152.

Anuntagool, N., Naigowit, P., Petkanchanapong, V., Aramsri, P., Panichakul, T., and Sirisinha, S. (2000a). Monoclonal antibody-based rapid identification of *Burkholderia pseudomallei* in blood culture fluid from patients with community-acquired septicaemia. *J. Med. Microbiol.* 49:1075–1078.

Anuntagool, N., Rugdech, P., and Sirisinha, S. (1993). Identification of specific antigens of *Pseudomonas pseudomallei* and evaluation of their efficacies for diagnosis of melioidosis. *J. Clin. Microbiol.* 31:1232–1236.

Anuntagool, N., and Sirisinha, S. (2002). Antigenic relatedness between *Burkholderia pseudomallei* and *Burkholderia mallei*. *Microbiol. Immunol.* 46:143–150.

Arun, S., Neubauer, H., Gürel, A., Ayyildiz, G., Kusçu, B., Yesildere, T., Meyer, H., and Hermanns, W. (1999). Equine glanders in Turkey. *Vet. Rec.* 144:255–258.

Ashdown, L.R. (1979a). An improved screening technique for isolation of *Pseudomonas pseudomallei* from clinical specimens. *Pathology* 11:293–297.

Ashdown, L.R. (1979b). Identification of *Pseudomonas pseudomallei* in the clinical laboratory. *J. Clin. Pathol.* 32:500–504.

Ashdown, L.R. (1979c). Nosocomial infection due to *Pseudomonas pseudomallei*: two cases and an epidemiologic study. *Rev. Infect. Dis.* 1:891–894.

Ashdown, L.R. (1981). Relationship and significance of specific immunoglobulin M antibody response in clinical and subclinical melioidosis. *J. Clin Microbiol.* 14:361–364.
Ashdown, L.R. (1987). Indirect haemagglutination test for melioidosis. *Med. J. Aust.* 147: 364–365.
Ashdown, L.R. (1992a). Rapid differentiation of *Pseudomonas pseudomallei* from *Pseudomonas cepacia. Lett. Appl. Microbiol.*14:203–205.
Ashdown, L.R. (1992b). Serial C-reactive protein levels as an aid to the management of melioidosis. *Am. J. Trop. Med. Hyg.* 46:151–157.
Ashdown, L.R., Johnson, R.W., Koehler, J.M., and Cooney, C.A. (1989). Enzyme-linked immunosorbent assay for the diagnosis of clinical and sub-clinical melioidosis. *J. Infect. Dis.* 160:253–260.
Ashdown, L.R., and Koehler, J.M. (1990). Production of hemolysin and other extracellular enzymes by clinical isolates of *Pseudomonas pseudomallei. J. Clin. Microbiol.* 28:2331–2334.
Atkins, T., Prior, R., Mack, K., Russell, P., Nelson, M., Oyston, P.C.F., Dougan, G., and Titball, R.W. (2002b). A mutant of *Burkholderia pseudomallei*, auxotrophic in the branched chain amino acid biosynthetic pathway, is attenuated and protective in a murine model of melioidosis. *Infect. Immun.* 70:5290–5294.
Atkins, T., Prior, R., Mack, K., Russell, P., Nelson, M., Prior, J., Ellis, J., Oyston, P.C.F., Dougan, G., and Titball, R.W. (2002a). Characterisation of an acapsular mutant of *Burkholderia pseudomallei* identified by signature tagged mutagenesis. *J. Med. Microbiol.* 51:539–547.
Barnes, J.L., Ulett, G.C., Ketheesan, N., Clair, T., Summers, P.M., and Hirst, R.G. (2001). Induction of multiple chemokine and colony-stimulating factor genes in experimental *Burkholderia pseudomallei* infection. *Immunol. Cell. Biol.* 79:490–501.
Bauernfiend, A., Roller, C., Meyer, D., Jungwirth, R., and Schneider, I. (1998). Molecular procedure for rapid detection of *Burkholderia mallei* and *Burkholderia pseudomallei. J. Clin. Microbiol.* 36: 2737–2741.
Bernstein, J.M., and Carling, E.R. (1909). Observations on human glanders with a study of six cases and a discussion of the methods of diagnosis. *Br. Med. J.* i:319–325.
Blanc, G., and Baltazard, M. (1941). Transmission du bacilli de Whitmore par la puce du rat *Xenopsylla cheopis. Compt. Rend. Acad. Sci.* 213:541–543.
Blancou, J. (1994). Early methods for the surveillance and control of glanders in Europe. *Rev. Sci. Tech. Off. Int. Epiz.* 13:545–557.
Boerner, P. (1882). A preliminary report on work by the Imperial Health Care Office leading to discovery of the glanders bacillus. *Deutsche Med. Wochenschr.* 52:707–708.
Bojtzov, V., and Geissler, E. (1999). Military biology in the USSR, 1920–45. In: Geissler, E., Moon, J.E. van C. (eds.), *SIPRI Chemical and Biological Warfare Studies. 18. Biological and Toxin Weapons: Research, Development and Use from the Middle Ages to 1945*. Oxford University Press, Oxford, pp. 153–167.
Breitbach, K., Rottner, K., Klocke, S., Rohde, M., Jenzora, A., Wehland, J., and Steinmetz, I. (2003). Actin-based motility of *Burkholderia pseudomallei* involves the Arp 2/3 complex, but not N-WASP and Ena/VASP proteins. *Cell. Microbiol.* 5:385–393.
Brett, P.J., Burtnick, M.N., and Woods, D.E. (2003). The *wbiA* locus is required for the 2-*O*-acetylation of lipopolysaccharides expressed by *Burkholderia pseudomallei* and *Burkholderia thailandensis. FEMS Microbiol. Lett.* 218:323–328.
Brett, P.J., Mah, D.C.W., and Woods, D.E. (1994). Isolation and characterization of *Pseudomonas pseudomallei* flagellin proteins. *Infect. Immun.* 62:1914–1919.
Brett, P.J., DeShazer, D., and Woods, D.E. (1998). *Burkholderia thailandensis* sp. nov., a *Burkholderia pseudomallei*-like species. *Int. J. Syst. Bacteriol.* 48: 317–320.
Brett, P.J., and Woods, D.E. (1996). Structural and immunological characterization of *Burkholderia pseudomallei* O-polysaccharide-flagellin protein conjugates. *Infect. Immun.* 64:2824–2828.

Brook, M.D., Currie, B., and Desmarchelier, P.M. (1997). Isolation and identification of *Burkholderia pseudomallei* from soil using selective culture techniques and the polymerase chain reaction. *J. Appl. Microbiol.* 82:589–596.

Brown, N.F., and Beacham, I.R. (2000). Cloning and analysis of genomic differences unique to *Burkholderia pseudomallei* by comparison with *B. thailandensis*. *J. Med. Microbiol.* 49:993–1001.

Brown, N.F., Boddey, J.A., Flegg, C.A., and Beacham, I.F. (2002). Adherence of *Burkholderia pseudomallei* cells to cultured human epithelial cell lines is regulated by growth temperature. *Infect. Immun.* 70:974–980.

Brown, N.F., Lew, A.E., and Beacham, I.R. (2000). Identification of new transposable genetic elements in *Burkholderia pseudomallei* using subtractive hybridisation. *FEMS Microbiol. Lett.* 183:73–79.

Bryan, L.E., Wong, S., Woods, D.E., Dance, D.A.B., and Chaowagul, W. (1994). Passive protection of diabetic rats with antisera specific for the polysaccharide portion of the lipopolysaccharide isolated from *Pseudomonas pseudomallei*. *Can. J. Infect. Dis.* 5:170–178.

Burtnick, M.N., Brett, P.J., and Woods, D.E. (2002). Molecular and physical characterization of *Burkholderia mallei* O antigens. *J. Bacteriol.* 184:849–852.

Burtnick, M.N., and Woods, D.E. (1999). Isolation of polymyxin B-susceptible mutants of *Burkholderia pseudomallei* and molecular characterization of genetic loci involved in polymyxin B resistance. *Antimicrob. Agents Chemother.* 43:2648–2656.

Charuchaimontri, C., Suputtamongkol, Y., Nilakul, C., Chaowagul, W., Chetchotisakd P., Lertpatanasuwun, N., Intaranongpai, S., Brett, P.J., and Woods, D.E. (1999). Antilipopolysaccharide II: an antibody protective against fatal melioidosis. *Clin. Infect. Dis.* 29:813–818.

Chaowagul, W., Simpson, A.J.H., Suputtamongkol, Y., Smith, M.D., Angus, B.J., and White, N.J. (1999a). A comparison of chloramphenicol, trimethoprim-sulfamethoxazole, and doxycycline with doxycycline alone as maintenance therapy for melioidosis. *Clin. Infect. Dis.* 29:375–380.

Chaowagul, W., Simpson, A.J.H., Suputtamongkol, Y., and White, N.J. (1999b). Empirical cephalosporin treatment of melioidosis. *Clin. Infect. Dis.* 28:1328.

Chaowagul, W., Suputtamongkol, Y., Dance, D.A.B., Rajchanuvong, A., Pattara-arechachai, J., and White, N.J. (1993). Relapse in melioidosis: incidence and risk factors. *J. Infect. Dis.* 168:1181–1185.

Chaowagul, W., Suputtamongkol, Y., Smith, M.D., and White, N.J. (1997). Oral fluoroquinolones for maintenance treatment of melioidosis. *Trans. R. Soc. Trop. Med. Hyg.* 91:599–601.

Chaowagul, W., White, N.J., Dance, D.A.B., Wattanagoon, Y., Naigowit, P., Davis, T.M.E., Looareesuwan, S., and Pitakwatchara, P. (1989). Melioidosis: a major cause of community-acquired septicemia in northeastern Thailand. *J. Infect. Dis.* 159:890–899.

Chen, Y.S., Chen, S.C., Kao, C.M., and Chen, Y.L. (2003). Effects of soil pH, temperature and water content on the growth of *Burkholderia pseudomallei*. *Folia Microbiol.* 48:253–256.

Cheng, A.C., Stephens, D.P., Anstey, N.M., and Currie B.J. (2004). Adjunctive granulocyte colony-stimulating factor for treatment of septic shock due to melioidosis. *Clin. Infect. Dis.* 38:32–37.

Chenthamarakshan, V., Vadivelu, J., and Putucheary, S.D. (2001). Detection of immunoglobulins M and G using culture filtrate antigen of *Burkholderia pseudomallei*. *Diagn. Microbiol. Infect. Dis.* 39:1–7.

Chetchotisakd, P., Chaowagul, W., Mootsikapun, P., Budhsarawong, D., and Thinkamrop, B. (2001a). Maintenance therapy of melioidosis with ciprofloxacin plus azithromycin compared with cotrimoxazole plus doxycycline. *Am. J. Trop. Med. Hyg.* 64:24–27.

Chetchotisakd, P., Porramitikul, S., Mootsikapun, P., Anunnatsiri, S., and Thinkamrop, B. (2001b). Randomized, double-blind, controlled study of cefoperazone-sulbactam plus cotrimoxazole versus ceftazidime plus cotrimoxazole for the treatment of severe melioidosis. *Clin. Infect. Dis.* 33:29–34.

Chierakul, W., Rajanuwong, A., Wuthiekanun, V., Teerawattanasook, N., Gasiprong, M., Simpson, A., Chaowagul, W., White, N.J. (2004). The changing pattern of bloodstream infections associated with the rise in HIV prevalence in northeastern Thailand. *Trans. R. Soc. Trop. Med. Hyg.* 98: 678–686.

Chua, K.L., Chan, Y.Y., and Gan, Y.H. (2003). Flagella are virulence determinants of *Burkholderia pseudomallei*. *Infect. Immun.* 71:1622–1629.

Cottew, G.S. (1950). Melioidosis in sheep in Queensland. *Aust. J. Exp. Biol. Med. Sci.* 28:677–683.

Currie, B. (1995). *Pseudomonas pseudomallei*-insulin interaction. *Infect. Immun.* 63:3745.

Currie, B.J., Fisher, D.A., Anstey, N.M., and Jacups, S.P. (2000a). Melioidosis: acute and chronic disease, relapse and re-activation. *Trans. R. Soc. Trop. Med. Hyg.* 94:301–304.

Currie, B.J., Fisher, D.A., Howard, D.M., Burrow, J.N.C., Lo, D., Selva-nayagam, S., Anstey, N.M., Huffam, S.E., Snelling, P.L., Marks, P.J., Stephens, D.P., Lum, G.D., Jacups, S.P., and Krause, V.L. (2000b). Endemic melioidosis in tropical northern Australia: a 10-year prospective study and review of the literature. *Clin. Infect. Dis.* 31:981–986.

Currie, B.J., Fisher, D.A., Howard, D.M., Burrow, J.N.C., Selvanayagam, S., Snelling, P.L., Anstey, N.M., and Mayo, M.J. (2000c). The epidemiology of melioidosis in Australia and Papua New Guinea. *Acta Trop.* 74:121–127.

Currie, B.J., and Jacups, S.P. (2003). Intensity of rainfall and severity of melioidosis, Australia. *Emerg. Infect. Dis.* 9:1538–1542.

Currie, B.J., Mayo, M., Anstey, N.M., Donohoe, P., Haase, A., and Kemp, D.J. (2001). A cluster of melioidosis cases from an endemic region is clonal and linked to the water supply using molecular typing of *Burkholderia pseudomallei* isolates. *Am. J. Trop. Med. Hyg.* 65:177–179.

Currie, B., Smith-Vaughan, H., Golledge, C., Buller, N., Sriprakash, K.S., and Kemp, D.J. (1994). *Pseudomonas pseudomallei* isolates collected over 25 years from a nontropical endemic focus show clonality on the basis of ribotyping. *Epidemiol. Infect.* 113:307–312.

Cuzzubbo, A.J., Chenthamarakshan, V., Vadivelu, J., Putucheary, S.D., Rowland, D., and Devine, P.L. (2000). Evaluation of a new commercially available immunoglobulin M and immunoglobulin G immunochromatographic test for diagnosis of melioidosis infection. *J. Clin. Microbiol.* 38: 1670–1671.

Dance, D.A.B. (1990). Melioidosis. *Rev. Med. Microbiol.* 1:143–150.

Dance, D.A.B. (1991). Melioidosis: the tip of the iceberg? *Clin. Microbiol. Rev.* 4:52–60.

Dance, D.A.B. (2001). Melioidosis as an emerging global problem. *Acta Trop.* 74:115–119.

Dance, D.A.B., Davis, T.M.E., Wattanagoon, Y., Chaowagul, W., Saiphan, P., Looareesuwan, S., Wuthiekanun, V., and White, N.J. (1989a). Acute suppurative parotitis caused by *Pseudomonas pseudomallei* in children. *J. Infect. Dis.* 159:654–660.

Dance, D.A.B., Wuthiekanun, V., Chaowagul, W., and White N.J. (1989b). The antimicrobial susceptibility of *Pseudomonas pseudomallei*. Emergence of resistance *in vitro* and during treatment. *J. Antimicrob. Chemother.* 24:295–309.

Dance, D.A.B., Wuthiekanun, V., Chaowagul, W., and White, N.J. (1989c). Interactions *in vitro* between agents used to treat melioidosis. *J. Antimicrob. Chemother.*24:311–316.

Dance, D.A.B., Wuthiekanun, V., Naigowit, P., and White, N.J. (1989d). Identification of *Pseudomonas pseudomallei* in clinical practice: use of simple screening tests and API 2ONE. *J. Clin. Pathol.* 42: 645–648.

Dance, D.A.B., King, C., Aucken, H., Knott, C.D., West, P.G., and Pitt, T.L. (1992). An outbreak of melioidosis in imported primates in Britain. *Vet. Rec.* 130:525–529.

Dance, D.A.B., Smith, M.D., Aucken, H.M., and Pitt, T.L. (1999). Imported melioidosis in England and Wales. *Lancet* 353:208.

Dannenberg, A.M., and Scott, E.M. (1958). Melioidosis: pathogenesis and immunity in mice and hamsters. I. Studies with virulent strains of *Malleomyces pseudomallei*. *J. Exp. Med.* 107:153–166.

Dejsirilert, S., Butraporn, R., Chiewsilp, D., Kondo, E., and Kanai, K. (1989). High activity of acid phosphatase of *Pseudomonas pseudomallei* as a possible attribute relating to its pathogenicity. *Jap. J. Med. Sci. Biol.* 42:39–49.

Dejsirilert, S., Kondo, E., Chiewsilp, D., and Kanai, K. (1991). Growth and survival of *Pseudomonas pseudomallei* in acidic environments. *Jap. J. Med. Sci. Biol.* 44:63–74.

Derbyshire, J.B. (2002). The eradication of glanders in Canada. *Can. Vet. J.* 43:722–726.

Desakorn, V., Smith, M.D., Wuthiekanun, V., Dance, D.A.B., Aucken, H., Suntharasamai, P., Rajchanuvong, A., and White, N.J. (1994). Detection of *Pseudomonas pseudomallei* antigen in urine for the diagnosis of melioidosis. *Am. J. Trop. Med. Hyg.* 51:627–633.

DeShazer, D., Brett, P.J., Burtnick, M.N., and Woods, D.E. (1999). Molecular characterization of genetic loci required for secretion of exoproducts in *Burkholderia pseudomallei*. *J. Bacteriol.* 181: 4661–4664.

DeShazer, D., Brett, P.J., Carlyon, R., and Woods, D.E. (1997). Mutagenesis of *Burkholderia pseudomallei* with Tn5-OT182: isolation of motility mutants and molecular characterization of the flagellin structural gene. *J. Bacteriol.* 179:2116–2125.

DeShazer, D., Brett, P.J., and Woods, D.E. (1998). The type II O-antigenic polysaccharide moiety of *Burkholderia pseudomallei* lipopolysaccharide is required for serum resistance and virulence. *Mol. Microbiol.* 30:1081–1100.

DeShazer, D., Waag, D.M., Fritz, D.L., and Woods, D.E. (2001). Identification of a *Burkholderia mallei* polysaccharide gene cluster by subtractive hybridisation and demonstration that the encoded capsule is an essential virulence determinant. *Microb. Pathogenesis* 30:253–269.

DeShazer, D., and Woods, D.E. (1999). Animal models of melioidosis. In: Zek, O., Sande, M. (eds.), *Handbook of Animal Models of Infection*. Academic Press, pp. 199–203.

Desmarchelier, P.M., Dance, D.A.B., Chaowagul, W., Suputtamongkol, Y., White, N.J., and Pitt, T.L. (1993). Relationships among *Pseudomonas pseudomallei* isolates from patients with recurrent melioidosis. *J. Clin. Microbiol.* 31:1592–1596.

Dharakul, T., and Songsivilai, S. (1996). Recent developments in the laboratory diagnosis of melioidosis. *J. Infect. Dis. Antimicrob. Agents* 13:77–80.

Dharakul, T., Songsivilai, S, Anuntagool, N., Chaowagul, W., Wongbunnate, S., Intachote, P., and Sirisinha, S. (1997). Diagnostic value of an antibody enzyme-linked immunosorbent assay using affinity-purified antigen in an area endemic for melioidosis. *Am. J. Trop. Med. Hyg.* 56:418–423.

Dharakul, T., Songsivilai, S., Smithikarn, S., Thepthai, C., and Leelaporn, A. (1999). Rapid identification of *Burkholderia pseudomallei* in blood cultures using lipopolysaccharide-specific monoclonal antibody. *Am. J. Trop. Med. Hyg.* 61:658–662.

Dharakul, T., Songsivilai, S., Viriyachitra, S., Luangwedchakarn, V., Tassaneetritap, S., and Chaowagul, W. (1996). Detection of *Burkholderia pseudomallei* DNA in patients with septicemic melioidosis. *J. Clin. Microbiol.* 34:609–614.

Dharakul, T., Vejbaesya, S., Chaowagul, W., Luangtrakool, P., Stephens, H.A.F., and Songsivilai, S. (1998). HLA-DR and –DQ associations with melioidosis. *Hum. Immunol.* 59:580–586.

Dhiensiri, T., Puapairoj, S., and Susaengrat, W. (1988). Pulmonary melioidosis: clinical-radiologic correlation in 183 cases from northeastern Thailand. *Radiology* 166:711–715.

Dodin, A., and Fournier, J. (1970). Antigènes précipitants et agglutinants de *Pseudomonas pseudomallei* (B. de Whitmore). II. Mise en évidence d'antigènes précipitants communs à *Yersinia pestis* et *Pseudomonas pseudomallei*. *Ann. Inst. Pasteur* (Paris) 119:738–744.

Dorman, S.E., Gill, V.J., Gallin, J.I., and Holland, S.M. (1998). *Burkholderia pseudomallei* infection in a Puerto Rican patient with chronic granulomatous disease: case report and review of occurrences in the Americas. *Clin. Infect. Dis.* 26:889–894.

Doyle, A.C. (1913). The adventure of the dying detective. *The Strand Magazine* xivi:79.

Egan, M., and Gordon, D.L. (1996). *Burkholderia pseudomallei* inactivates complement and is ingested but not killed by polymorphonuclear leukocytes. *Infect. Immun.* 64:4952–4959.

Embi, N., Devarajoo, D., Mohamed, R., and Ismail, G. (1993). An ELISA-disc procedure for antibodies to *Pseudomonas pseudomallei*: application for serological study of melioidosis in an endemic area. *World J. Microbiol. Biotechnol.* 9:91–96.

Everett, E.D., and Nelson, R.A. (1975). Pulmonary melioidosis. Observations in thirty-nine cases. *Am. Rev. Respir. Dis.*112:331–340.

Faa, A.G., and Holt, P.J. (2002). Melioidosis in the Torres Strait islands of Far North Queensland. *Commun. Dis. Intell.* 26:279–283.

Friedland, J.S., Suputtamongkol, Y., Remick, D.G., Chaowagul, W., Strieter, R.M., Kunkel, S.L., White, N.J., and Griffin, G.E. (1992). Prolonged elevations of interleukin-8 and interleukin-6 concentrations in plasma and of leukocyte interleukin-8 mRNA levels during septicemic and localized *Pseudomonas pseudomallei* infection. *Infect. Immun.* 60:2402–2408.

Fritz, D.L., Vogel, P., Brown, D.R., DeShazer, D.E., and Waag, M. (2000). Mouse model of sublethal and lethal intraperitoneal glanders (*Burkholderia mallei*). *Vet. Pathol.* 37:626–636.

Fukuhara, H., Ishimine, T, Futenma, M., and Saito, A. (1995). Efficacy of antibiotics against extracellular and intracellular *Burkholderia pseudomallei*, and their therapeutic effects on experimental pneumonia in mice. *Jap. J. Trop. Med. Hyg.* 23:1–7.

Gaiger, S.H. (1913). Glanders in man. *J. Comp. Pathol. Therap.* XXVI:223–236.

Gan, Y.-H., Chua, K.L., Chua, H.H., Liu, B., Hii, C.S., Chong, H.L, and Tan, P. (2002). Characterization of *Burkholderia pseudomallei* infection and identification of novel virulence factors using a *Caenorhabditis elegans* host system. *Mol. Microbiol.* 44:1185–1197.

Gauthier, Y.P., Thibault, F.M., Paucod, J.C., and Vidal, D.R. (2000). Protease production by *Burkholderia pseudomallei* and virulence in mice. *Acta Trop.* 74:215–220.

Gee, J.E., Sacchi, C.T., Glass, M.B., De, B.K., Weyant, R.S., Levett, P.N., Whitney, A.M., Hoffmaster, A.R., and Popovic, T. (2003). Use of 16S rRNA gene sequencing for rapid identification and differentiation of *Burkholderia pseudomallei* and *B. mallei*. *J. Clin. Microbiol.* 41:4647–4654.

Godfrey, A.J., Wong, S., Dance, D.A.B., Chaowagul, W., and Bryan, L.E. (1991). *Pseudomonas pseudomallei* resistance to β-lactamase antibiotics due to alterations in the chromosomally encoded β-lactamase. *Antimicrob. Agents Chemother.* 35:1635–1640.

Godoy, D., Randle, G., Simpson, A.J., Aanensen. D.M., Pitt, T.L., Kinoshita, R., and Spratt, B.G. (2003). Multilocus sequence typing and evolutionary relationships among the causative agents of melioidosis and glanders, *Burkholderia pseudomallei* and *Burkholderia mallei*. *J. Clin. Microbiol.* 41:2068–2079.

Goshorn, R.K. (1987). Recrudescent pulmonary melioidosis. A case report involving the so-called 'Vietnamese Time Bomb.' *Indiana Med.* 80:247–249.

Gotoh, N., White, N.J., Chaowagul, W., and Woods, D.E. (1994). Isolation and characterization of the outer membrane proteins of *Burkholderia* (*Pseudomonas*) *pseudomallei*. *Microbiology* 140: 797–805.

Green, R.N., and Tuffnell, P.G. (1968). Laboratory acquired melioidosis. *Am. J. Med.* 44:599–605.

Guard, R.W., Khafagi, F.A., Brigden, M.C., and Ashdown, L.R. (1984). Melioidosis in far North Queensland. A clinical and epidemiological review of twenty cases. *Am. J. Trop. Med. Hyg.* 33: 467–473.

Haase, A., Brennan, M., Barrett, S., Wood, Y., Huffam, S., O'Brien, D., and Currie, B. (1998). Evaluation of PCR for diagnosis of melioidosis. *J. Clin. Microbiol.* 36:1039–1041.

Haase, A., Janzen, J., Barrett, S., and Currie, B. (1997). Toxin production by *Burkholderia pseudomallei* strains and correlation with severity of melioidosis. *J. Med. Microbiol.* 46:557–563.

Haase, A., Smith-Vaughan, H., Melder, A., Wood, Y., Janmaat, A., Gilfedder, J., Kemp, D., and Currie, B. (1995). Subdivision of *Burkholderia pseudomallei* ribotypes into multiple types by random amplified polymorphic DNA analysis provides new insights into epidemiology. *J. Clin. Microbiol.* 33:1687–1690.

Hagen, R.M., Gauthier, Y.P., Sprague, L.D., Vidal, D.R., Zysk, G., Finke, E.-J., and Neubauer, H. (2002). Strategies for PCR based detection of *Burkholderia pseudomallei* DNA in paraffin wax embedded tissues. *Mol. Pathol.* 55:398–400.

Harley, V.S., Dance, D.A.B., Drasar, B.S., and Tovey, G. (1998b). Effects of *Burkholderia pseudomallei* and other *Burkholderia* species on eukaryotic cells in tissue culture. *Microbios* 96:71–93

Harley, V.S., Dance, D.A.B., Tovey, G., McCrossan, M.V., and Drasar, B.S. (1998a). An ultrastructural study of the phagocytosis of *Burkholderia pseudomallei*. *Microbios* 94:35–45.

Harris, S. (1999). The Japanese biological warfare programme: an overview. In: Geissler, E., Moon, and J.E. van C. (eds.), *SIPRI Chemical and Biological Warfare Studies. 18. Biological and Toxin Weapons: Research, Development and Use from the Middle Ages to 1945*. Oxford University Press, Oxford, pp. 127–152.

Häußler, S., Nimtz, M., Domke, T., Wray, V., and Steinmetz, I. (1998). Purification and characterization of a cytotoxic exolipid of *Burkholderia pseudomallei. Infect. Immun.* 66:1588–1593.

Häußler, S., Rohde, M., and Steinmetz, I. (1999). Highly resistant *Burkholderia pseudomallei* small colony variants isolated in vitro and in experimental melioidosis. *Med. Microbiol. Immunol.* (Berlin) 188:91–97.

Häußler, S., Rohde, M., von Neuhoff, N. Nimtz, M., and Steinmetz, I. (2003). Structural and functional cellular changes induced by *Burkholderia pseudomallei* rhamnolipid. *Infect. Immun.* 71: 2970–2975.

Heckly, R.J., and Nigg, C. (1958). Toxins of *Pseudomonas pseudomallei*. II. Characterization. *J. Bacteriol.* 76:427–436.

Heine, H.S., England, M.J., Waag, D.M., and Byrne, W.R. (2001). In vitro antibiotic susceptibilities of *Burkholderia mallei* (causative agent of glanders) determined by broth microdilution and e-test. *Antimicrob. Agents Chemother.* 45: 2119–2121.

Heng, B.H., Goh, K.T., Yap, E.H., Loh, H., and Yeo, M. (1998). Epidemiological surveillance of melioidosis in Singapore. *Ann. Acad. Med. Singapore* 27:478–484.

Hicks, C.L., Kinoshita, R., and Ladds, P.W. (2000). Pathology of melioidosis in captive marine animals. *Aust. Vet. J.* 78:193–195.

Ho, M., Schollaardt, T., Smith, M.D., Perry, M.B., Brett, P.J., Chaowagul, W., and Bryan, L.E. (1997). Specificity and functional activity of anti-*Burkholderia pseudomallei* polysaccharide antibodies. *Infect. Immun.* 65:3648–3653.

Ho, P.L., Cheung, T.K.M., Kinoshita, R., Tse, C.W.S., Yuen, K.Y., and Chau, P.Y. (2002a). Activity of five fluoroquinolones against 71 isolates of *Burkholderia pseudomallei. J. Antimicrob. Chemother.* 49:1042–1044.

Ho, P.L., Cheung, T.K.M., Yam, W.C., and Yuen, K.Y. (2002b). Characterization of a laboratory-generated variant of BPS β-lactamase from *Burkholderia pseudomallei* that hydrolyses ceftazidime. *J. Antimicrob. Chemother.* 50:723–726.

Holden, M.S., Titball, R.W., Peacock, S.J., et al. (2004). Genomic plasticity of the causative agent of melioidosis, Burkholderia pseudomallei. *Proc. Nat. Acad. Sci.* 101:14240–14245.

Holmes, B., Pinning, C.A., and Dawson, C.A. (1986). A probability matrix for the identification of gram-negative, aerobic, non-fermentative bacteria that grow on nutrient agar. *J. Gen. Microbiol.* 132:1827–1842.

Hoppe, I., Brenneke, B., Rohde, M., Kreft, A., Haußler, S., Regnzerowski, A., and Steinmetz, I. (1999). Characterisation of a murine model of melioidosis: characterisation of different strains of mice. *Infect. Immun.* 67:2891–2900.

Howard, K., and Inglis, T.J.J. (2003). Novel selective medium for isolation of *Burkholderia pseudomallei. J. Clin. Microbiol.* 41:3312–3316.

Howe, C. (1950). Glanders. In: Christian, H.A. *Oxford Medicine*, Vol. 5, pp. 185–202. Oxford University Press, Oxford.

Howe, C., and Miller, W.R. (1947). Human glanders: report of six cases. *Ann. Intern. Med.* 26: 93–115.

Howe, C., Sampath, A., and Spotnitz, M. (1971). The pseudomallei group: a review. *J. Infect. Dis.* 124:598–606.

Hunting, W. (1908). *Glanders. A clinical treatise*. H&W Brown, London.

Inglis, T.J.J., Chiang, D., Lee, G.S.H., and Lim, C.K. (1998). Potential misidentification of *Burkholderia pseudomallei* by API 20NE. *Pathology* 30:62–64.

Inglis, T.J.J., Garrow, S.C., Henderson, M., Clair, A., Sampson, J., O'Reilly, L., and Cameron, B. (2000a). *Burkholderia pseudomallei* traced to water treatment plant in Australia. *Emerg. Infect. Dis.* 6:56–59.

Inglis T.J.J., Rigby, P., Robertson, T.A., Dutton, N.S., Henderson, M., and Chang, B.J. (2000b). Interaction between *Burkholderia pseudomallei* and *Acanthamoeba* species results in coiling phagocytosis, endamebic bacterial survival, and escape. *Infect. Immun.* 68:1681–1686.

Inglis, T.J.J., Mee, B.J., and Chang, B.J. (2001). The environmental microbiology of melioidosis. *Rev. Med. Microbiol.* 12:13–20.

Inglis, T.J.J., O'Reilly, L.O., Foster, N., Clair, A., and Sampson, J. (2002). Comparison of rapid, automated ribotyping and DNA macrorestriction analysis of *Burkholderia pseudomallei*. *J. Clin. Microbiol.* 40:3198–3203.

Inglis, T.J.J., Robertson, T., Woods, D.E. Dutton, N., and Chang, B.J. (2003). Flagellum-mediated adhesion by *Burkholderia pseudomallei* precedes invasion of *Acanthamoeba astronyxis*. *Infect. Immun.* 71:2280–2282.

Ismail, G Embi, M.N., Omar, O., Razak, N., Allen, J.C., and Smith, J.C. (1987). A competitive immunosorbent assay for detection of *Pseudomonas pseudomallei* exotoxin. *J. Med. Microbiol.* 23:353–357.

Isshiki, Y., Matsuura, M., Dejsirilert, S., Ezaki, T., and Kawahara, K. (2001). Separation of 6-deoxyheptane from a smooth-type lipopolysaccharide preparation of *Burkholderia pseudomallei*. *FEMS Microbiol. Lett.* 199:21–25.

Jeddeloh, J.A., Fritz, D.L., Waag, D.M., Hartings, J.M., and Andrews, G.P. (2003). Biodefense-driven murine model of pneumonic melioidosis. *Infect. Immun.* 71:584–587.

Jenney, A.W.J., Lum, G., Fisher, D.A., and Currie, B.J. (2001). Antibiotic susceptibility of *Burkholderia pseudomallei* from tropical northern Australia and implications for therapy of melioidosis. *Int. J. Antimicrob. Agents* 17:109–113.

Jones, A.L., Beveridge, T.J., and Woods, D.E. (1996). Intracellular survival of *Burkholderia pseudomallei*. *Infect. Immun.* 64:782–790.

Jones, A.L., DeShazer, D., and Woods, D.E. (1997). Identification and characterization of a two-component regulatory system involved in invasion of eukaryotic cells and heavy-metal resistance in *Burkholderia pseudomallei*. *Infect. Immun.* 65:4972–4977.

Jones, S.M., Ellis, J.F., Russell, P., Griffin, K.F., and Oyston, P.C.F. (2002). Passive protection against *Burkholderia pseudomallei* infection in mice by monoclonal antibodies against capsular polysaccharide, lipopolysaccharide or proteins. *J. Med. Microbiol.* 51:1055–1062.

Kanai, K., Suzuki, Y., Kondo, E., Maejima, Y., Miyamoto, D., Suzuki, T., and Kurata, T. (1997). Specific binding of *Burkholderia pseudomallei* cells and their cell-surface acid phosphatase to gangliotetraosylceramide (asialo GM1) and gangliotriaosylceramide (asialo GM2). *Southeast Asian J. Trop. Med. Hyg.* 28:781–790.

Kanaphun, P., Thirawattanasuk, N., Suputtamongkol, Y., Naigowit, P., Dance, D.A.B., Smith, M.D., and White, N.J. (1993). Serology and carriage of *Pseudomonas pseudomallei:* a prospective study in 1000 hospitalized children in Northeast Thailand. *J. Infect. Dis.* 167:230–233.

Kawahara, K., Dejsirilert, S., and Ezaki, T. (1998). Characterization of three capsular polysaccharides produced by *Burkholderia pseudomallei*. *FEMS Microbiol. Lett.* 169:283–287.

Kenny, D.J., Russell, P., Rogers, D., Eley, S.M., and Titball, R.W. (1999). In vitro susceptibilities of *Burkholderia mallei* in comparison to those of other pathogenic *Burkholderia* spp. *Antimicrob. Agents Chemother.* 43:2773–2775.

Kespichayawattana, W., Rattanachetkul, S., Wanun, T., Utaisincharoen, P., and Sirisinha, S. (2000). *Burkholderia pseudomallei* induces cell fusion and actin-associated membrane protrusion: a possible mechanism for cell-to-cell spreading. *Infect. Immun.* 68:5377–5384.

Ketheesan, N., Barnes, J.L, Ulett, G.C., VanGessel, H.J., Norton, R.E., Hirst, R.G., and LaBrooy, J.T. (2002). Demonstration of a cell-mediated immune response in melioidosis. *J. Infect. Dis.* 186: 286–289.

Khupulsup, K., and Petchclai, B. (1986). Application of indirect haemagglutination test and indirect fluorescent antibody test for IgM antibody for diagnosis of melioidosis in Thailand. *Am. J. Trop. Med. Hyg.* 35:366–369.

Klein, G.C. (1980). Cross-reaction to *Legionella pneumophila* antigen in sera with elevated titers to *Pseudomonas pseudomallei*. *J. Clin. Microbiol.* 11:27–29.

Knirel, Y.A., Paramonov, N.A., Shashkov, A.S., Kochetkov, N.K., Yarullin, R.G., Farber, S.M., and Efremenko, V.I. (1992). Structure of the polysaccharide chains of *Pseudomonas pseudomallei* lipopolysaccharides. *Carbohydr. Res.* 233:185–193.

Koh, C.L., Fong, M.Y., Pang, T., Parasakthi, N., and Putucheary, S.D. (1989). Detection, purification and partial characterization of plasmid DNA in *Pseudomonas pseudomallei* from Malaysia. *Trop. Biomed.* 6:141–143.

Koh, H.T., Ng, L.S.Y., Sng, L.-H., Wang, G.C.Y., and Lin, R.V.T.P. (2003). Automated identification systems and *Burkholderia pseudomallei*. *J. Clin. Microbiol.* 41:1809.

Kondo, E., Kurata, T., Naigowit, P., and Kanai, K. (1996). Evolution of cell surface acid phosphatase of *Burkholderia pseudomallei*. *Southeast Asian J. Trop. Med. Public Health* 27:592–599.

Korbrisate, S., Suwanasai, N., Leelaporn, A., Ezaki, T., Kawamura, Y., and Sarasombath, S. (1999). Cloning and characterization of a nonhemolytic phospholipase C gene from *Burkholderia pseudomallei*. *J. Clin. Microbiol.* 37:3742–3745.

Kosuwon, W., Saengnipanthkul, S., Mahaisavariya, B., and Laupattarakasem, W. (1993). Musculoskeletal melioidosis. *J. Bone Joint Surg.* 75-A:1811–1815.

Kunakorn, M., Boonma, P., Khupulsup, K., and Petchclai, B. (1990). Enzyme-linked immunosorbent assay for immunoglobulin M specific antibody for the diagnosis of melioidosis. *J. Clin. Microbiol.* 28:1249–1253.

Kunakorn, M., Jayanetra, P., and Tanphaichitra, D. (1991a). Man-to-man transmission of melioidosis. *Lancet* 337:1290–1291.

Kunakorn, M., Petchclai, B., Khupulsup, K., and Naigowit, P. (1991b). Gold blot for detection of immunoglobulin M (IgM)- and IgG-specific antibodies for rapid serodiagnosis of melioidosis. *J. Clin. Microbiol.* 29:2065–2067.

Kunakorn, M., and Markham, R.B. (1995). Clinically practical seminested PCR for *Burkholderia pseudomallei* quantitated by enzyme immunoassay with and without solution hybridization. *J. Clin. Microbiol.* 33:2131–2135.

Kunakorn, M., Raksakait, K., Sethaudom, C., Sermswan, R., and Dharakul, T. (2000). Comparison of three PCR primer sets for diagnosis of septicemic melioidosis. *Acta Trop.* 74:247–251.

Lauw, F.N., Simpson, A.J.H., Hack, C.E., Prins, J.M., Wolbink, A.M., van Deventer, S.J.H., Chaowagul, W., White, N.J., and van der Poll, T. (2000a). Soluble granzymes are released during human endotoxaemia and in patients with severe infection due to Gram-negative bacteria. *J. Infect. Dis.* 182:206–213.

Lauw, F.N., Simpson, A.J.H., Prins, J.M., Smith, M.D., Kurimoto, M., van Deventer, S.J.H., Speelman, P., Chaowagul, W., White, N.J., and van der Poll, T. (1999). Elevated plasma concentrations of interferon (IFN)-γ and the IFN-γ-inducing cytokines interleukin (IL)-18, IL-12, and IL-15 in severe melioidosis. *J. Infect. Dis.* 180:1878–1885.

Lauw, F.N., Simpson, A.J.H., Prins, J.M., van Deventer, S.J.H., Chaowagul, W., White, N.J., and van der Poll, T. (2000b). The CXC chemokines gamma interferon (IFN-γ)-inducible protein 10 and monokine induced by IFN-γ are released during severe melioidosis. *Infect. Immun.* 68:3888–3893.

Leakey, A.K., Ulett, G.C., and Hirst, R.G. (1998). BALB/c and C57BL/6 mice infected with virulent *Burkholderia pseudomallei* provide contrasting animal models for the acute and chronic forms of melioidosis. *Microb. Pathogenesis* 24:269–275.

Lee, M.-A., and Liu, Y. (2000). Sequencing and characterization of a novel serine metalloprotease from *Burkholderia pseudomallei*. *FEMS Microbiol. Lett.* 192:67–72.

Leelarasamee, A. (1997). Diagnostic value of indirect haemagglutination test in melioidosis. *J. Infect. Dis. Antimicrob. Agents* 14:57–59.

Leelarasamee, A., and Bovornkitti, S. (1989). Melioidosis: review and update. *Rev. Infect. Dis.* 11: 413–425.

Leelarasamee, A., Trakulsomboon, S., Kusum, M., and Dejsirilert, S. (1997). Isolation rates of *Burkholderia pseudomallei* among the four regions of Thailand. *Southeast Asian J. Trop. Med. Public Health* 28:107–112.

Lertmemongkolchai, G., Manmontri, W., Leelayuwat, C., Romphruk, A., and Waropastrakul, S. (1991). Immunoblot analysis to demonstrate antigenic variability of clinical isolated *Pseudomonas pseudomallei*. *Asian Pacific J. Allerg. Immunol.* 9:5–8.

Lertpatanasuwan, N., Sermsri, K., Petkasaem, A., Trakulsomboon, S., Thamlikitkul, V., and Suputtamongkol, Y. (1999). Arabinose-positive *Burkholderia pseudomallei* infection in humans: case report. *Clin. Infect. Dis.* 28:927–928.

Lever, M.S., Nelson, M., Ireland, P.I., Stagg, A.J., Beedham, R.J., Hall, G.A., Knight, G., and Titball, R.W. (2003). Experimental aerogenic *Burkholderia mallei* (glanders) infection in the BALB/c mouse. *J. Med. Microbiol.* 52:1109–1115.

Lew, A., and Desmarchelier, P.M. (1993). Molecular typing of *Pseudomonas pseudomallei*: restriction fragment length polymorphism of rRNA genes. *J. Clin. Microbiol.* 31:533–539.

Lew, A., and Desmarchelier, P.M. (1994). Detection of *Pseudomonas pseudomallei* by PCR and hybridization. *J. Clin. Microbiol.* 32:1326–1332.

Levi, M.I. (1960). Current studies of melioidosis and certain tasks for scientific investigation. *Zh. Mikrobiol. Epidemiol. Immunobiol.* 31:133–139.

Liu, P.V. (1957). Survey of haemolysin production among species of pseudomonads. *J. Bacteriol.* 74:718–727.

Liu, B., Koo, G.C., Yap, E.H., Chua, K.L., and Gan, Y.-H. (2002). Model of differential susceptibility to mucosal *Burkholderia pseudomallei* infection. *Infect. Immun.* 70:504–511.

Livermore, D.M., Chau, P.Y., Wong, A.I., and Leung, Y.K. (1987). β-Lactamase of *Pseudomonas pseudomallei* and its contribution to antibiotic resistance. *J. Antimicrob. Chemother.* 20: 313–321.

Lopez, J., Copps, J., Wilhemsen, C., Moore, R., Kubay, J., St-Jacques, M., Halayko, S., Kranendonk, C., Toback, S., DeShazer, D., Fritz, D.L., Tom, M., and Woods, D.E. (2003). Characterization of experimental equine glanders. *Microb. Infect.* 5:1125–1131.

Loprasert, S., Sallabhan, R., Whangsuk, W., and Mongkolsuk, M. (2000). Characterization and mutagenesis of *fur* gene from *Burkholderia pseudomallei.* Gene 254:129–137.

Low Choy, J., Mayo, M., Janmaat, A., and Currie, B.J. (2000). Animal melioidosis in Australia. *Acta Trop.* 74:153–158.

Lowe, P., Engler, C., and Norton, R. (2002). Comparison of automated and nonautomated systems for identification of *Burkholderia pseudomallei*. *J. Clin. Microbiol.* 40:4625–4627.

Lumbiganon, P., Pengsaa, K., Puapermpoonsiri, S., and Puapairoj, A. (1988). Neonatal melioidosis: a report of 5 cases. *Pediatr. Infect. Dis. J.* 7:634–636.

Lumbiganon, P., and Viengnondha. (1994). Clinical manifestations of melioidosis in children. *Pediatr. Infect. Dis.* 14:136–140.

Mack, K., and Titball, R.W. (1996). Transformation of *Burkholderia pseudomallei* by electroporation. *Anal. Biochem.* 242:73–76.

Mack, K., and Titball, R.W. (1998). The detection of insertion sequences within the human pathogen *Burkholderia pseudomallei* which have been identified previously in *Burkholderia cepacia*. *FEMS Microbiol. Lett.* 162:69–74.

MacKnight, K., Chow, D., See, B., and Vedros, N. (1990). Melioidosis in a macaroni penguin *Eudyptes chrysolophus*. *Dis. Aquat. Org.* 9:105–107.

Markovitz, A. (1979). Inoculation by bronchoscopy. *West. Med. J.* 131:550.

Masoud, H., Ho, M., Schollaardt, T., and Perry, M.B. (1997). Characterization of the capsular polysaccharide of *Burkholderia* (*Pseudomonas*) *pseudomallei* 304b. *J. Bacteriol.* 179:5663–5669.

Matsuura, M., Kawahara, K., Ezaki, T., and Nakano, M. (1996). Biological activities of lipopolysaccharide of *Burkholderia* (*Pseudomonas*) *pseudomallei*. *FEMS Microbiol. Lett.* 137:79–83.

Mays, E.E., and Ricketts, E.A. (1975). Melioidosis: recrudescence associated with bronchogenic carcinoma twenty-six years following initial geographic exposure. *Chest* 68:261–263.

McCormick, J.B., Sexton, D.J., McMurray, J.G., Carey, E., Hayes, P., and Feldman, R.A. (1975). Human-to-human transmission of *Pseudomonas pseudomallei*. *Ann. Intern. Med.* 83:512–513.

McCormick, J.B., Weaver, R.E., Hayes, P.S., Boyce, J.M., and Feldman, R.A. (1977). Wound infection by an indigenous *Pseudomonas pseudomallei*-like organism isolated from the soil: case report and epidemiologic study. *J. Infect. Dis.* 135:103–107.

Merianos, A., Patel, M., Lane, J.M., Noonan, C.N., Sharrock, D., Mock, P.A., and Currie, B. (1993). The 1990–1991 outbreak of melioidosis in the Northern Territory of Australia: epidemiology and environmental studies. *Southeast Asian J. Trop. Med. Public Health* 24:425–435.

Miller, R., and Clinger, D. (1961). Melioidosis pathogenesis in rabbits. I. In vivo studies in the rabbit ear chamber. *Arch. Pathol.* 71:629–634.

Miller, W.R., Pannell, L., Cravitz, L., Tanner, W.A., and Ingalls, M.S. (1948a). Studies on certain biological characteristics of *Malleomyces mallei* and *Malleomyces pseudomallei*. I. Morphology, cultivation, viability, and isolation from contaminated specimens. *J. Bacteriol.* 55:115–126.

Miller, W.R., Pannell, L., Cravitz, L., Tanner, W.A., and Rosebury, T. (1948b). Studies on certain biological characteristics of *Malleomyces mallei* and *Malleomyces pseudomallei*. II. Virulence and infectivity for animals. *J. Bacteriol.* 55:127–135.

Miralles, I.S., Maciel, M.C.A, Angelo, M.R.F., Gondini, M.M., Frota, L.H.F., Reis, C.M.F., Hofer, E. (2004). Burkholderia pseudomallei: A case report of a human infection in Ceará, Brazil. *Rev. Inst. Med. Trop. S. Paulo.* 46:51–54.

Miyagi, K., Kawakami, K., and Saito, A. (1997). Role of reactive nitrogen and oxygen intermediates in gamma interferon-stimulated murine macrophage bactericidal activity against *Burkholderia pseudomallei*. *Infect. Immun.* 65:4108–4113.

Mohamed, R Nathan, S., Embi, N., Razak, N., and Ismail, G. (1989). Inhibition of macromolecular synthesis in cultured macrophages by *Pseudomonas pseudomallei* exotoxin. *Microbiol. Immunol.* 33:811–820.

Mollaret, H.H. (1988). "L'affaire du Jardin des Plantes" ou comment la mélioïdose fit son apparition en France. *Mad. Mal. Infect.* 18:643–654.

Moon, J.E., and van C. (1999). US biological warfare planning and preparedness: the dilemmas of policy. In: Geissler, E., and Moon, J.E. van C. (eds.), *SIPRI Chemical and Biological Warfare Studies. 18. Biological and Toxin Weapons: Research, Development and Use from the Middle Ages to 1945.* Oxford University Press, Oxford, pp. 215–254.

Moore, R.A., DeShazer, D., Reckseidler, S., Weissman, A., and Woods, D.E. (1999). Efflux-mediated aminoglycoside and macrolide resistance in *Burkholderia pseudomallei*. *Antimicrob. Agents Chemother.* 43:465–470.

Nachiangmai, N., Patamasucon, P., Tipayamonthein, B., Kongpon, A., and Nakaviroj, S. (1985). *Pseudomonas pseudomallei* in southern Thailand. *Southeast Asian J. Trop. Med. Public Health.* 16:83–87.

Naigowit, P., Kurata, T., Wangroongsub, P., Petkanjanapong, V., Kondo, E., and Kanai, K. (1993). Application of indirect immunofluorescence microscopy to colony identification of *Pseudomonas pseudomallei*. *Asian Pacific J. Allerg. Immunol.* 11:149–154.

Narita, M., Loganathan, P., Hussain, A., Jamaluddin, A., and Joseph, P.G. (1982). Pathological changes in goats experimentally inoculated with *Pseudomonas pseudomallei*. *Nat. Inst. Animal Health Q.* 22:170–179.

Ngauy, V., Lemeshev, Y., Sadkowski, L., Crawford, G. (2005). Cutaneous melioidosis in a man who was taken as a prisoner of war by the Japanese during World War II. *J. Clin. Microbiol.* 43:970–972.

Nierman, W.C., DeShazer, D., Kim, H.S., et al. (2004). Structural flexibility in the Burkholderia mallei genome. *Proc. Nat. Acad. Sci.* 101:14246–14251.

Nimtz, M., Wray, T., Domke, B., Brenneke, B., Häussler, S., and Steinmetz, I. (1997). Structure of an acidic exopolysaccharide of *Burkholderia pseudomallei*. *Eur. J. Biochem.* 250:608–616.

Niumsup, P., and Wuthiekanun, W. (2002). Cloning of the class D β-lactamase gene from *Burkholderia pseudomallei* and studies on its expression in ceftazidime-susceptible and resistant strains. *J. Antimicrob. Chemother.* 50:445–455.

Norton, R., Roberts, B., Freeman, M., Wilson, M., Ashurst-Smith, C., Lock, W., Brookes, D., and La Brooy, J. (1998). Characterisation and molecular typing of *Burkholderia pseudomallei*: are disease presentations of melioidosis clonally related? *FEMS Immunol. Med. Microbiol.* 20:37–44.

Nuntayanuwat, S., Dharakul, T., Chaowagul, W., and Songsivilai, S. (1999). Polymorphism in the promoter region of tumor necrosis factor-alpha gene is associated with severe melioidosis. *Hum. Immunol.* 60:979–983.

O'Carroll, M.R., Kidd, T.J., Coulter, C., Smith, H.V., Rose, B.R, Harbour, C., and Bell, S.C. (2003). *Burkholderia pseudomallei*: another emerging pathogen in cystic fibrosis. *Thorax* 58:1087–1091.

Office International des Épizooties (2003). http://www.oie.int.

O'Quinn, A.L., Wiegand, E.M., and Jeddeloh, J.A. (2001). *Burkholderia pseudomallei* kills the nematode *Caenorhabditis elegans* using an endotoxin-mediated paralysis. *Cell Microbiol.* 3:381–393.

Perry, M.B., MacLean, L.L., Schollaardt, T., Bryan, L.E., and Ho, M. (1995). Structural characterization of the lipopolysaccharide O antigens of *Burkholderia pseudomallei*. *Infect. Immun.* 63: 3348–3352.

Petkanjanapong, V., Naigowit, P., Kondo, E., and Kanai, K. (1992). Use of endotoxin antigens in enzyme-linked immunosorbent assay for the diagnosis of *P. pseudomallei* infections (melioidosis). *Asian Pacific J. Allerg. Immunol.* 10:145–150.

Phung, L.V., Han, Y., Oka, S., Hotta, H., Smith, M.D., Theeparakun, P., Yabuuchi, E., and Yano, I. (1995). Enzyme-linked immunosorbent assay (ELISA) using a glycolipid antigen for the serodiagnosis of melioidosis. *FEMS Immunol. Med. Microbiol.* 12:259–264.

Pilourias, P., Ulett, G.C., Ashurst-Smith, C., Hirst, R.G., and Norton, R.E. (2002). A comparison of antibiotic susceptibility testing methods for cotrimoxazole with *Burkholderia pseudomallei*. *Int. J. Antimicrob. Agents* 19:427–429.

Pitt, T.L., Aucken, H., and Dance, D.A.B. (1992). Homogeneity of lipopolysaccharide antigens in *Pseudomonas pseudomallei*. *J. Infect.* 25:139–146.

Pitt, T.L., Trakulsomboon, S., and Dance, D.A.B. (2000). Molecular phylogeny of *Burkholderia pseudomallei*. *Acta Trop.* 74:181–185.

Poe, R.H., Vassallo, C.L., and Domm, B.M. (1971). Melioidosis: the remarkable imitator. *Am. Rev. Respir. Dis.* 104:427–431.

Pongsunk, S., Ekpo, P., and Dharakul, T. (1996). Production of specific monoclonal antibodies to *Burkholderia pseudomallei* and their diagnostic application. *Asian Pacific J. Allerg. Immunol.* 14: 43–47.

Pongsunk, S., Thirawattanasuk, N., Piyasangthong, N., and Ekpo, P. (1999). Rapid identification of *Burkholderia pseudomallei* in blood cultures by a monoclonal antibody assay. *J. Clin. Microbiol.* 37:3662–3667.

Powell, K., Ulett, G, Hirst, R., and Norton, R. (2003). G-CSF immunotherapy for treatment of acute disseminated murine melioidosis. *FEMS Microbiol. Lett.* 224:315–318.

Pruksachartvuthi, S., Aswapokee, N., and Thankerngpol, K. (1990). Survival of *Pseudomonas pseudomallei* in human phagocytes. *J. Med. Microbiol.* 31:109–114.

Punyagupta, S. (1989). Melioidosis: review of 686 cases and presentation of a new clinical classification. In: *Melioidosis*. Bangkok Medical Publisher, Bangkok, pp. 217–229.

Putucheary, S.D., Parasakthi, N., and Lee, M.K. (1992). Septicaemic melioidosis: a review of 50 cases from Malaysia. *Trans. R. Soc. Trop. Med. Hyg.* 86:683–685.

Puthucheary, S., Vadivelu, J., Ce-Cile, C., Kum-Thong, W., and Ismail, G. (1996). Short report: electron microscopic demonstration of extracellular structure of *Burkholderia pseudomallei*. *Am. J. Trop. Med. Hyg.* 54:313–314.

Puthucheary, S., Vadivelu, J., Wong, K.T., and Ong, G.S.Y. (2001). Acute respiratory failure in melioidosis. *Singapore Med. J.* 42:117–121.

Rajchanuvong, A., Chaowagul, W., Suputtamongkol, Y., Smith, M.D., Dance, D.A.B., and White, N.J. (1995). A prospective comparison of co-amoxiclav and the combination of chloramphenicol, doxycycline, and co-trimoxazole for the oral maintenance treatment of melioidosis. *Trans. R. Soc. Trop. Med. Hyg.* 89:546–549.

Rattanathongkom, A., Sermswan, R.W., and Wongratanacheewin, S. (1997). Detection of *Burkholderia pseudomallei* in blood samples using polymerase chain reaction. *Mol. Cell. Probes* 11: 25–31.

Reckseidler, S.L., DeShazer, D., Sokol, P.A., and Woods, D.E. (2001). Detection of bacterial virulence genes by subtractive hybridization: identification of capsular polysaccharide of *Burkholderia pseudomallei* as a major virulence determinant. *Infect. Immun.* 69:34–44.

Rode, J.W., and Webling, D.D'A. (1981). Melioidosis in the Northern Territory of Australia. *Med. J. Australia* 1:181–184.

Rogul, M., Brendle, J.J., Haapala, D.K., and Alexander, A.D. (1970). Nucleic acid similarities among *Pseudomonas pseudomallei, Pseudomonas multivorans*, and *Actinobacillus mallei. J. Bacteriol.* 101:827–835.

Rotz, L.D., Khan, A.S., Lillibridge, S.R., Ostroff, S.M., and Hughes, J.M. (2002). Public health assessment of potential biological terrorism agents. *Emerg. Infect. Dis.* 8:225–230.

Rugdech, P., Anuntagool, N., and Sirisinha, S. (1995). Monoclonal antibodies to *Pseudomonas pseudomallei* and their potential for diagnosis of melioidosis. *Am. J. Trop. Med. Hyg.* 52: 231–235.

Russell, P., Eley, S.M., Ellis, J., Green, M., Bell, D.L., Kenny, D.J., and Titball, R.W. (2000). Comparison of efficacy of ciprofloxacin and doxycycline against experimental melioidosis and glanders. *J. Antimicrob. Chemother.* 45:813–818.

Samosornsuk, N., Lulitanond, A., Saenla, N., Anuntagool, N., Wongratanacheewin, S., and Sirisinha, S. (1999). Short report: evaluation of a monoclonal antibody-based latex agglutination test for rapid diagnosis of septicemic melioidosis. *Am. J. Trop. Med. Hyg.* 61:735–737.

Samuel, M, and Ti, Y. (2003). Interventions for treating melioidosis (Cochrane Methodology Review). In: *The Cochrane Library*, Issue 4. Chichester, UK: John Wiley & Sons, Ltd.

Santanirand, P., Harley, V.S., Dance, D.A.B., Drasar, B.S., and Bancroft, G.J. (1999). Obligatory role of gamma interferon for host survival in a murine model of infection with *Burkholderia pseudomallei. Infect. Immun.* 67:3593–3600.

Schlech, W.F., Turchik, J.B., Westlake, R.E., Klein, G.C., Band, J.D., and Weaver, R.E. (1981). Laboratory-acquired infection with *Pseudomonas pseudomallei* (melioidosis). *N. Engl. J. Med.* 305: 1133–1135.

Sermswan, R.W., Wongratanacheewin, S., Anuntagool, N., and Sirisinha, S. (2000). Comparison of the polymerase chain reaction and serological tests for diagnosis of septicemic melioidosis. *Am. J. Trop. Med. Hyg.* 63:146–149.

Sexton, M.M., Goebel, L.A., Godfrey, A.J., Chaowagul, W., White, N.J., and Woods, D.E. (1993). Ribotype analysis of *Pseudomonas pseudomallei* isolates. *J. Clin. Microbiol.* 31:238–243.

Sexton, M.M., Jones, A.L., Chaowagul, W., and Woods, D.E. (1994). Purification and characterization of a protease from *Pseudomonas pseudomallei. Can. J. Microbiol.* 40:903–910.

Simpson, A.J.H., Suputtamongkol, Y., Smith, M.D., Angus, B.J., Rajanuwong, A., Wuthiekanun, V., Howe, P.A., Walsh, A.L., Chaowagul, W., and White, N.J. (1999). Comparison of imipenem and ceftazidime as therapy for severe melioidosis. *Clin. Infect. Dis.* 29:381–387.

Simpson, A.J.H., Opal, S.M., Angus, B.J., Prins, J.M., Palardy, J.E., Parejo, N.A., Chaowagul, W., and White, N.J. (2000a). Differential antibiotic-induced endotoxin release in severe melioidosis. *J. Infect. Dis.* 181:1014–1019.

Simpson, A.J.H., Smith, M.D., Weverling, G.J., Suputtamongkol, Y., Angus, B.J., Chaowagul, W., White, N.J., van Deventer, S.J.H., and Prins, J.M. (2000b). Prognostic value of cytokine concentrations (tumour necrosis factor, interleukin 6 and interleukin 10) and clinical parameters in severe melioidosis. *J. Infect. Dis.* 181:621–625.

Simpson, A.J.H., Newton, P.J., Chierakul, W., Chaowagul, W., and White, N.J. (2003). Diabetes mellitus, insulin, and melioidosis in Thailand. *Clin. Infect. Dis.* 36:e71–72.

Simpson, A.J.H., and Wuthiekanun, V. (2000). Interaction of insulin with *Burkholderia pseudomallei* may be caused by a preservative. *J. Clin Pathol.* 53:159–160.

Sirisinha, S. (1991). Diagnostic value of serological tests for melioidosis in an endemic area. *Asian Pacific J. Allerg. Immunol.* 9:1–3.

Smith, M.D., Wuthiekanun, V., Walsh, A.L., and Pitt, T.L. (1993). Latex agglutination test for identification of *Pseudomonas pseudomallei*. *J. Clin. Pathol.* 46:374–375.

Smith, M.D., Wuthiekanun, V., Walsh, A.L., and White, N.J. (1994). Susceptibility of *Pseudomonas pseudomallei* to some newer β-lactam antibiotics and antibiotic combinations using time-kill studies. *J. Antimicrob. Chemother.* 33:145–149.

Smith, M.D., Angus, B.J., Wuthiekanun, V., and White, N.J. (1997). Arabinose assimilation defines a nonvirulent biotype of *Burkholderia pseudomallei*. *Infect Immun.* 65:4319–4321.

Smith, M.D., Suputtamongkol, Y., Chaowagul, W., Assicot, M., Bohoun, C., Petitjean, S., and White, N.J. (1995a). Elevated serum procalcitonin levels in patients with melioidosis. *Clin. Infect. Dis.* 20:641–645.

Smith, M.D., Wuthiekanun, V., Walsh, A.L., Teerawattanasook, N., Desakorn, V. Suputtamongkol, Y., Pitt, T.L., and White, N.J. (1995b). Latex agglutination for rapid detection of *Pseudomonas pseudomallei* antigen in urine of patients with melioidosis. *J. Clin. Pathol.* 48:174–176.

Smith, M.D., Wuthiekanun, V., Walsh, A.L., and White, N.J. (1996). *In vitro* action of carbapenem antibiotics against β-lactamase susceptible and resistant strains of *Burkholderia pseudomallei*. *J. Antimicrob. Chemother.* 37:611–615.

Sodeman, W.A. (1994). Sherlock Holmes and tropical medicine: a centennial appraisal. *Am. J. Trop. Med. Hyg.* 50:99–101.

Sookpranee, M., Boonma, P., Susaengrat, W., Bhuripanyo, K., and Punyagupta, S. (1992). Multicenter prospective randomised trial comparing ceftazidime plus co-trimoxazole with chloramphenicol plus doxycycline and co-trimoxazole for treatment of severe melioidosis. *Antimicrob. Agents. Chemother.* 36:158–162.

Sprague, L.D., Zysk, G., Hagen, R.M., Meyer, H., Ellis, J., Anuntagool, N., Gauthier, Y., and Neubauer, H. (2002). A possible pitfall in the identification of *Burkholderia mallei* using molecular identification systems based on the sequence of the flagellin *fliC* gene. *FEMS Immunol. Med. Microbiol.* 34:231–236.

Srinavasan. A., Kraus, C.N, DeShazer, D. Becker, P.M., Dick, J.D., Spacek, L, Bartlett, J.G., Byrne, W.R., and Thomas, D.L. (2001). Glanders in a military research microbiologist. *N. Engl. J. Med.* 345:256–258.

Stanton, A.T., and Fletcher, W. (1932). Studies from the Institute for Medical Research, Federated Malay States. 21. Melioidosis. John Bale & Sons and Danielson Ltd., London.

Stanton, A.T., Fletcher, W., and Symonds, S.L. (1927). Melioidosis in a horse. *J. Hyg.* (Cambridge) 26:33–35.

Steinmetz, I., Reganzerowski, A., Brenneke, B., Häussler, S., Simpson, A., and White, N.J. (1999). Rapid identification of *Burkholderia pseudomallei* by latex agglutination based on an exopolysaccharide-specific monoclonal antibody. *J. Clin. Microbiol.* 37:225–228.

Steinmetz, I., Rohde, M., and Brenneke, B. (1995). Purification and characterization of an exopolysaccharide of *Burkholderia* (*Pseudomonas*) *pseudomallei*. *Infect. Immun.* 63:3959–3965.

Stevens, M.P., Wood, M.W., Taylor, L.A., Monaghan, P., Hawes, P., Jones, P.W., Wallis, T.S., and Galyov, E.E. (2002). An Inv/Mxi-Spa-like type III protein secretion system in *Burkholderia*

pseudomallei modulates intracellular behaviour of the pathogen. *Mol. Microbiol.* 46:649–659.
Stevens, M.P., Friebel, A., Taylor, L.A., Wood, M.W., Brown, P.J., Hardt, W.D., and Galyov, E.E. (2003). A *Burkholderia pseudomallei* type III secreted protein, BopE, facilitates bacterial invasion of epithelial cells and exhibits guanine nucleotide exchange factor activity. *J. Bacteriol.* 185: 4992–4996.
Strauss, J.M., Groves, M.G., Mariappan, M, and Ellison, D.W. (1969). Melioidosis in Malaysia. II. Distribution of *Pseudomonas pseudomallei* in soil and surface water. *Am. J. Trop. Med. Hyg.* 18:698–702.
Sulaiman, S., Othman, M.Z., and Aziz, A.H. (2000). Isolations of enteric pathogens from synanthropic flies trapped in downtown Kuala Lumpur. *J. Vector. Ecol.* 25:90–93.
Suputtamongkol, Y., Chaowagul, W., Chetchotisakd, P., Lertpatanasawun, N., Intaranongpai, S., Ruchutrakool, T., Budhsarawong, D., Mootsikapun, P., Wuthiekanun, V., Teerawattanasook, N., and Lulitanond, A. (1999). Risk factors for melioidosis and bacteremic melioidosis. *Clin. Infect. Dis.* 29:408–413.
Suputtamongkol, Y., Hall, A.J., Dance, D.A.B., Chaowagul, W., Rajchanuvong, A., Smith, M.D., and White, N.J. (1994a). The epidemiology of melioidosis in Ubon Ratchatani, northeast Thailand. *Int. J. Epidemiol.* 23:1082–1090.
Suputtamongkol, Y., Kwiatkowski, D., Dance, D.A.B., Chaowagul, W., and White, N.J. (1992). Tumor necrosis factor in septicemic melioidosis. *J. Infect. Dis.* 165:561–564.
Suputtamongkol, Y., Rajchanuvong, A., Chaowagul, W., Dance, D.A.B., Smith, M.D., Wuthiekanun, V., Walsh, A.L., Pukrittayakamee, S., and White, N.J. (1994b). Ceftazidime vs. amoxycillin/clavulanate in the treatment of severe melioidosis. *Clin. Infect. Dis.* 19:846–853.
Sura, T., Smith, M.D., Cowan, A.L. Walsh, A.L., White, N.J., and Krishna, S. (1997). Polymerase chain reaction for the detection of *Burkholderia pseudomallei*. *Diagn. Microbiol. Infect. Dis.* 29: 121–127.
Thomas, A.D. (1983). Evaluation of the API 20E and Microbact 24E systems for the identification of *Pseudomonas pseudomallei*. *Vet. Microbiol.* 8:611–615.
Thomas, A.D., and Forbes-Faulkner, J.C. (1981). Persistence of *Pseudomonas pseudomallei* in soil. *Aust. Vet. J.* 57:535–536.
Thomas, A.D., Forbes-Faulkner, J.C., Norton, J.H., and Trueman, K.F. (1988). Clinical and pathological observations on goats experimentally infected with *Pseudomonas pseudomallei*. *Aust. Vet. J.* 65:43–46.
Tong, S., Yang, S., Lu, Z., and He, W. (1996). Laboratory investigation of ecological factors influencing the environmental presence of *Burkholderia pseudomallei*. *Microbiol. Immunol.* 40:451–453.
Tribuddharat, C., Moore, R.A., Baker, P., and Woods, D.E. (2003). *Burkholderia pseudomallei* class A β-lactamase mutations that confer selective resistance against ceftazidime or clavulanic acid inhibition. *Antimicrob. Agents Chemother.* 47:2082–2087.
Tungpradabkul, S., Wajanarogana, S., Tunpiboonsak, S., and Panyim, S. (1999). PCR-RFLP analysis of the flagellin sequences for identification of *Burkholderia pseudomallei* and *Burkholderia cepacia* from clinical isolates. *Mol. Cell. Probes.* 13:99–105.
Ulett, G.C., Hirst, R., Bowden, B., Powell, K., and Norton, R. (2003). A comparison of antibiotic regimens in the treatment of acute melioidosis in a mouse model. *J. Antimicrob. Chemother.* 51: 77–81.
Ulett, G.C., Ketheesan, N., and Hirst, R.G. (2000). Cytokine gene expression in innately susceptible BALB/c mice and relatively resistant C57BL/6 mice during infection with virulent *Burkholderia pseudomallei*. *Infect. Immun.* 68:2034–2042.
Utaisincharoen, P., Kespichayawattana, W., Anuntagool, N., Chaisuriya, P., Pichyangkul, S., and Sirisinha, S. (2003). CpG ODN enhances uptake of bacteria by mouse macrophages. *Clin. Exp. Immunol.* 132:70–75.

Utaisincharoen, P., Tangthawornchaikul, N., Kespichayawattana, W., Chaisuriya, P., and Sirisinha, S. (2001). *Burkholderia pseudomallei* interferes with inducible nitric oxide synthase (iNOS) production: a possible mechanism of evading macrophage killing. *Microbiol. Immunol.* 45:307–313.

Vadivelu, J., and Putucheary, S.D. (2000). Diagnostic and prognostic value of an immunofluorescent assay for melioidosis. *Am. J. Trop. Med. Hyg.* 62:297–300.

Vadivelu, J., Putucheary, S.D., Gendeh, G.S., and Parasakthi, N. (1995). Serodiagnosis of melioidosis in Malaysia. *Singapore Med. J.* 36:299–302.

Vasu, C., Vadivelu, J., and Putucheary, S.D. (2003). The humoral response in melioidosis patients during therapy. *Infection* 31:24–30.

Vatcharapreechasakul, T., Suputtamongkol, Y., Dance, D.A.B., Chaowagul, W., and White, N.J. (1992). *Pseudomonas pseudomallei* liver abscesses: a clinical, laboratory, and ultrasonographic study. *Clin. Infect. Dis.* 14:412–417.

Vesselinova, A., Najdenski, H., Nikolova, S., and Kussovski, V. (1996). Experimental melioidosis in hens. *J. Vet. Med.* 43:371–378.

Vora, S.K. (2002). Sherlock Holmes and a biological weapon. *J. R. Soc. Med.* 95:101–103.

Vorachit, M., Lam, K., Jayanetra, P., and Costerton, J.W. (1993). Resistance of *Pseudomonas pseudomallei* growing as a biofilm on silastic discs to ceftazidime and co-trimoxazole. *Antimicrob. Agents Chemother.* 37:2000–2002.

Vorachit, M., Lam, K., Jayanetra, P., and Costerton, J.W. (1995). Electron microscopy study of the mode of growth of *Pseudomonas pseudomallei in vitro* and *in vivo*. *J. Trop. Med. Hyg.* 98: 379–391.

Vuddhakul, V., Tharavichitkul, P., Na-Ngam, N., Jitsurong, S., Kunthawa, B., Noimay, P., Noimay, P., Binla, A., and Thamlikitkul, V. (1999). Epidemiology of *Burkholderia pseudomallei* in Thailand. *Am. J. Trop. Med. Hyg.* 60:458–461.

Walsh, A.L., Smith, M.D., Wuthiekanun V., Suputtamongkol, Y., Chaowagul, W., Dance, D.A.B., Angus, B., and White, N.J. (1995a). Prognostic significance of quantitative bacteraemia in septicemic melioidosis. *Clin. Infect. Dis.* 21:1498–1500.

Walsh, A.L., Smith, M.D., Wuthiekanun, V., Suputtamongkol, Y., Desakorn, V., Chaowagul, W., and White, N.J. (1994). Immunofluorescence microscopy for the rapid diagnosis of melioidosis. *J. Clin. Pathol.* 47:377–399.

Walsh, A.L., Smith, M.D., Wuthiekanun, V., and White, N.J. (1995b). Postantibiotic effects and *Burkholderia* (*Pseudomonas*) *pseudomallei*: evaluation of current treatment. *Antimicrob. Agents Chemother.* 39:2356–2358.

Walsh, A.L., Wuthiekanun, V., Smith, M.D. Suputtamongkol, Y., and White, N.J. (1995c). Selective broths for the isolation of *Pseudomonas pseudomallei* from clinical samples. *Trans. R. Soc. Trop. Med. Hyg.* 89:124.

Warawa, J., and Woods, D.E. (2002). Melioidosis vaccines. *Expert. Rev. Vaccines* 1:477–482.

Wheelis, M. (1999). Biological sabotage in World War I. In: Geissler, E., Moon, J.E. van C. (eds.), *SIPRI Chemical and Biological Warfare Studies. 18. Biological and Toxin Weapons: Research, Development and Use from the Middle Ages to 1945*. Oxford University Press, Oxford, pp. 35–62.

White, N.J. (2003). Melioidosis. *Lancet* 361:1715–1722.

White, N.J., Dance, D.A.B., Chaowagul, W., Wattanagoon, Y., Wuthiekanun, V., and Pitakwatchara, N. (1989). Halving of mortality of melioidosis by ceftazidime. *Lancet* ii:697–700.

Whitmore, A. (1913). An account of a glanders-like disease occurring in Rangoon. *J. Hyg.* XIII:1–34.

Winstanley, C., Hales, B.A., and Hart, C.A. (1999). Evidence for the presence in *Burkholderia pseudomallei* of a type III secretion system associated gene cluster. *J. Med. Microbiol.* 48: 649–656.

Winstanley, C., and Hart, C.A. (2000). Presence of type III secretion genes in *Burkholderia pseudomallei* correlates with ara$^-$ phenotypes. *J. Clin. Microbiol.* 38:883–885.

Wong, K.T., Puthucheary, S.D., and Vadivelu, J. (1995). The histopathology of human melioidosis. *Histopathology* 26:51–55.

Wong, K.T., Vadivelu, J., Puthucheary, S.D., and Tan, K.L. (1996). An immunohistochemical method for the diagnosis of melioidosis. *Pathology* 28:188–191.

Wongprompitak, P., Thepthai, C., Songsivilai, S., and Dharakul, T. (2001). *Burkholderia pseudomallei*-specific recombinant protein and its potential in the diagnosis of melioidosis. *Asian Pacific J. Allerg. Immunol.* 19:37–41.

Wongratanacheewin, S., Sermswan, R.W, Anuntagool, N., and Sirisinha, S. (2001). Retrospective study on the diagnostic value of IgG ELISA, dot immunoassay and indirect haemagglutination in septicemic melioidosis. *Asian Pacific J. Allerg. Immunol.* 19:129–133.

Wongratanacheewin, S., Tattawasart, U., Lulitanond, V., Wongwajana, S., Sermswan, R.W., Sookpranee, M., and Nuntirooj, K. (1993). Characterization of *Pseudomonas pseudomallei* antigens by SDS-polyacrylamide gel electrophoresis and western blot. *Southeast Asian J. Trop. Med. Pub. Health* 24: 107–113.

Woo, P.C.Y., Woo, G.K.S., Lau, S.K.P., Wong, S.S.Y., and Yuen, K.Y. (2002). Single gene target bacterial identification: *groEL* gene sequencing for discriminating clinical isolates of *Burkholderia pseudomallei* and *Burkholderia thailandensis*. *Diagn. Microbiol. Infect. Dis.* 44:143–149.

Woods, D.E. (2002). The use of animal infection models to study the pathogenesis of melioidosis and glanders. *Trends Microbiol.* 10:483–484.

Woods, M.L., Currie, B.J., Howard, D.M., Tierney, A., Watson, A., Anstey, N.M., Philpott, J., Asche, V., and Withnall, K. (1992). Neurological melioidosis: seven cases from the Northern Territory of Australia. *Clin. Infect. Dis.* 15:163–169.

Woods, D.E., DeShazer, D., Moore, R.A., Brett, P.J., Burtnick, M.J., Reckseidler, S.L., and Senkiew, M.D. (1999). Current studies on the pathogenesis of melioidosis. *Microb. Infect.* 2:157–162.

Woods, D.E., Jeddeloh, J.A., Fritz, D.L., and DeShazer, D. (2002). *Burkholderia thailandensis* E125 harbors a temperate bacteriophage specific for *Burkholderia mallei*. *J. Bacteriol.* 184:4003–4017.

Woods, D.E., Jones, A.L., and Hill, P.J. (1993). Interaction of insulin with *Pseudomonas pseudomallei*. *Infect. Immun.* 61:4045–4050.

Wuthiekanun, V., Dance, D.A.B., Wattanagoon, Y., Supputtamongkol, Y., Chaowagul, W., and White, N.J. (1990). The use of selective media for the isolation of *Pseudomonas pseudomallei* in clinical practice. *J. Med. Microbiol.* 33:121–126.

Wuthiekanun, V., Anuntagool, N., White, N.J., and Sirisinha, S. (2002). Short report: a rapid method for the differentiation of *Burkholderia pseudomallei* and *Burkholderia thailandensis*. *Am. J. Trop. Med. Hyg.* 66:759–761.

Wuthiekanun, V., Smith, M.D., Dance, D.A.B., Walsh, A.L., Pitt, T.L., and White, N.J. (1996). Biochemical characteristics of clinical and environmental isolates of *Burkholderia pseudomallei*. *J. Med. Microbiol.* 45:408–412.

Wuthiekanun, V., Smith, M.D., Dance D.A., and White, N.J. (1995a). Isolation of *Pseudomonas pseudomallei* from soil in north-eastern Thailand. *Trans. R. Soc. Trop. Med. Hygiene* 89:41–43.

Wuthiekanun, V., Smith, M.D., and White, N.J. (1995b). Survival of *Burkholderia pseudomallei* in the absence of nutrients. *Trans. R. Soc. Trop. Med. Hyg.* 89:491.

Wuthiekanun, V., Suputtamongkol, Y., Simpson, A.J.H., Kanaphun, P., and White, N.J. (2001). Value of throat swab in the diagnosis of melioidosis. *J. Clin. Microbiol.* 39:3801–3802.

Yabuuchi, E., Kosako, Y., Oyaizu, H., Yano, I, Hotta, H., Hashimoto, Y., Ezaki, T., and Arakawa, M. (1992). Proposal of *Burkholderia* gen. nov. and transfer of seven species of the genus *Pseudomonas* homology group II to the new genus, with the type species *Burkholderia cepacia* (Palleroni and Holmes 1981) comb. nov. *Microbiol. Immunol.* 36:1251–1275.

Yang, H., Chaowagul, W., and Sokol, P.A. (1991). Siderophore production by *Pseudomonas pseudomallei*. *Infect. Immun.* 59:776–780.

Yang, H., Kooi, C.D., and Sokol, P.A. (1993). Ability of *Pseudomonas pseudomallei* malleobactin to acquire transferrin-bound, lactoferrin-bound, and cell-derived iron. *Infect. Immun.* 61:656–62.

5

Smallpox as a Weapon for Bioterrorism

J. Michael Lane, M.D., and Lila Summer

1. INTRODUCTION

Smallpox, the only disease ever eradicated, is one of the six pathogens considered a serious threat for biological terrorism (Henderson *et al.*, 1999; Mahy, 2003; Whitley, 2003). Smallpox has several attributes that make it a potential threat. It can be grown in large amounts. It spreads via the respiratory route. It has a 30% mortality rate. The potential for an attack using smallpox motivated President Bush to call for phased vaccination of a substantial number of American health care and public health workers (Grabenstein and Winkenwerder, 2003; Stevenson and Stolberg, 2002). Following September 11, 2001, the United States rebuilt its supplies of vaccine and Vaccinia Immune Globulin (VIG), expanded the network of laboratories capable of testing for variola virus, and engaged in a broad education campaign to help health care workers and the general public understand the disease (Centers for Disease Control and Prevention, 2003a). This chapter summarizes the scientific and theoretical bases for use of smallpox as a bioweapon and options for preparation for defense against it.

2. VIROLOGY

Variola major, the virus that causes smallpox, is an orthopox. *Variola minor*, its less pathogenic cousin, has little theoretical expectation for use as a bioweapon. V. *major* is a large DNA virus (350 × 270 nm), with one of the most complex genomes of human viruses (180 kbp double-stranded DNA). The genome of several strains of *V. major* has been completely sequenced, but the functions of the genes have not all been elucidated (Moss, 2001).

Bioterrorism and Infectious Agents
Edited by Fong and Alibek, Springer Science+Business Media, Inc., New York, 2005

Like most orthopox viruses, variola is host specific. Humans are the only natural hosts. Experimental infection of small numbers of monkeys with large intravenous doses of virus has been accomplished (LeDuc and Jahrling, 2001). The virus grows well on many tissue cultures and on the chorioallantoic membrane of embryonated chicken eggs. Vaccinia virus, a close cousin of variola, is routinely grown in many laboratories, and can be lyophilized to ensure stability to heat. The same techniques could be used with variola.

The genetics of vaccinia are well known, in part because vaccinia strains have been proposed as carrier viruses for genes from other agents, including HIV (Moss, 2001; Smith *et al.*, 2002). This work shows that the genes of orthopox viruses are amenable to deletions and additions. Researchers working with ectromelia (mousepox), another close cousin of variola, have created a very virulent strain, able to escape the protective effect of prior immunization with vaccinia (Born *et al.*, 2000; Jackson *et al.*, 2001). Recent research suggests that smallpox virus could be recreated by synthesizing long strands of DNA, thus enhancing its availability for bioterrorism. The viral genome has been deposited in public databases, making such work possible although time-consuming and highly technical (Wade, 2005).

Variola virus is fairly hardy in the environment if protected from heat and ultraviolet light. However, it is easy to kill with standard hospital disinfectants or ultraviolet light (Fenner *et al.*, 1988). There is little information about its ability to survive when aerosolized; by analogy to vaccinia virus, it probably could survive for an hour or more if not in direct sunlight (Harper, 1961; Thomas, 1974).

In summary, variola has several virologic attributes that make it attractive as a terrorist weapon. It is easy to grow. It can be lyophilized to protect it from heat. It can be aerosolized. Its genome is large and theoretically amendable to modification.

3. PATHOGENESIS

The pathogenesis of smallpox is believed to resemble that of mousepox (ectromelia), which was elucidated by Fenner and colleagues during the 1950s and 1960s. Subsequent work with rabbitpox and monkeypox, orthopoxes that also cause species-specific systemic disease, have refined our understanding of pathogenesis (Fenner *et al.*, 1988). Infection is via the respiratory route. During the first 4 or 5 days after infection, the virus multiplies in the epithelium of the upper respiratory tract. It is then released into the bloodstream in a primary (asymptomatic) viremia. This viremia is cleared by the reticuloendothelial system, where the virus again multiplies. About 8 or 9 days after the initial infection, the reticuloendothelial cells release a secondary viremia, which is probably cell-associated. This is a massive viremia that causes an intense and prostrating prodrome, with fever, myalgias, and other symptoms of a vigorous viremia. The secondary viremia is cleared toward the end of the prodrome, when the leukocyte-associated virus becomes localized in small blood vessels in the dermis and upper respiratory epithelium.

The skin lesions evolve in a stately and characteristic way (see section on "Clinical Disease"). The histopathology is characteristic. The lesions are full of virus. The fluid from vesicles and pustules and scabs is infectious, and virus can be isolated from scabs (Breman and Henderson, 2002; Fenner *et al.*, 1988). The main source of natural infection is the secretions from the upper respiratory tract, where lesions quickly break down and excrete virus because the epithelium in the nose and throat lacks the firm keratinized layer that seals in the virus in the lesions on the skin. Patients are therefore infectious from about 12–24 hours before the initial faint macular rash appears on the skin, and remain infectious throughout the course of the rash (Breman and Henderson, 2002; Dixon, 1962; Fenner *et al.*, 1988).

The clinical illness and fatality rate roughly parallel the density of the skin lesions. When lesions are sparse, cases are unlikely to die and probably are not efficient transmitters. However, their mobility may allow them to have enough social interaction to result in transmission. Vaccine-modified smallpox can be very mild and nonfatal, although if vaccination was more than 20 years prior to exposure, the fatality rate is not trivial and patients can still transmit the disease. As lesions become denser and confluent, the fatality rate increases, the amount of virus in the respiratory secretions increases, and patients are more infectious (Fenner *et al.*, 1988; Mack, 1972; Rao, 1972). Hemorrhagic smallpox has a fatality rate of nearly 100%, and patients are highly infectious. About 1–5% of unvaccinated patients with *V. major* get hemorrhagic smallpox, probably with disseminated intravascular coagulation (Bray and Buller, 2004; McKenzie *et al.*, 1965; Rao, 1972). They are usually very sick, usually unable to get out of bed and thus may not transmit efficiently. The clinical presentation (from mild to discrete to confluent to hemorrhagic) is a function of the host response, not the virus. The clinical types do not breed true, in that transmission from any patient can give rise to any of the clinical presentations, and the virus is the same.

The immune response in smallpox includes both cell-mediated immunity and production of neutralizing antibodies. Both probably appear about 6 days after the onset of the rash (Fenner *et al.*, 1988). Immunity is essentially life-long in recovered patients, although very rare cases of second infections have been reported several decades after initial infections. The immune response in hemorrhagic smallpox is probably poor, contributing to the very bad outcome.

In summary, smallpox causes an acute illness with a devastating prodrome, with virus primarily transmitted via respiratory secretions. The fatality rate roughly parallels the density of the skin lesions and hence the intensity of the preceding viremia.

4. CLINICAL DISEASE

The clinical spectrum of variola major has been well described (Breman and Henderson, 2002; Dixon, 1962; Fenner *et al.*, 1988; Rao, 1972). The illness starts with a dramatic prodrome, with high fever and signs and symptoms indicative of massive viremia. The patient usually improves somewhat when the viremia is cleared, although the fever does not return to normal. As the fever decreases (about 2–4 days after the onset of the prodrome), the characteristic rash becomes evident. During the first day or two of the rash, it may be impossible to distinguish from measles or many other viral exanthms. On dark skin, the rash may not be apparent on the first day or two, being simply faint macules. If the mouth is carefully examined, an enanthem can be detected.

The lesions of smallpox have a predilection for the cooler parts of the body and are most dense on the face and peripheral extremities. Lesions are present on the palms and/or soles in the majority of cases.

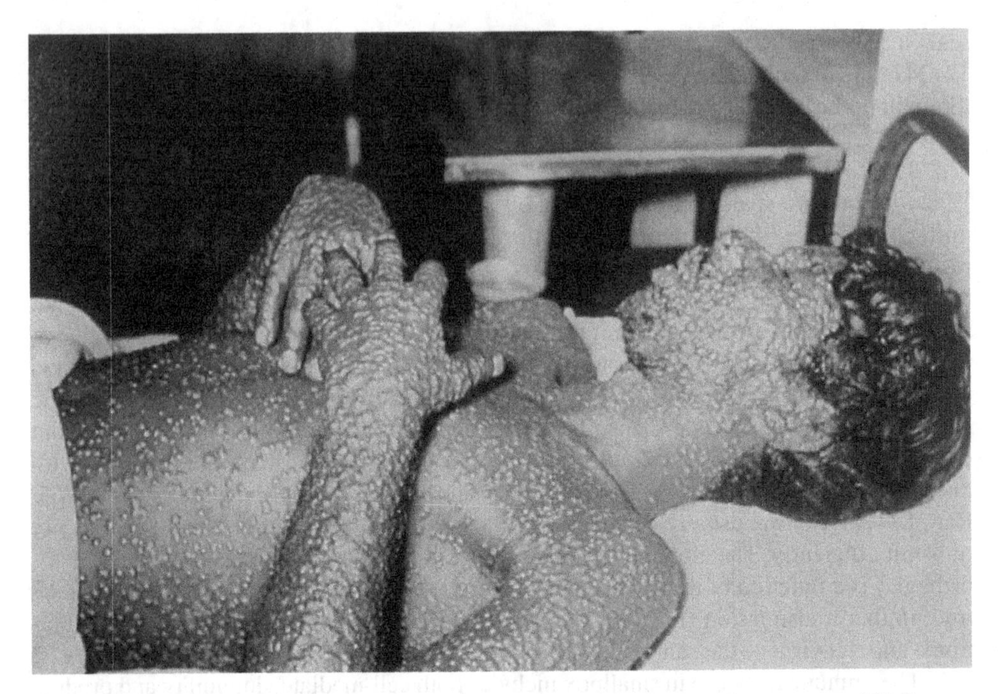

Figure 5.1. Typical semiconfluent smallpox on the seventh or eighth day. Photo courtesy of CDC.

The individual lesions undergo a slow and predictable evolution. Excellent photographs can be found at the Centers for Disease Control and Prevention (CDC) website (Centers for Disease Control and Prevention, 2002a; World Health Organization, 2004). By about the 3rd day, the macules become papular, and the papules progress to fluid-filled vesicles by about the 5th day. These vesicles become large, hard, tense pustules by about the seventh or eighth day. Figure 5.1 is a typical patient with semiconfluent smallpox on about the 7th day of rash. The pustules are "in" the skin, not just "on" the skin. They are deep-seated and feel like dry garbanzo beans, usually nearly 1 cm in diameter.

About the 8th or 9th day, the lesions begin to dry up and umbilicate. By about 2 weeks after the onset of the rash, lesions are scabbing. About 3 weeks after onset, the scabs begin to separate, leaving pitted and depigmented scars.

The causes of death from smallpox are not well elucidated. Massive viral toxemia probably causes a sepsis cascade. Cardiovascular shock may be part of the agonal syndrome. In hemorrhagic cases, disseminated intravascular coagulation probably occurs. Antibacterial agents are not helpful. Loss of fluid and proteins from the exudative rash probably contribute to death. Modern medical care might reduce the fatality rate, but there is no way to prove that contention (Bray and Buller, 2004; Breman and Henderson, 2002; Fenner *et al.*, 1988; Koplan and Foster, 1979; Rao, 1972).

In summary, smallpox produces a serious and prostrating clinical disease. The characteristic pustular rash is easy to diagnose if smallpox is known to be circulating. There is no proven therapy. No data exist to show whether modern supportive care could reduce the death rate.

5. DIAGNOSIS

When smallpox is known to be circulating, the clinical presentation and characteristic rash make diagnosis fairly easy. Diagnosis can be difficult when smallpox is not high on the index of suspicion. Initial cases after a covert bioterrorist attack will probably be missed, at least until the 4th or 5th day of the rash. Transmission may have already taken place by this time. For that reason, but also as good clinical and public health practice, patients with undiagnosed rashes accompanied by fever should always be isolated until the diagnosis is established.

The CDC has produced a diagnostic algorithm based on experience with the differential diagnosis of suspect smallpox cases. This algorithm is Figure 5.2 and can also be found at the CDC website (Centers for Disease Control and Prevention, 2002b; Seward *et al.*, 2004). Most cases initially considered suspect smallpox turn out to be chickenpox, disseminated herpes simplex, secondary syphilis, or drug eruptions.

If this algorithm indicates that a patient is high risk to be smallpox, local and national public health authorities should be immediately notified by telephone, and laboratory specimens taken for polymerase chain reaction (PCR), electron photomicroscopy (EM), and viral culture. A network of laboratories around the United States can do real-time PCR on very short notice (Centers for Disease Control and Prevention, 2003b). State health department laboratories are or know of the nearest one of these Laboratory Resource Network laboratories. Instructions for obtaining, handling, and shipping specimens can be found at the CDC website (Centers for Disease Control and Prevention, 2003b).

Rapid diagnosis requires a sophisticated viral diagnostic laboratory. Rapid testing is best done using EM to identify actual virions and real-time PCR assays to detect viral DNA. These tests are highly sensitive if adequate specimens are provided to the laboratory (Kulesh *et al.*, 2004; Ropp *et al.*, 1995; Sofi *et al.*, 2003). PCR is very specific, whereas most mammalian orthopoxviruses, including vaccinia and monkeypox, have the same brick-like viral structure and cannot be reliably differentiated from variola with EM. Public health action should be initiated by either a positive PCR or EM test, but confirmation by viral culture should also be attempted.

Laboratory tests require adequate specimens. Copious specimens *MUST* be provided if smallpox is seriously suspected (i.e., if the clinical picture fits the "high-risk" category in the algorithm, or if intelligence or communications from terrorists suggests an attack). At least six lesions should be unroofed. The roof tissue and the pus should be placed in separate sterile vials. EM grids should be touched to the base of the opened lesions. Pus can be dried directly onto plastic slides. Punch biopsies of several lesions should be taken, with half put into formalin fixative and half sent unfixed. Shipping and handling of specimens are important; guidance should be sought from CDC (Centers for Disease Control and Prevention, 2003b).

Efforts are currently underway to detect smallpox virus in the environment, including in air distribution systems in large buildings. These involve filtration of large volumes of air and testing the material from the filters with PCR (NBC10 News, 2003). These techniques are conceptually encouraging, but have unknown sensitivity and specificity.

In summary, clinical diagnosis is easy when smallpox is suspected and the rash is fully developed. Sophisticated laboratory tests are available to diagnose suspect cases

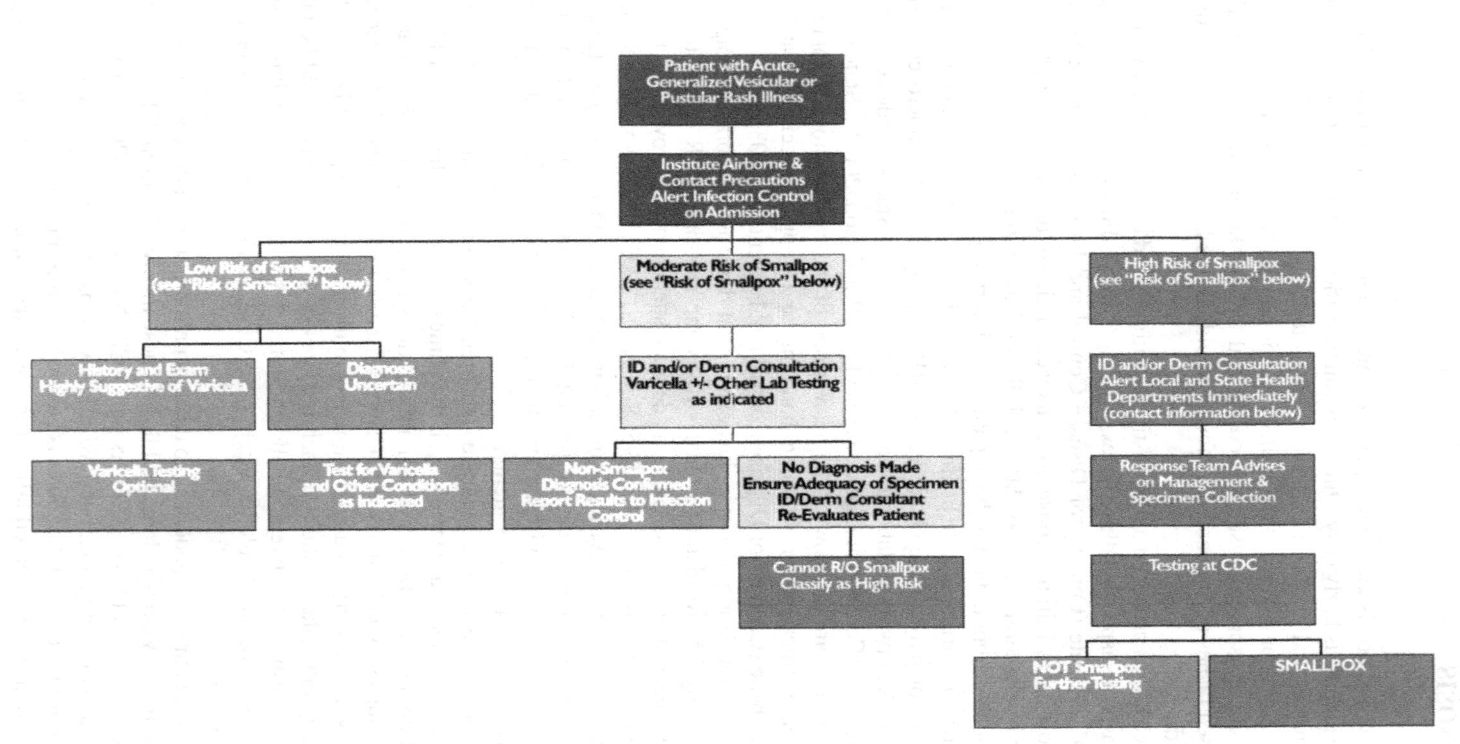

Figure 5.2. Diagnostic algorithm for rash illness. Courtesy of CDC.

rapidly and accurately. Laboratory specimens must be adequate in volume, and carefully handled. Suspect smallpox is a public health emergency and proper public health authorities must be immediately informed.

6. EPIDEMIOLOGY

There is no nonhuman host for smallpox, and there are no subclinical carriers. Once patients have recovered, they are immune and cannot transmit the infection (Fenner *et al.*, 1988). Since there have been no cases since 1977, any new cases must be the result of bioterrorism, or a highly unlikely escape from one of the two official laboratories.

The epidemiology of naturally occurring smallpox has been well studied. The work of Downie *et al.* (1965), Rao *et al.* (1968), Heiner *et al.* (1971), Mack (1972), Mukherjee *et al.* (1974), Sommer and Foster (1974), and the recent demographic analysis by Gani and Leach (2001) provides a good picture of the occurrence and spread of the disease. Smallpox does not ordinarily spread rapidly. Transmission requires prolonged face-to-face contact, such as that which occurs among family members or caregivers. Transmission is most efficient when the index patient is less than 6 feet from the recipient, so that the large-droplet respiratory secretions can be inhaled (Downie *et al.*, 1965; Mack, 1972; Sarkar *et al.*, 1973). Since virus is not secreted from the respiratory tract until the end of the prodrome, patients are usually bedridden when they become infectious and usually do not transmit the disease widely.

Smallpox has been documented to spread from bedding containing pus and scabs, and from dead bodies (Dixon, 1962; Fenner *et al.*, 1988; Hopkins *et al.*, 1971). Modern hospital systems for handling infectious wastes and cremation of bodies should eliminate such transmission (Centers for Disease Control and Prevention, 2003c).

Under natural conditions, smallpox is a highly seasonal disease. Transmission is most efficient in cool dry seasons, possibly because respiratory droplets evaporate quickly in dry conditions – creating small droplets that remain suspended in the air. Although the vast majority of smallpox is acquired by prolonged face-to-face contact, a well-documented outbreak proves that true aerosolization does occur. A patient whose smallpox had not been diagnosed was hospitalized in Meschede, Germany, in 1972. He was coughing vigorously, which probably contributed to creating an aerosol. Patients and visitors on a floor above his room became infected with smallpox (Wehrle *et al.*, 1970). Modern hospital infection control practices should keep such transmission from occurring in hospitals, but the outbreak proves that aerosols can be dangerous.

6.1. Surveillance and Containment Strategy

These epidemiological observations led the World Health Organization in 1968 to switch tactics from mass vaccination to the "surveillance and containment" method. This method is also sometimes called "ring vaccination". C.W. Dixon describes, in his once-definitive textbook on smallpox (1962), how he used "expanding ring" vaccination to control a smallpox epidemic in North Africa after World War II. Vaccine supplies and workers were in short supply, so mass vaccination was impractical. The identification of patients and vaccination of their contacts was his first priority. The next was vaccinating people living in tents surrounding the infected family. Finally, an entire infected village was vaccinated if time and vaccine supply permitted (Dixon, 1948). This concept formed the basis of the surveillance and containment method, after Foege and his colleagues made similar observations about the ease of controlling smallpox by vaccinating close contacts in West Africa in 1967–1968 (Foege *et al.*, 1971; Foege *et al.*, 1975). Surveillance and containment methods work well, even in large and geographically extensive outbreaks, such as that which occurred among several tens of millions of people in the Ganges floodplain in 1974 (Fenner *et al.*, 1988). It is the method that eradicated smallpox when 150 years of mass vaccination failed.

Surveillance and containment consists of five steps. The first step is to identify and report cases. The diagnostic algorithm developed by CDC (Figure 5.2), and the widespread availability of PCR lab tests expedite this process. The second step is to isolate the patient(s). The profound illness and fear of transmission make isolation readily acceptable to the patients and the public. Legal authority exists for isolation if necessary (Centers for Disease Control and Prevention, 2002c). The third step is to identify contacts, the persons who might have had prolonged face-to-face interactions with the patients during the time they were clearly ill. Often patients themselves are dead or moribund and cannot be interviewed. Their family members should be considered contacts, and one or more of them can usually provide information about other possible contacts. Contacts are usually easy to find; they want to be vaccinated and will seek health officials once the patient has been diagnosed.

The fourth step is to vaccinate the contacts. Vaccination prevents smallpox from developing if performed within 3 or 4 days of exposure (Massoudi *et al.*, 2003, Kennedy *et al.*, 2004). The copious supply of bifurcated needles and lyophilized vaccine now available means that vaccination can be rapid and efficient. The rate-limiting factor will be the paperwork demanded by modern medicolegal systems. The contacts are placed under fever surveillance, with temperatures taken twice a day. If they become febrile, and therefore possibly prodromal, they are immediately isolated before they become capable of transmitting the virus.

The fifth step is to vaccinate people who have been or might be associates of one of the first-ring contacts, particularly if the process is not initiated until several days after the index patient(s) became infectious. There is ample time to vaccinate these "second-ring" contacts, because they have not yet been exposed to actual illness. Most new cases detected after initiation of containment activities will be known contacts vaccinated in the incubation period and can be promptly isolated. In the unlikely event that a case develops in a missed contact, the containment process is promptly restarted.

In summary, under natural conditions, smallpox does not spread rapidly. Transmission is by prolonged face-to-face contact. True aerosol spread can occur, but is rare. Surveillance and containment tactics should quickly control outbreaks, even large ones resulting from widespread bioterrorist activity.

7. PATIENT MANAGEMENT AND INFECTION CONTROL

Patients suspected of having smallpox must be immediately isolated under full contact and airborne precautions (Centers for Disease Control and Prevention, 2003d). Suspect smallpox is a public health emergency, and public health authorities should be immediately notified. Smallpox outbreaks in Europe and the United States had a high degree of nosocomial transmission (Mack, 1972; Mack, 2003). If the diagnosis is suspected, modern infection control procedures and effective isolation should virtually eliminate nosocomial transmission. Early in a bioterrorist attack, if smallpox is not high on most clinicians' index of suspicion, cases may be missed and transmit the disease before being effectively isolated. Most hospitals have procedures to isolate patients with fever and an undiagnosed rash to reduce nosocomial transmission of measles and chickenpox, but these procedures may be difficult to implement in a large, busy city emergency ward.

Only recently vaccinated personnel should be allowed to attend patients. If the health facility does not have such personnel prevaccinated and designated as clinical team members, anyone entering the room must be vaccinated with fresh vaccine and vigorous technique, and should also wear a properly fitting N-95 respirator. Personnel should be vaccinated even if they claim to have had a recent successful vaccination; if they are immune, such vaccination carries no risk, and it will eliminate the possibility of an error in the vaccination history.

Supportive care is the basis of the clinical management of smallpox (Breman and Henderson, 2002; Fenner *et al.*, 1988). Adequate food and fluids must be provided in a clean environment. If patients are obtunded, intravenous hydration with monitoring of electrolytes is important, although intravenous access may be difficult through the edematous and pock-laden skin.

Smallpox is disfiguring. Older texts suggested removing mirrors from patients' rooms (Dixon, 1962).

There is controversy about the optimal locus for medical care for smallpox patients. There is an ethical imperative to provide the best care possible, but modern hospitals have many immunosuppressed patients (HIV, cancer therapy, transplants, etc.) who would do poorly if they contracted smallpox or had to be vaccinated. Facilities with good isolation possibilities – such as motels, older tuberculosis, mental, or Veterans Administration hospitals, or mobile hospitals, such as those available from the Federal Emergency Management Authority or the military – might make adequate smallpox hospitals. Medical care can be brought to such facilities, and isolation of the entire facility may be possible. (Centers for Disease Control and Prevention, 2003d)

Dead bodies, bedding, and wastes are potentially infectious (Fenner *et al.*, 1988; Hopkins *et al.*, 1971). Bodies of patients dying from smallpox should be cremated. Modern handling of infectious bedding and wastes will kill the variola virus. Rooms where patients have been cared for should be disinfected with any standard hospital disinfectant. Personnel handling bodies, linen, or wastes must be vaccinated.

In summary, careful consideration should be given to the locus of medical care during an outbreak because smallpox is transmissible nosocomially. Scrupulous adherence to infection control guidelines must be maintained, including judicious disposal of medical wastes and dead bodies. Only recently vaccinated personnel should provide care to patients.

8. POTENTIAL AS A BIOWEAPON

No historical evidence exists that smallpox was an effective bioweapon. Over several centuries of colonial settlement in North and South America, anecdotes, diaries, and public letters expressed intent to use smallpox against indigenous people. Much like the current discussions about the potential use of smallpox as a weapon, what has been written into historical texts and some medical journals may have been fueled more by fear than plausibility.

The correspondence between the British General Jeffery Amherst and his colonels about infecting hostile Indians with smallpox is prolifically referenced in recent journals. The general made this suggestion in July 1763, as strategy against tribes near Fort Pitt, Pennsylvania, involved in Pontiac's Rebellion: "Could it not be contrived to send a small pox among the disaffected tribes of Indians?" A week later, his colonel replied in the only mention of it ever made again: "I will try to inoculate the ___ with Some Blankets that may fall in their Hands, and take care not to get the disease myself" (Knollenberg, 1954). Several weeks before that communication William Trent, an Indian trader at Fort Pitt, suspicious of two Delaware Indian visitors, wrote in his personal journal, "We gave them two Blankets and a Handkerchief out of the Small Pox Hospital. I hope it will have the desired effect." Intent is clear, but the epidemiological record shows that smallpox was raging among the tribes the previous spring, weeks before these documents were written (Knollenberg, 1954). More reasonably, the seasonal nature of smallpox caused subsequent outbreaks rather than blankets. The disease, particularly devastating to American-Indians, had been endemic among them for over a century.

During the American Revolution, George Washington may have believed that British soldiers infected fleeing citizens by using variolation, despite the fact that variolation was a procedure normally used to prevent smallpox. In an 1811 record, a council of Indian chiefs faced a Pacific Fur Company trader, who called himself "the smallpox chief," and threatened to uncork a bottle of the virus if the council decided to attack; fear of smallpox may have averted war. Oral history relates the use of scabs in blankets, linen, clothing, and virus-contaminated tobacco, spanning activities throughout North America and Brazil (Knollenberg, 1954; Wheelis, 1999).

Smallpox virus currently exists legally in only two laboratories: the CDC in Atlanta and at the State Research Center for Virology and Biotechnology in the Novosibirsk region of Russia. Possession of smallpox virus in any place other than these two laboratories is illegal by international convention.

A former Deputy Director of the Soviet Union's bioweapons program has written that, during the cold war, their laboratories produced smallpox in large amounts, and made efforts to adapt it for loading into intercontinental missiles (Alibek, 1999). Scientists defecting from the former Soviet Union, or leaving Russia seeking work in other nations, may have illegally carried stocks of the virus to "rogue" nations (Alibek, 1999; Gellman, 2002; Mangold *et al.*, 1998; Warrick, 2002). There is no publicly accessible proof that such defectors actually transported smallpox out of Russia, but no way of disproving that they did. Allegations have been made that Iraq, Iran, North Korea, and France may have stocks of the virus; these allegations have neither been proved nor disproved (Gellman, 2002; Johnson *et al.*, 2003). During the middle of 2003, a team of top American scientists found no physical or anecdotal evidence to suggest that Iraq was producing smallpox or had stocks in its possession. (Linzer, 2003).

Smallpox can be grown in large quantities and lyophilized for stability. Large amounts can be stored in relatively small containers. The virus would be relatively stable in an aerosol if protected from heat and ultraviolet light (Harper, 1961). The infectious dose may be fairly small (Fenner *et al.*, 1988). An outbreak of 10 cases occurred in Aralsk, Kazakhstan, in 1971 and was kept secret by the Soviets. The index case was allegedly on a boat on the Aral Sea when she was apparently infected (Enserink, 2002; Zelicoff, 2003) A recent claim suggests that an aerosol released from a bioweapons installation on an island in the Aral Sea infected her while the boat passed close to the island. The fact that only one of several workers on the boat was infected makes this distant mode of infection improbable. The subsequent additional cases, experiencing close face-to-face contact, were consistent with natural spread of the disease (Henderson, 2003).

"Dark Winter," a widely publicized political exercise intended to educate public health and government leaders about biological terrorism, used smallpox in its script (O'Toole and Inglesby, 2001). The narrative exaggerated transmission rates, and many leaders and media took the fictitious results literally. BBC Television followed with a docudrama, "Smallpox 2002," in which a crazed terrorist infected himself and spread smallpox through casual and biologically impossible contact (BBC2 England, 2002). The epidemiology of smallpox renders these fictitious accounts highly improbable. Such zeal to alert us to general biosecurity issues obscures many specific aspects of an attack using smallpox.

While no official statements identify specific modes of spread that terrorists might choose, unofficial speculations abound. A fear-inducing hoax, or virus sprayed into a building's air circulation system, or the use of nebulizers to infect thousands at a large airport, are all realistic scenarios (Bozzette *et al.*, 2003). The concept of a volunteer suicidal terrorist who walks around a busy mall, or a big city subway, is unrealistic because smallpox renders people so sick that they are rarely mobile, and they are so visibly sick that they would be avoided by the general public (Piller, 2002). Given the respiratory portal of entry of the virus, a spray or aerosol may be the likely method of introduction.

A large aerosol spray from a light airplane, such as a crop duster outfitted to release lyophilized smallpox virus over a public event such as a political rally or sports competition, is technically feasible. The Federal Aviation Administration, or any developed country's aviation security system, would quickly impound such an unauthorized airplane. If smallpox virus is found, the containment efforts, beginning with a public announcement that attendees of the relevant event should get vaccinated, would abort the attack.

If smallpox virus is introduced into the air circulation of a large building, large numbers of respiratory infections could take place. Several large cities are currently using air filtration systems and doing daily laboratory tests for smallpox, anthrax, and other biohazards. The U.S. government's "Biowatch" program deploys devices to sample air throughout 31 cities and operates in open air, but not closed arenas (NBC10 News, 2003). The sensitivity and specificity of these tests in realistic situations have not been determined, but they are theoretically very sensitive because they use real-time PCR.

Some terrorist groups are associated with the illegal drug trade (Hastert, 2002). Cocaine is sniffed into the upper respiratory tract, so contamination of cocaine with smallpox virus could seed the infection in a widespread population. Many drug users do not readily access the health care system when they are ill and might not be quickly diagnosed, and thus transmit the virus.

Terrorists might spray virus in an airport bus outside the target nation, and infect passengers bound for widespread destinations within that nation. Multiple simultaneous cases would thus occur in many cities, with resultant panic.

Large numbers of smallpox cases are not necessary for an effective bioterrorist attack. A small number of cases could generate public panic. Terrorists could put a solution of smallpox virus into hand-held atomizers, and station volunteers outside of such places as entertainment theme parks, military installations, and critical industries. The virus could be sprayed directly into the faces of persons leaving such facilities, under the guise of marketing a new perfume, etc. Half a dozen cases, with obvious connections to well-known institutions, could invoke panic and social disruption, such as that which followed the 17 anthrax cases in the United States in 2001.

Terrorists with access to a modern virus laboratory might genetically modify smallpox in ways similar to the published manipulations of ectromelia (Jackson *et al.*, 2001; Roos, 2003). Such a strain could not be tested for pathogenicity because there is no animal model for human infection. Genetically altered strains might pose problems of transmission; alteration of pathogenicity might have unknown effects on the transmissibility of the virus. Experienced intelligence observers feel that terrorists would avoid creating a strain with enhanced virulence. Such strains could devastate developing countries with poor public health systems, and a widespread outbreak would quickly spread to such countries (Johnson *et al.*, 2003). Natural smallpox could similarly boomerang. Terrorists with the ability to manufacture it would realize that an effective attack might cause widespread disease in nations harboring their colleagues. Many such nations have poor public health systems and little vaccine, and would be more devastated than the nation initially attacked (Johnson *et al.*, 2003; Oxford, 2003).

In summary, smallpox virus may exist in "rogue" nations. Scenarios involving aerosol spread of smallpox are technically feasible, but have limitations. Even small numbers of cases might suffice to create considerable panic.

9. PREVENTION

The most important aspect of prevention of a bioterrorist attack using smallpox is reliable intelligence. The location of the virus and the abilities and intent of its possessors drive preventive efforts. Allegations that Iran has weaponized smallpox must be viewed through our knowledge that these allegations have been made by an Iranian opposition group (Warwick, 2002). The claims that France, Iraq, North Korea, and Yemen have the virus await firm data (Gellman, 2002). Unproven statements in an internationally politicized climate must be weighed with evidence-based intelligence. Reliable intelligence, the key factor, is an unknown.

Public health efforts, and the extensive publicity about them, have direct preventive value for averting an attack or a resulting major epidemic of smallpox, and can be emulated in many nations. The announced British strategy is to stockpile vaccine and train a cohort of health workers who can identify and isolate smallpox cases and start surveillance and containment activities (Oxford, 2003). In the months after September 11, 2001, the United States increased its vaccine supply from 15 million doses to about 340 million doses. The supply of VIG increased more than 10-fold. The number of laboratories capable of doing real-time PCR for orthopoxviruses increased from two to at least 50. Courses for clinicians, public health officials, and others have been widely held. Several new websites have been created to assist professionals with the diagnosis and management of smallpox, and acquaint them with all aspects of vaccination. Posters and training materials have been widely distributed. Nearly 40,000 front-line medical personnel have been vaccinated (Centers for Disease Control and Prevention, 2003f).

In the absence of solid information about the risk of a bioterrorist attack using smallpox, nations face a policy dilemma. If vaccination carried no risk, there would be no dilemma; widespread vaccination would be reinstated. However, vaccinia is a pathogenic live virus. Vaccination carries well-known risks, particularly in populations that are largely naive to the virus, and in which large numbers of persons are immunosuppressed by HIV, posttransplant therapy, cancer chemotherapy, and steroid medication (Lane and Goldstein, 2003).

The potentially fatal complications of vaccination have recently been reviewed (Centers for Disease Control and Prevention, 2003e; Fulginiti *et al.,* 2003; Lane and Goldstein, 2003). Postvaccinial encephalitis occurs about four or five times per million vaccinees, but carries a 25% fatality rate. Progressive vaccinia is rare, but in our highly immunosuppressed population could be more common than in the past. It may be fatal in 20% or more of cases. Eczema vaccinatum is more common, and the background prevalence of eczema (atopic dermatitis) is about three times higher today than in the 1960s when studies of the frequency of eczema vaccinatum were performed. It has a fatality rate of about 1%. Figure 5.3 is the face of an eczematous woman who acquired vaccinia from her child. Recently, myocarditis has been found to occur about once in every in 18,000 vaccinees (Arness *et al.*, 2004; Eckart *et al.*, 2004; Halsell *et al.*, 2003). Postvaccinial myocarditis has an unknown fatality rate, but deaths have been reported in the past (Dalgaard, 1957; Ferry, 1977; Finlay-Jones, 1964). In addition to these serious complications, a number of vaccinees suffer from "robust" major reactions, with fever, pain, and inflammation at the site; about 1% develop inconsequential, but sometimes, unsightly rashes (Frey *et al.*, 2002; Grabenstein and Winkenwerder, 2003).

Vaccinia is transmissible by direct contact. Transmission is rare and generally only to very close contacts such as family members sharing a bed (Grabenstein and Winkenwerder, 2003; Neff *et al.*, 2002). The presence of immunosuppressed patients on the wards of major hospitals has kept many medical personnel from accepting vaccination. Careful management of the vaccination site and scrupulous hand hygiene can minimize transmission (Lane and Fulginiti, 2003).

Vaccination is not the only means of preventing smallpox. Rigorous isolation of patients coupled with quarantine of contacts until their incubation period is over will prevent transmission (Bozzette *et al.*, 2003; Eichner, 2003; Eubank *et al.*, 2004; Mack, 2003; Meltzer *et al.*, 2001). The prolonged incubation period (12 days) provides time to allow public health action and education of the population once initial cases are recognized.

9.1 Vaccination Policy

In the absence of circulating smallpox, but the presence of a theoretical threat of a bioterrorist attack, there are several options for the use of smallpox vaccine. In 1980, the United States government's Advisory Committee on Immunization Practices (ACIP) recommended that vaccinia be reserved for laboratory personnel who work with orthopoxviruses capable of infecting humans (e.g., variola, vaccinia, and monkeypox) (Centers for Disease Control and Prevention, 1980). These guidelines were expanded to include animal handlers working with animals infected with orthopoxviruses.

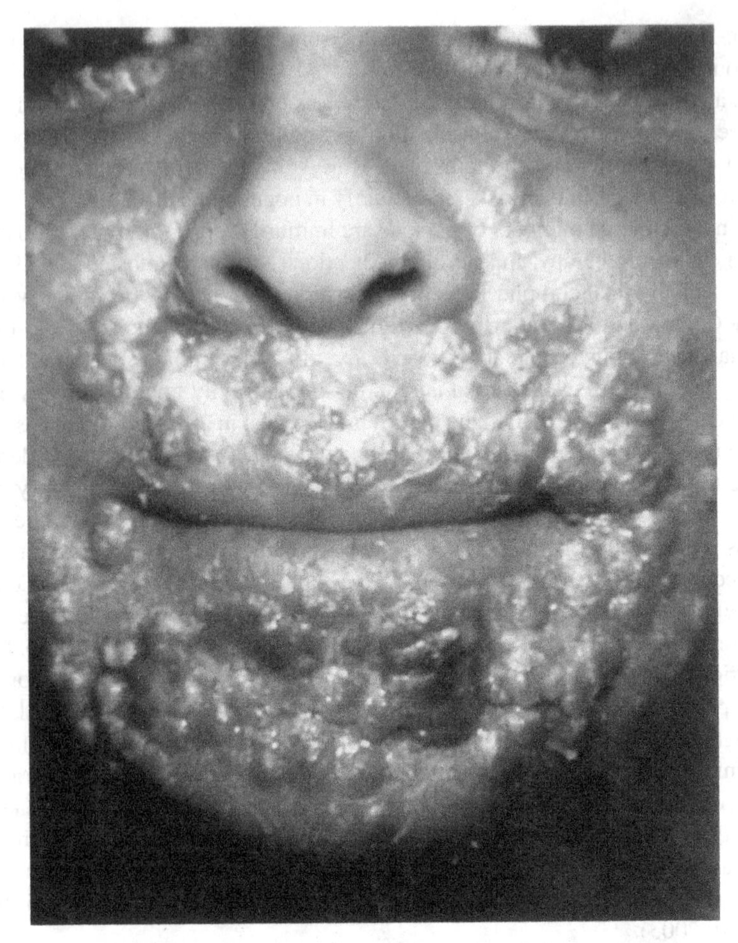

Figure 5.3. Woman with eczema vaccinatum. Photo courtesy of CDC.

After September 11, 2001, when smallpox vaccine in the United States was limited to 15 million doses, the ACIP formed a Smallpox Working Group, which held public hearings and studied scientific data and political opinions. Many public health and academic officials suggested that vaccine be made available to limited numbers of workers who would form response teams in the event of a terrorist attack. In June 2002, the ACIP recommended that vaccine be made available to selected state and local officials with responsibility for protection of public health (Centers for Disease Control and Prevention 2002d). This was consistent with CDC's plan for containment of a smallpox attack and coincided with proof that the 15 million doses of vaccine could be diluted 1:5 (Centers for Disease Control and Prevention, 2002d; Frey *et al.,* 2002), that additional vaccine was on hand, and that production of new vaccine was well under way.

Vaccine supplies are now more than adequate to immunize the entire U.S. population, and suggestions have been made to resume vaccination more widely (Bicknell, 2002; DeRugy and Pena, 2002). Public announcements suggested vaccinating considerable numbers of health care workers, particularly in big city emergency wards and other areas where initially undiagnosed smallpox cases might seek medical care. Many states sought to revaccinate predominantly older workers who had a past history of vaccination, because the residual immunity in such revaccinees is not trivial, and they are less likely to experience adverse events after vaccination (Hammarlund *et al.*, 2003). The number of health care workers in big city hospitals is large, and they turn over rapidly. Many health care workers vaccinated under such a policy are in frequent contact with patients with HIV, status postorgan transplants, atopic dermatitis, etc. Given the recent data (Frey *et al.*, 2002) that adults given primary vaccination frequently need to take one or more days off work because of fever and malaise, many hospitals expressed concern about the effects of widespread vaccination (Connolly, 2003; Gettleman, 2002). Other "first responders" – including policemen, firemen, ambulance workers, and other emergency medical personnel – should be considered for preevent vaccination (Bush, 2002; Centers for Disease Control and Prevention, 2003g; Stevenson and Altman, 2002). In part reflecting these concerns, in August 2003 the Institute of Medicine recommended shifting emphasis away from vaccinations to focus on measures to improve coordination and quicken response time to any public health threat (Olson, 2003).

Several writers suggest that individuals have the right to decide for themselves about the risks and benefits of vaccination, particularly given the current copious vaccine supply (Bicknell, 2002; DeRugy and Pena, 2002). The morbidity and mortality from vaccinia, the problems of screening potential vaccinees and their household members for contraindications, and the efficacy of surveillance and containment methods for controlling smallpox outbreaks, argue against such a policy (Bozzette *et al.*, 2003; Eichner, 2003; Eubank *et al.*, 2004; Lane and Goldstein, 2003; Meltzer *et al.*, 2001).

The United States stopped routine vaccination in 1972. It could be resumed if the threat of smallpox becomes considerable. Only in a scenario where smallpox becomes widespread would it be wise to resume mass vaccination. Such a scenario would inevitably include inadvertent exportation of the disease to other nations, including those with poor fiscal and public health resources and no vaccine. Such resumption of endemicity of smallpox would be a human tragedy of untold proportions. Thus, developed nations must stockpile vaccine and maintain preventive public health efforts for rapid containment of terrorist-induced outbreaks.

In summary, intelligence about the existence of smallpox in the hands of terrorist groups, and their intent for it use, is the best means of preventing an attack. Readiness to control an outbreak resulting from an attack entails a high index of suspicion among clinicians, a good network of diagnostic laboratory capabilities, and a plan for use of surveillance and isolation techniques to quickly contain outbreaks. Vaccination policies need periodic review, because the risks and benefits of vaccination of various medical and law enforcement groups are controversial. Resumption of widespread vaccination is dangerous and unnecessary.

10. FUTURE DIRECTIONS

The United States has rebuilt its vaccine supply, trained diagnostic laboratories, enhanced the availability of clinical and epidemiologic information, and completed a detailed plan for handling a smallpox attack. What should be done to reduce the threat of smallpox further? There are scientific and political efforts currently underway to prepare for an outbreak, to reduce its likelihood, and to enhance the safety of vaccines.

The intelligence community and the new Department of Homeland Security have increased the antiterrorism budget and the amount of attention paid to possible bioterrorist threats. Presumably much of this work is classified, but rumors of nations that may still have (illegal) stocks of variola virus are under investigation.

Rapid detection of variola virus in the environment would be useful. Air filtration systems are being improved so that huge volumes of air inside large buildings, such as commercial or government offices, can be routinely and rapidly filtered. The filters are then tested by PCR for the presence of variola virus DNA (Fackelmann, 2003; NBC10News, 2003). The sensitivity and specificity of such systems cannot be calculated, because no real smallpox has been found or can be injected into these systems. Given the high sensitivity of PCR, they should be good at detecting variola.

A test that could find virus in the throats of patients early in the prodrome could help identify patients who need immediate isolation in the event of an outbreak. Such tests have been developed, again using PCR techniques, and are being explored using material taken from experimentally infected monkeys (Jahrling *et al.*, 2004).

Antiviral compounds, particularly ones derived from cidofovir and its analogues, are being investigated. A nontoxic antiviral with good activity against variola, or indeed against the serious complications of vaccinia, would change the risk-benefit calculations for both pre- and postevent vaccination programs. Compounds that are more soluble and thus more bioavailable than cidofovir are currently being screened for activity against variola (Bray and Roy, 2004; Langbein, 2003; Nyets and DeClercq, 2003; Smee and Sidwell, 2003).

A vaccine that provides protection against variola but is less pathogenic than vaccinia would be useful. Several candidate strains derived from vaccinia are being actively investigated. Modified Vaccinia Ankara, NYVAC, and LC 16m8 are prominent among these attenuated strains (Bonnilla-Guerrero and Poland, 2003; Drexler *et al.*, 2003; Earl *et al.*, 2004; Henderson, Borio, and Lane, 2003; Lane and Goldstein, 2003; McCurdy *et al.*, 2004).

Ultimately, the United States, Russia, and the World Health Organization will have to decide whether retaining the existing legal stocks of variola virus is worthwhile. As long as the virus exists in any laboratory, terrorist groups may claim that one motivation for retaining the virus is its potential as a bioweapon. If the stocks are destroyed, this argument no longer exists. International cooperation under the aegis of the World Health Organization is needed to affirm the illegality of biowarfare using variola, and maximize efforts to find and destroy illegal stockpiles (Breman *et al.*, 2003).

References

Alibek, K. (1999). *Biohazard.* New York: Random House, Inc.

Arness, M.K., Eckart, R.E., Love, S.S., Atwood, J.E., Wells, T.S., Engler, R.J., Collins, L.C., Ludwig, S.L., Riddle, J.R., Grabenstein, J.D., Tornberg, D.N. (2004) Myopericarditis following smallpox vaccination. *American Journal of Epidemiology.* 160:642–51.

BBC2 England. (2002). *Smallpox 2002.* Produced by Simon Chinn, in association with the Learning Channel and Granada International, written by Daniel Percival.

Bicknell, W.J. (2002). The case for voluntary smallpox vaccination. *N. Engl. J. Med.* 346:1323–1325.

Bonnilla-Guerrero, R., and Poland, G. (2003). Smallpox vaccines: current and future. *J. Lab. Clin. Med.* 142:252–257.

Born, T.L., Morrison, L.A., Esteban, D.J., VandenBos, T., Thebeau, L.G. Chen, M., Spriggs, M.K., Sims, J.E., Buller, R.M. A poxvirus protein that binds to and inactivates IL-18 and inhibits NK cell response. (2000) J. Immunology 164:3246–54.

Bozzette, S.A., Boer, R., Bhatnagar, V., Brower, J.L., Keeler, E.B., Morton, S.C., and Stoto, M.A. (2003). A model for a smallpox-vaccination policy. *N. Engl. J. Med.* 348:416–425.

Bray, M., Buller, M.(2004) Looking back at smallpox. Clin Infect Dis 38:882–9.

Bray, M., Roy, C.J. (2004) Antiviral prophylaxis of smallpox. *Journal of Antimicrobial Chemotherapy.* 54:1–5.

Breman, J.G., Arita, I., and Fenner, F. (2003). Preventing the return of smallpox. *N. Engl. J. Med.* 348:463–466.

Breman, J.G., and Henderson, D.A. (2002). Diagnosis and management of smallpox. *N. Engl. J. Med.* 346:1300–1308.

Broad, W.J. (2002). U.S. to vaccinate 500,000 workers against smallpox. *The New York Times*, July 7, 2002, Section A1.

Bush, G.W. (2002). Protecting Americans: Smallpox Vaccination Program. http://www.whitehouse.gov/news/releases/2002/12/20021213-1.html.

Cassimatis, D.C., Atwood, J.E., Engler, R.M., Linz, P.E., Grabenstein, J.D., Vernalis, M.N. (2004) Smallpox vaccination and myopericarditis: a clinical review. *Journal of the American College of Cardiology.* 43:1503–10.

Centers for Disease Control. (1980). Smallpox vaccine: recommendations of the Immunization Practices Advisory Committee (ACIP). *MMWR* 29:417–420.

Centers for Disease Control and Prevention (2002a). Smallpox images. http://www.bt.cdc.gov/agent/smallpox/smallpox-images/index.asp.

Centers for Disease Control and Prevention. (2002b). Diagnostic algorithm for rash illnesses. http://www.bt.cdc.gov/agent/smallpox/diagnosis/evalposter.asp.

Centers for Disease Control and Prevention. (2002c). Smallpox response plan and guidelines. Guide C. Legal and quarantine authority. http://www.bt.cdc.gov/agent/smallpox/response-plan/files/guide-c-part-2.pdf.

Centers for Disease Control and Prevention (2002d). CDC telebriefing transcript smallpox vaccine June 20, 2002. http://www.cdc.gov/od/oc/media/transcripts/t020620.htm.

Centers for Disease Control and Prevention (2003a). Smallpox response plan and guidelines (version 3.0). http://www.bt.cdc.gov/agent/smallpox/response-plan/index.asp.

Centers for Disease Control and Prevention (2003b). Smallpox response plan and guidelines (version 3.0). Guide D—Specimen collection and transport. http://www.bt.cdc.gov/agent/smallpox/response-plan/files/guide-d.doc.

Centers for Disease Control and Prevention (2003c). Smallpox response plan and guidelines (version 3.0). Guide F—Environmental control of smallpox virus. http://www.bt.cdc.gov/agent/smallpox/response-plan/guide-f.doc.

Centers for Disease Control and Prevention (2003d). Smallpox response plan and guidelines (version 3.0). Annex 1. Medical management of smallpox patients and vaccination complications. http://www.bt.cdc.gov/agent/smallpox/response-plan/annex-1.doc.

Centers for Disease Control and Prevention. (2003e). Smallpox vaccination and adverse events. Guidelines for clinicians. *MMWR* 52(RR04):RR04:1–28.

Centers for Disease Control and Prevention. (2003f). Smallpox vaccination program status by state. http://www.cdc.gov/od/oc/media/spvaccin.htm.

Centers for Disease Control and Prevention. (2003g). Recommendations for using smallpox vaccine in a pre-event vaccination program: supplemental recommendations of the Advisory Committee on Immunization practices (ACIP) and the Healthcare Infection Control Practices Advisory Committee (HICPAC). *MMWR* 52(RR-7).

Connolly C. (2003). Bush inoculation plan near standstill. Medical professionals cite possible side effects, uncertainty of threat. *The Washington Post* Feb 24, Section A6.

Cummings, J.F., Polhemus, M.E., Hawkes, C., Klote, M., Ludwig, G.V., Wortmann, G. (2004) Lack of vaccinia viremia after smallpox vaccination. (2004) *Clinical Infectious Diseases*. 38:456–8.

Dalgaard, J.B. (1957). Fatal myocarditis following smallpox vaccination. *Am. Heart J.* 54:156–157.

Damon, I., and Regnery, R. (2003). Orthopoxviruses. In: Marrey, P.C., and Baron, E.J. (eds.), *Manual of Clinical Microbiology*, 8th ed. American Society of Microbiology Press, Washington, D.C.

DeRugy, V., and Pena, C.V. (2002). Responding to the threat of smallpox bioterrorism. *Policy Anal.* 434:1–16.

Dixon, C.W. (1948). Smallpox in Tripolitania, 1946; an epidemiological and clinical study of 500 cases, including trials of penicillin treatment. *J Hyg.* 46:35–77.

Dixon, C.W. (1962). *Smallpox*. Little Brown and Company, Boston.

Downie, A.W., Meikeljohn, M., St. Vincent, L., Rao, A.R., Sundara Babu, B.V., and Kempe, C.H. (1965). The recovery of smallpox virus from patients and their environment in a smallpox hospital. *Bull. WHO* 33:615–622.

Drexler, I, Staib, C., Kastenmuller, W., Stevanoivic, S., Schmidt, B., Lemonnier, F.A., Rammensee, H.-G., Busch, D.H., Bernhard, H., Erfle, V., and Sutter, G. (2003). Identification of vaccinia virus epitope-specific HLA-A*0201-restricted T cells and comparative analysis of smallpox vaccines . *Proc. Soc.* NH. *Acad. Sci.* 100:217–222.

Earl, P.L., Americo, J.L., Wyatt, L.S., Eller, L.A., Whitbeck, J.C., Cohen, G.H., Eisenberg, R.J., Hartmann, C.J., Jackson, D.L., Kulesh, D.A., Martinez, M.J., Miller, D.M., Mucker, E.M., Shamblin, J.D., Zwiers, S.H., Huggins, J.W., Jahrling, P.B., Moss, B. (2004) Immunogenicity of a highly attenuated MVA smallpox vaccine and protection against monkeypox. *Nature*. 428:182–5.

Eckart, R.E., Love, S.S., Atwood, J.E., Arness. M.K., Cassimatis, D.C., Campbell, C.L., Boyd, S.Y., Murphy, J.G., Swerdlow, D.L., Collins, L.C., Riddle, J.R., Tornberg, D.N., Grabenstein, J.D., Engler, R.J. Department of Defense Smallpox Vaccination Clinical Evaluation Team. (2004) Incidence and follow-up of inflammatory cardiac complications after smallpox vaccination. *Journal of the American College of Cardiology*. 44:201–5.

Eichner, M. (2003). Case isolation and contact tracing can prevent the spread of smallpox. *Am. J. Epidemiol.* 158:118–128.

Enserink, M. (2002). Did bioweapons test cause a deadly smallpox outbreak? *Science* 296:2116–2117.

Eubank, S., Guclu, H., Kumar, V.S., Marathe, M.V., Srinivasan, A., Toroczkai, Z., Wang, N. (2004) Modelling disease outbreaks in realistic urban social networks. *Nature*. 429:180–4.

Fackelmann, K. (2003). Handheld bioterror detectors in works. *USA Today*, July 10, 2003, p. 6D.

Fenner, F., Henderson. D.A., Arita, I., Jezek, Z., and Ladnyi, I.D. (1988). *Smallpox and Its Eradication*. WHO, Geneva.

Ferry, B.J. (1977). Adverse reactions after smallpox vaccination. *Med. J. Australia* 2:180–183.

Finlay-Jones, L.R. (1964). Fatal myocarditis after vaccination against smallpox: report of a case. *N. Engl. J. Med.* 270:41–42.

Foege, W.H., Lane, J.M., and Millar, J.D. (1971). Selective epidemiologic control in smallpox eradication. *Am. J. Epidemiol.* 94:311–315.

Foege WH, Millar JD, Henderson DA. (1975) Smallpox eradication in West and Central Africa. Bull. WHO 52: 209–222.

Frey, S.E., Couch, R.B., Tacket, C.O., Treanor, J.J., Wolff, M., and Newman, F.K. (2002). Clinical responses to undiluted and diluted smallpox vaccine. *N. Engl. J. Med.* 346:1265–1274.

Fulginiti, V.A., Papier, A., Lane, J.M., Neff, J.M., and Henderson, D.A. (2003). Smallpox vaccination (vaccinia): a review. Part I. Background, vaccination technique, normal vaccination and revaccination and expected normal reactions. *Clin. Infect. Dis.* 37:241–250.

Fulginiti, V.A., Papier, A., Lane, J.M., Neff, J.M., and Henderson, D.A. (2003). Smallpox vaccination (vaccinia): a review. Part II. Adverse events. *Clin. Infect. Dis.* 37:251–271.

Gani, R., Leach, S. (2001). Transmission potential of smallpox in contemporary populations. *Nature* 414:748–751.

Garde, V., Harper, D., Fairchok, M.P. (2004) Tertiary contact vaccinia in a breastfeeding infant. *JAMA*. 291:725–7.

Gellman, B. (2002). Four nations thought to possess smallpox. *The Washington Post*, November 5, 2002, A1.

Gettleman, J. (2002). Two hospitals refuse to join Bush's plan for smallpox. *The NewYork Times*, December 19, 2002, A16.

Grabenstein, J.D., and Winkenwerder, W. (2003). US military smallpox vaccination program experience. *J.A.M.A.* 289:3278–3282.

Halsell, J.S., Riddle, J.R., Atwood, J.E., Gardner, P., Shope, R., Poland, G.A., Gray, G.C., Ostroff, S., Eckart, R.E., Hospenthal, D.R., Gibson, R.L., Grabenstein, J.D., Arness, M.K., and Tornbery, D.N. (2003). Myopericarditis following smallpox vaccination among vaccinia-naïve US military personnel. *J.A.M.A.* 289:3283–3289.

Hammarlund, E., Lewis, M.W., Hansen, S.G., Strelow, L.I., Nelson, J.A., Sexton, G.J., Hanifin, J.M., and Slifka, M.K. (2003). Duration of antiviral immunity after smallpox vaccination. Nature Med. (online), August 17, 2003. http://www.nature.com/naturemedicine/.

Harper, G.J. (1961). Airborne micro-organisms: survival test with four viruses. *J. Hyg.* 59:479–486.

Hastert, J.D. (2002). House Speaker Hastert introduces the Speaker's Task Force for a Drug Free America. http://speaker.house.gov/library/drugs/010921task.asp.

Heiner, G.G., Fatima, N., and McCrumb, F.R. (1971). A study of intrafamilial transmission of smallpox. *Am.J. Epidemiol.* 94:316–326.

Henderson, D.A. (2003). Commentary on Dr Alan Zelikoff's epidemiological analysis of the Aralsk outbreak (N3). *Crit. Rev. Microbiol.* 29:169–170.

Henderson, D.A., Borio, L.L., and Lane, J. M. (2003). Smallpox and vaccinia. In: Plotkin and Orenstein (eds.), *Vaccines,* 4th ed. W.B. Saunders Company, New York.

Henderson, D.A., Inglesby, T.V., Bartlett, J.G., Ascher. M.S., Eitzen, E., Jahrling, P.B., Hauer, J., Layton, M., McDade, J., Osterholm, M.T., O'Toole. T., Parker, G., Perl, T., Russell, P,K., and Tonat, K. (1999). Smallpox as a biological weapon: medical and public health management. Working Group on Civilian Biodefense. *J.A.M.A.* 281:2127–2137 [review].

Hepburn, M.J., Dooley, D.P., Murray, C.K., Hospenthal, D.R., Hill, B.L., Nauschuetz, W.N., Davis, K.A.m Crouch, H.K., McAllister, C.K.(2004) Frequency of vaccinia virus isolation on semipermeable versus nonocclusive dressings covering smallpox vaccination sites in hospital personnel. *American Journal of Infection Control.* 32:126–30.

Hopkins, D.R., Lane, J.M., Cummings, E.C., and Millar, J.D. (1971). Two funeral-associated smallpox outbreaks in Sierra Leone. *Am. J. Epidemiol.* 94:341–347.

Jackson, R.J., Ramsay, A.J., Christensen, C.D., Beaton, S., Hall, D.F., and Ramshaw, I.A. (2001). Expression of mouse interleukin-4 by a recombinant ectromelia virus suppresses cytolytic lymphocyte responses and overcomes genetic resistance to mousepox. *J. Virol.* 75:1205–1210.

Jahrling, P.B., Hensley, L.E., Martinez, M.J., LeDuc, J.W., Rubins, K.H., Relman, D.A., Huggins, J.W., (2004) Exploring the potential of variola virus infection of cynomolgus macaques as a model for human smallpox. Proc National Acad Sci. 101:15197–204.

Johnson, K., Locy, T., and Manning, A. (2003). Terrorists' most likely weapon here? Bombs. *USA Today,* May 16, 2003, 1A.

Kennedy, J.S., Frey, S.E., Yan, L., Rothman, A.L., Cruz, J., Newman, F.K., Orphin, L., Belshe, R.B., Ennis. F.A., (2004) Induction of human T cell-mediated immune responses after primary and secondary smallpox vaccination. *Journal of Infectious Diseases.* 190:1286–94.

Knollenberg, B. (1954). General Amherst and germ warfare. *Mississippi Valley Hist. Rev.* 41:489–494.

Koplan, J.P., and Foster, S.O. (1979). Smallpox: clinical types, causes of death, and treatment. *J. Infect. Dis.* 140:440–441.

Kulesh, D.A., Baker, R.O., Loveless, B.M., Norwood, D., Zwiers, S.H., Mucker, E., Hartmann, C., Herrera, R., Miller, D., Christensen, D., Wasieloski, L.P Jr., Huggins, J., Jahrling, P.B. (2004) Smallpox and pan-orthopox virus detection by real-time 3'-minor groove binder TaqMan assays on the roche LightCycler and the Cepheid smart Cycler platforms. *Journal of Clinical Microbiology.* 42:601–7.

Lane, J.M., and Goldstein, J. (2003). Evaluation of 21ist century risks of smallpox vaccination and policy options. *Ann. Intern. Med.* 138:488–493.

Lane, J.M., and Fulginiti, V.A. (2003). Transmission of vaccinia virus, and rationale for measures for prevention. *Clin. Infect. Dis.* 37:281–284.

Langbein, B. (2003). New compounds show promise as smallpox drugs. *Reuters,* July 31, 2003.

LeDuc, J.W., Damon, I., Relman, D.A., Huggins, J., and Jahrling, P.B. (2002). Smallpox research activities: U.S. interagency collaboration, 2001. *Emerg. Infect. Dis.* 8:743–745.

LeDuc J.W., and Jahrling, P.B. (2001). Strenthening national preparedness for smallpox; an update. *Emerg. Infect. Dis.* 7:155–157.

Legrand, J., Viboud, C., Boelle, P.Y., Valleron, A.J., Flahault, A. (2004) Modelling responses to a smallpox epidemic taking into account uncertainty. *Epidemiology & Infection.* 13:19–25.

Linzer, D. (2003). U.S.team in Iraq finds no smallpox. *The Atlanta Journal-Constitution.* September 19, 2003, A14.

Mack, T.M. (1972). Smallpox in Europe, 1950–1971. *J. Infect. Dis.* 125:161–169.

Mack, T.M. (2003). A different view of smallpox and vaccination. *N. Engl. J. Med.* 348:460–463.

Mack, T.M., Thomas, D.B., Ali, A., and Khan, M.M. (1972). Epidemiology of smallpox in West Pakistan I: acquired immunity and the distribution of disease. *Am. J. Epidemiol.* 95:157–168.

Mahy, B.W.J. (2003). An overview on the use of a viral pathogen as a bioterrorism agent: why smallpox? *Antiviral Res.* 57:1–5.

Mangold, T., Gilmore J., and Molloy P. (1998). *Plague War*, PBS Videotape, Alexandria, VA.

McCurdy, L.H., Larkin, B.D., Martin, J.E., Graham, B.S. (2004) Modified vaccinia Ankara: potential as an alternative smallpox vaccine.*Clinical Infectious Diseases.* 38:1749–53.

McKenzie, P.J., Githens, J.H., Harwood, M.E., Roberts, J.F., Rao, A.R., and Kempe, C.H. (1965). Haemorrhagic smallpox. 2. Specific bleeding and coagulation studies. *Bull. WHO* 33:773–782.

Meltzer, M.I., Damon, I., LeDuc, J.W., and Millar, J.D. (2001). Modeling potential responses to smallpox as a bioterrorist weapon. *Emerg. Infect. Dis.* 7:959–969.

Moss, B. (2001). Poxviridae: the viruses and their replication. In: Knipe, D.M., and Howley, P.M. (eds), *Fields Virology,* 4th ed., chap. 84. Lippincott Williams & Wilkins, Philadelphia.

Mukherjee, M.K., Sarkar, J.K., and Mitra, A.S. (1974). Pattern of intrafamilial transmission of smallpox in Calcutta, India. *Bull. WHO* 51:215–219.

NBC10News. (July 10, 2003). Government monitoring Philadelphia air for biotoxins. Sensors check for anthrax, smallpox, in 31 cities. http://www.nbc10.com/2323887/detail.html.

Neff, J.M., Lane, J.M., Fulginiti, V.A., and Henderson, D.A. (2002). Contact vaccinia—transmission of vaccinia from smallpox vaccination. *J.A.M.A.* 288:1901–1905.

Neyts, J., and De Clercq E. (2003). Therapy and short-term prophylaxis of poxvirus infections: historical background and perspective. *Antiviral Res.* 57:25–33.

Olson, E. (2003). Panel urges shift of focus in preparing for smallpox. *The New York Times*, August 13, 2003, A15.

O'Toole, T., and Inglesby, T. (2001). Shining light on dark winter. Baltimore: Johns Hopkins Center for Civilian Defense Strategies. Accessed July 10, 2003 at www.hopkins-biodefense.org/lessons.html.

Oxford, J. (2003). Smallpox bioterrorism unlikely: populations are easily protected. *Basic Notes,* March 25, 2003.

Piller, C. (2002). Smallpox strike called unlikely. *The Los Angeles Times,* December 13, 2002.
Rao, A.R., Jacob, E.S., Kamalakshi, S., Appaswamy, S., and Bradbury. (1968). Epidemiological studies of smallpox. A study of intrafamilial transmission in a series of 254 infected families. *Ind. J. Med. Res.* 56:1826–1854.
Rao, A.R. (1972). *Smallpox.* The Kothari Book Depot, Bombay.
Roos, R. (2003). Scientists research antidotes to super mousepox virus. CIDRAP News. http://www.cidrap.umn.edu/cidrap/content/bt/smallpox/news/nov0603/mousepox.html
Ropp, S.L., Jin, Q., Knight, J.C. Massung, R.F., and Esposito, J.J. (1995). PCR strategy for identification and differentiation of smallpox and other orthopoxviruses. *J Clin. Microbiol.* 33:2069–2076.
Sarkar, J.K., Mitra, A,C., Mukherjee, M.K., and De, S.K. (1973). Virus excretion in smallpox. 2. Excretion in the throats of household contacts. *Bull. WHO* 48:523–527.
Schriewer J., Buller, R.M., Owens, G. (2004) Mouse models for studying orthopoxvirus respiratory infection. Methods in Molecular Biol. 264:289–308.
Seward, J.E, Galil, K, Damon, I, Norton, S.A, Rotz, L., Harpaz, R., Cono, J., Marin, M., Hutchins, S., Chaves, S.S., McCauley, M.M. (2004) Development and experience with an algorithm to evaluate suspected smallpox cases in the United States, 2002–2004 Clin Infect Dis 39:1477–84.
Smee, D.F., and Sidwell, R.W. (2003). A review of compounds exhibiting anti-orthopoxvirus activity in animal models. *Antiviral Res.* 57:41–52.
Smith, G.L., Vanderplasschen, A., and Law, M. (2002). The formation and function of extracellular enveloped vaccinia virus. *J Gen. Virol.* 83:2914–2931 [review].
Sofi, I.M., Kulesh, D.A., Saleh, S.S., Damon, I.K., Esposito, J.J., Schmaljohn, A.L., and Jahrling, P.B. (2003). Real-time PCR assay to detect smallpox virus. *J. Clin. Microbiol.* 41:3835–3839.
Sommer, A., and Foster, S.O. (1974). The 1972 Smallpox outbreak in Khulna municipality, Bangladesh. 1. Methodology and epidemiologic findings. *Am. J. Epidemiol.* 99:291–302.
Stevenson, R.W., and Stolberg, S.G. (2002). Bush lays out plan on smallpox shots; military is first. *The New York Times,* December 14, 2002, Section A1.
Stevenson, R.W., and Altman, L.K. (2002). Smallpox shots will start soon under Bush plan. *The New York Times*, December 12, 2002, A1.
Talbot, T.R., Stapleton, J.T., Brady, R.C., Winokur, P.L., Bernstein, D.I., Germanson, T., Yoder, S.M., Rock, M.T., Crowe, J.E. Jr., Edwards, K.M. (2004)Vaccination success rate and reaction profile with diluted and undiluted smallpox vaccine: a randomized controlled trial. *JAMA*. 292:1205–12.
Thomas, G. (1974). Air sampling of smallpox virus. *J. Hyg.* 73:1–7.
Thorpe, L.E., Mostashari, F., Karpati, A,M., Schwartz, S.P., Manning, S.E., Marx, M.A., Frieden, T.R.(2004) Mass smallpox vaccination and cardiac deaths, New York City, 1947. *Emerging Infectious Diseases.* 10:917–20.
Vrantseva, Y. (2002). US paper to face Russian lawsuit. *GAZETA, RU*, December 5, 2002.
Warrick, J. (2002). Iran said to be producing bioweapons. *The Washington Post*, May 15, 2002, A22.
Wehrle, P.F., Posch, J., Richter, K.H., and Henderson, D.A. (1970). An airborne outbreak of smallpox in a German hospital and its significance with respect to other outbreaks in Europe. *Bull. WHO* 4: 669–679.
Wheelis, M. (1999). Biological warfare before 1914. In: Geissler, E., and van Courtland Moon J.E. (eds.), *Biological and Toxin Weapons: Research and Use from the Middles Ages to 1945*. Oxford University Press, New York, pp. 8–34.
Whitley, R.J. (2003). Smallpox: a potential agent of bioterrorism. *Antiviral Res.* 57:7–12.
World Health Organization. (2005). http://www.who.int/emc/diseases/smallpox/slideset/index.htm
Zelicoff, A.P. (2003). An epidemiological analysis of the 1971 smallpox outbreak in Aralsk, Kazakhstan. *Crit. Rev. Microbiol.* 129:97–108.

6

Hemorrhagic Fever Viruses as Biological Weapons

Allison Groseth, Steven Jones,
Harvey Artsob, and Heinz Feldmann

1. INTRODUCTION

Biological agents have a number of attractive features for use as weapons. Not only do they have the potential to result in substantial morbidity and mortality, but also their use would result in fear and public panic. This may be sufficient to produce severe social and economic results disproportionate to the actual damage caused by the disease itself in terms of illness and death. These agents are also comparably easy and inexpensive to produce from only a very minute amount of starting material. Finally, as a result of the prolonged incubation times required for the appearance of symptoms, it is not only possible for an attack to be completed without being recognized, but also distribution of the disease over a large geographical region can occur if infected individuals travel following infection. In the face of an increased threat of terrorism, the potential for biological agents to be used as weapons has to be considered.

When determining the potential of an agent to bc used as a biological weapon, there are a number of factors that must be taken into consideration, including the ability of the agent to cause a high degree of morbidity and mortality based on a low infectious dose, environmental stability, the ability to undergo person-to-person transmission and be transmitted via aerosols, the availability of an effective vaccine, the potential to cause anxiety among health care workers and the public, suitability for large-scale production and a history of previous bioweapons research programs with the agent (Borio *et al.*, 2002; Centers for Disease Control and Prevention, 2000). Based on their properties with respect to a number of these criteria (Tables 6.1 and 6.2), one potentially attractive group of biological agents for weaponization are the viral hemorrhagic fever (VHF) agents. These viruses are part of four families: *Filoviridae*, *Arenaviridae*, *Bunyaviridae*, and *Flaviviridae*, which are grouped as VHFs based on their ability to cause a clinical illnesses associated with fever and bleeding diathesis (Peters and Zaki, 2002).

Bioterrorism and Infectious Agents
Edited by Fong and Alibek, Springer Science + Business Media, Inc., New York, 2005

Table 6.1
Transmission and Risk Factors for Use as Biological Weapons of Hemorrhagic Fever Viruses

Family	Genus	Agent	Animal reservoir	Arthropod vector	Human-to-human transmission	Agricultural risk	Vaccine[a]
Filoviridae	*Marburgvirus*	*Marburg virus*	Unknown	Unknown	+	–	–
	Ebolavirus	*Ebola virus*	Unknown	Unknown	+	–	–
Arenaviridae	*Arenavirus*	Lassa virus	Rodent (*Mastomys* sp.)	None	+	–	–
		Junin virus	Rodent (*Calomys* sp.)	None	+	–	Live, attenuated virus
		Machupo virus	Rodent (*Calomys* sp.)	None	+	–	–
		Guanarito virus	Rodent (*Sigmodon* sp. and *Zygodontomys* sp.)	None	+	–	–
		Sabia virus	Unknown	Unknown	+	–	–
Bunyaviridae	*Nairovirus*	Crimean-Congo Hemorrhagic Fever virus	Numerous bird, small and large vertebrate species	Tick (*Hyalomma* sp.)	+	+[b]	–
	Phlebovirus	Rift Valley Fever virus	Unknown	Mosquito (*Aedes* and *Culex* sp.)	–	+	Formalin-inactivated virus
	Hantavirus	Hantaan virus	Rodent (*Apodemus* sp.)	None	–	–	–
		Sin Nombre virus	Rodent (*Peromyscus* sp.)	None	–	–	–
		Andes virus	Rodent (*Oligoryzomys* sp.)	None	+	–	–
Flaviviridae	*Flavivirus*	Dengue virus	Humans and various other primate species	Mosquito (*Aedes* sp.)	–	–	–
		Yellow Fever virus	Numerous primate species	Mosquito (*Aedes* sp.)	–	–	Live, attenuated virus
		Omsk Hemorrhagic Fever virus	Rodent (*Ondatra* sp.)	Tick (*Dermacento* sp.)	–	–	–
		Kyasanur Forest Disease virus	Shrews and several primate species	Tick (*Haemaphysalis* sp.)	–	–	–

[a] Only vaccines licensed for use or that have been used experimentally in humans are listed.
[b] Although this virus does not cause significant morbidity in infected animals, the economic burden associated with infection make it an agricultural concern.

Table 6.2
Possible Mechanisms of Dissemination of Agents with a High Risk of Being Used as Bioweapons

Family	Genus	Agent	Method of dissemination				Probable prior weaponization
			Aerosol	Food[a]	Water[b]	Infected vector	
Filoviridae	*Marburgvirus*	*Marburg virus*	K (Belanov *et al.*, 1996)	P	U	N	Yes (Center for Nonproliferation Studies, 2000)
	Ebolavirus	*Ebola virus*	K (Jaax *et al.*, 1996)	P[c]	U	N	Yes (Center for Nonproliferation Studies, 2000)
Arenaviridae	*Arenavirus*	Lassa virus	K (Stephenson *et al.*, 1984)	K (ter Meulen *et al.*, 1996)	U	N	Yes (Center for Nonproliferation Studies, 2000)
		Junin virus	K (Kenyon *et al.*, 1992)	P	U	N	Yes (Center for Nonproliferation Studies, 2000)
		Machupo virus	P	P	U	N	Yes (Center for Nonproliferation Studies, 2000)
		Guanarito virus	P	P	U	N	No
		Sabia virus	K (Centers for Disease Control, 1994)	P	U	N	No
Bunyaviridae	*Nairovirus*	Crimean-Congo Hemorrhagic Fever virus	P	P	U	U	No
	Phlebovirus	Rift Valley Fever virus	K (Brown *et al.*, 1981)	P	U	P	Yes (Center for Nonproliferation Studies, 2000)

K, known mechanism of dissemination (natural or experimental transmission); P, possible mechanism for dissemination; U, unlikely to be a successful mechanism of dissemination; N, not a relevant dissemination mechanism.

[a] Food would not represent a potential source of infection if properly cooked; however, foods consumed raw or contaminated following cooking could be a potential risk.

[b] Although we consider dissemination of these agents using a chlorinated city water supply to be highly unlikely, the possibility remains that successful dissemination could be achieved through contamination of a fresh water source (e.g., bottled water).

[c] Experimental infection of nonhuman primates by the oral route was documented (Jaax *et al.*, 1996).

VHF agents have been previously weaponized by the former Soviet Union and the United States, and possibly by North Korea as well (Alibek and Handelman, 1999; Center for Nonproliferation Studies, 2000; Miller *et al.*, 2002). The Soviet bioweapons program focused on the study of *Marburg virus* (MARV), *Ebola virus* (EBOV), Lassa virus (LASV), and the new world arenaviruses Junin (Argentinean Hemorrhagic fever) (JUNV) and Machupo (Bolivian Hemorrhagic fever) (MACV) (Alibek and Handelman, 1999; Center for Nonproliferation Studies, 2000; Miller *et al.*, 2002), and continued until 1992. In particular, Soviet researchers investigated and quantified the aerosol infectivity and stability of freeze-dried MARV (Bazhutin *et al.*, 1992). In contrast, American offensive bioweapons programs were primarily based on Yellow Fever and Rift Valley Fever viruses (RVFV) (Center for Nonproliferation Studies, 2000) until their termination in 1969. Research by U.S biodefense programs now focuses on the detection, identification, and treatment of these agents and is defensive in nature.

One important consideration with potential biological weapons is their ability to be obtained from a variety of sources. A number of the VHF agents could be obtained from infected humans or animals in endemic areas or during outbreak situations. In particular, it is believed that the Japanese cult Aum Shinrikyo unsuccessfully attempted to obtain EBOV during during the 1995 outbreak in Zaire (Global Proliferation of Weapons of Mass Destruction, 1996; Kaplan, 2000). In addition to natural sources, it may also be possible to obtain these agents from laboratories by a variety of means. First of all, it is possible that these agents may be held at institutions other than those that are officially recognized. For example, MARV was extensively distributed to laboratories worldwide following the initial outbreak in 1967. All existing stocks outside recognized institutions may or may not have been destroyed. Another potential concern relates to the breakup of the Soviet Union and its subsequent economic difficulties, which may have provided other nations or terrorist groups access to weapons, agents, and/or the personnel involved in former bioweapons programs. In particular, it is feared that criminal organizations may have stolen samples from laboratories in the former Soviet Union and could have then sold them to terrorist groups. Finally, it is possible that samples of virus might be obtained from a BSL 4 facility recognized to hold these agents by legal means. Although attempts are continuously being made to improve security at these facilities, it is problematic to control the minute quantities of material required to initiate cultures of replicating agents.

However, based on the properties mentioned previously, some of these agents present a greater risk of being used as bioweapons than others. In particular, we have omitted detailed description of the *Flaviviridae* based on a number of factors, including the low mortality and availability of effective vaccinations in the case of Yellow Fever, as well as geographical restriction and lack of any evidence demonstrating a capability for small particle aerosol dissemination in the case of Omsk Hemorrhagic Fever and Kyasanur Forest Disease viruses. Among the *Bunyaviridae*, we have also elected to forgo discussion of some members. In particular, members of the *Hantavirus* genus will not be covered since most of the hantaviruses are very difficult to grow in amounts necessary for weaponization and because early treatment with Ribavirin may limit the mortality associated with infection by agents causing hemorrhagic fever with renal symptoms. Although these limitations also apply to Crimean-Congo Hemorrhagic Fever (CCHF) virus, we have included this agent based on its high rate of nosocomial transmission, as well as its very pronounced bleeding signs, both of which make this agent a potentially effective agent for creating fear among both the general public and among health care workers. In addition, although this virus does not cause morbidity in agriculturally important animal species, it could still present a significant agricultural concern based on the potentially large economic burdens, that would be associated with animal infection. Despite its low mortality rate in African populations and the lack of human-to-human transmission, we have also included a discussion of RVFV. This decision was based on its previous study by the U.S. bioweapons program, as well as the belief that this agent may present a particularly serious threat to agriculture if introduced into nonendemic regions, such as North America, where the apparent presence of a suitable arthropod vector could lead to long-term establishment on the continent.

Based on their properties, different VHF agents might potentially be disseminated via different routes during a bioterrorist attack (Table 6.2). For all of the agents discussed in this article, aerosol transmission has either been naturally or experimentally observed, or can be considered very likely by analogy to closely related family members (Belanov *et al.*, 1996; Brown *et al.*, 1981; Centers for Disease Control, 1994; Jaax *et al.*, 1996; Kenyon *et al.*, 1992; Stephenson *et al.*, 1984). LASV has also been shown to be transmissible through contamination of food with rodent excreta or consumption of infected rodents (ter Meulen *et al.*, 1996). This has never been shown for any of the other agents, although experimental oral transmission of EBOV has been shown (Jaax *et al.*, 1996). But assuming a food supply that will not be further cooked could be contaminated, this route has to be considered a possibility. We consider it to be universally unlikely that any of these agents would survive in a chlorinated city water supply long enough to cause infection; however, we cannot exclude the possibility of transmission via this mechanism if a suitable fresh water source could be accessed. One such possibility might be the contamination of bottled water. Finally, in the cases of RVFV and CCHF, the introduction of infected arthropod vectors has to be considered. In the case of CCHF, we consider this to be relatively unlikely, since this agent is transmitted by tick species, which do not adjust well to new geographical areas, thus complicating their introduction to a nonendemic region. However, this limitation does not apply to the same extent to mosquito-borne diseases, such as RVFV and, therefore, this method of introduction has to be considered as a possibility, since the introduction into even a few animals could then result in secondary transmission of the disease.

2. EPIDEMIOLOGY

All the agents responsible for causing VHFs have proven or presumed animal reservoirs, although no reservoirs have been identified for the filoviruses and Sabia virus to date (Table 6.1). Infection with VHF agents occurs as a result of receiving a bite from an infected arthropod, exposure to infected rodent excreta, or contact with the carcass of an infected animal (LeDuc, 1989). With the exception of RVFV and the flaviviruses, subsequent person-to-person transmission to close contacts can occur, thus making community outbreaks, as well as nosocomial spread, a risk. In general, there is little knowledge regarding the transmission of these viruses, since outbreaks are sporadic and typically occur in areas lacking in adequate health care infrastructure (Figure 6.1). As a result, outbreaks are often well underway or even waning before data gathering can be initiated. In addition, it is difficult to determine the risks associated with specific modes of transmission, since

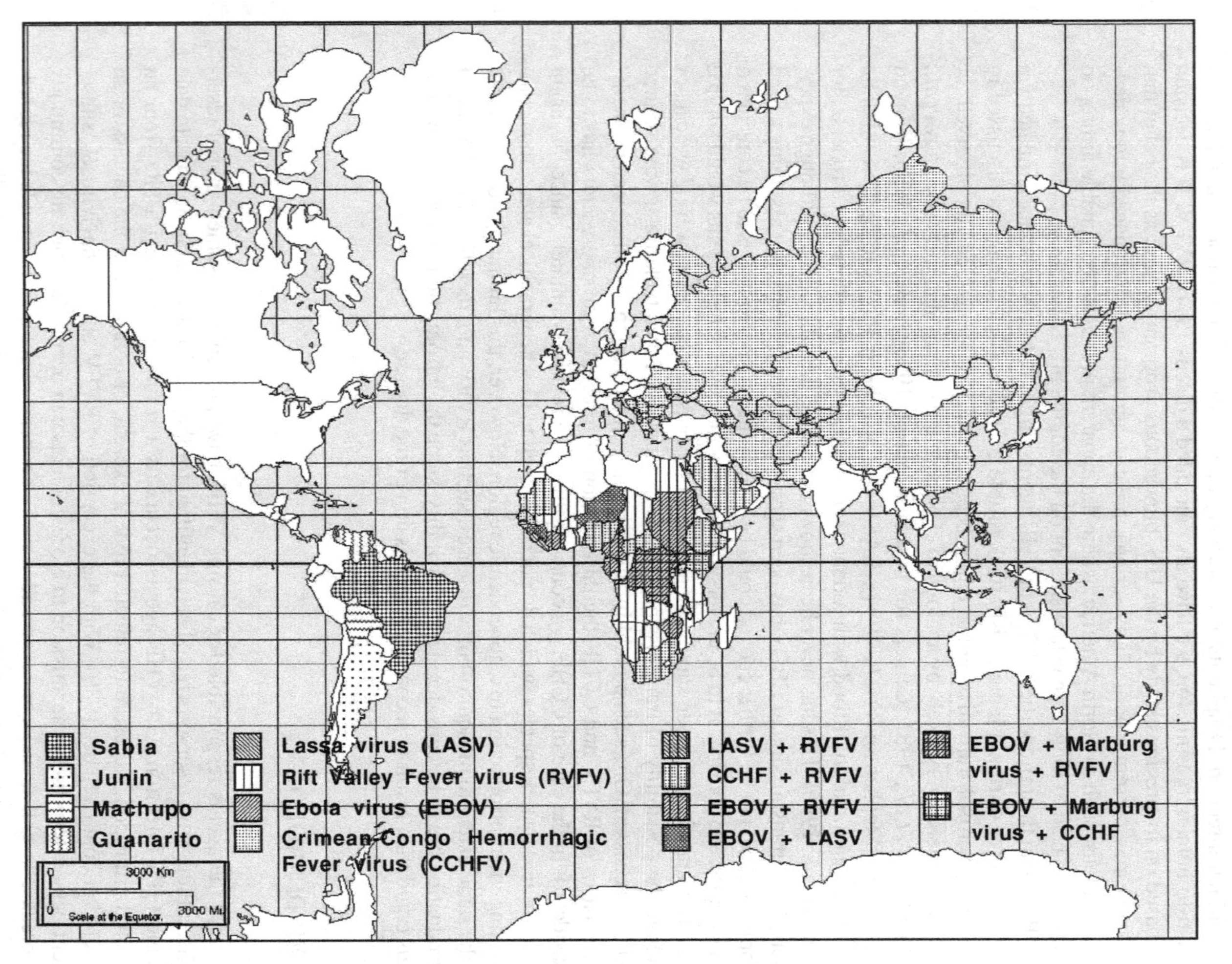

Figure 6.1. Geographical distribution of viral hemorrhagic fever agents with a high risk of being used as bioweapons.

patients often have multiple contacts, which involve different routes of potential exposure. However, it can be generally noted that percutaneous infections seem to be associated with the shortest incubation periods and the highest risk of mortality, whereas person-to-person airborne transmission appears to be relatively rare, although it is the only plausible explanation in some instances. Intentional dissemination of VHF agents as small particle aerosols would probably be highly effective as a weapon. However, there is little evidence that these agents are normally transmissible from human to human by the aerosol or droplet routes. This is in stark contrast to other biological agents, such as smallpox and plague, that are highly transmissible between humans by the inhalation route, and whose use as bioweapons would, consequently, result in numerous secondary infections and possibly much higher total mortalities despite lower case fatality rates.

2.1. *Filoviridae: Ebola* and *Marburg viruses*

Most cases of filovirus hemorrhagic fever occur in Africa, where infection results from contact with blood, secretions, or tissues from patients or nonhuman primates (Feldmann *et al.*, 2003). In particular, cases often result following injection with contaminated syringes and a number of cases have occurred as a result of accidental needle stick injury as well. The mortality rate for infections acquired by the percutaneous route is particularly high, and even low inocula can result in infection (World Health Organization, 1978). Transmission has been shown to occur through mucosal exposure in nonhuman primates (Jaax *et al.*, 1996; Simpson, 1969). Similarly, in humans, infection is thought to be possible through contact between contaminated hands and the mucosa or eyes, but this has never been directly shown (Colebunders and Borchert, 2000). Finally, there have been a number of cases in which transmission is suspected to have occurred via an airborne route (Centers for Disease Control and Prevention, 2001; Roels *et al.*, 1999). However, this does not appear to be a major contributing mechanism, since all epidemics to date have been successfully controlled using isolation techniques without specific airborne precautions.

Although MARV has been successfully isolated from healthy-looking monkeys prior to disease onset, there has never been any transmission documented prior to the onset of clinical symptoms (Simpson, 1969; Slenczka, 1999). Transmissibility of filoviruses increases during the course of infection and seems to be very rare during incubation, although a case was documented in which contact with a patient hours before the onset of symptoms resulted in transmission (Dowell *et al.*, 1995). Following convalescence, virus can persist for a short time in immunologically privileged sites. EBOV has been isolated from seminal fluid for up to 82 days after onset of symptoms and detected by reverse transcriptase-polymerase chain reaction (RT-PCR) for up to 101 days (Rodriguez *et al.*, 1999). Similarly, MARV was isolated from the seminal fluid of a patient up to 83 days after the onset of symptoms and resulted in infection of the patient's spouse (Martini, 1969; Slenczka, 1999). Similarly, virus can be isolated from liver biopsies and the anterior chamber of the eye 37 days or 12 weeks postonset of symptoms, respectively. This is despite clinical recovery and apparently normal immune function.

Following an incubation period, which typically lasts between 2 and 14 days, but may last as long as 21 days, infected individuals experience abrupt onset of fever (Table 6.3).

Table 6.3
Clinical Features of Viral Hemorrhagic Fevers Caused by Agents with a High Risk of Being Used as Bioweapons

Biological Agent	Distinguishing Clinical Features	Incubation Period (d)	Mortality (%)	Treatment
Ebola virus	Fever, severe prostration, maculopapular rash, bleeding, and disseminated intravascular coagulation common	2–21	50–90 [a]	Supportive
Marburg virus	Fever, myalgia, nonpruritic maculopapular rash, bleeding, and disseminated intravascular coagulation common	2–14	23–33 [b]	Supportive
Lassa virus	Gradual onset fever, nausea, abdominal pain, severe sore throat, conjunctivitis, ulceration of buccal mucosa, exudative pharyngitis, cervical lymphadenopathy, swelling of head and neck, pleural and pericardial effusions, and less commonly hemorrhages	5–21	15–20	Ribavirin, supportive
New World Arenaviruses	Gradual onset fever, myalgia, nausea, abdominal pain, conjunctivitis, generalized lymphadenopathy, petechiae, bleeding, and central nervous system dysfunctions	7–14	15–30	Ribavirin, supportive
Rift Valley Fever virus	Fever, headache, retro-orbital pain, photophobia, jaundice, and rarely hemorrhages	2–6	<1	Ribavirin, supportive
Crimean-Congo Hemorrhagic Fever virus	Fever, myalgia, petechial rash, echymoses, hematensis, melena, thrombocytopenia, leukopenia, hepatitis, and frequently jaundice	1–3 (tick bite) 4–6 (blood)	10–60	Ribavirin, supportive

[a] There are four different species of *Ebola virus*: *Zaire ebola virus* (ZEBOV), *Sudan ebola virus* (SEBOV), *Ivory Coast ebola virus* (ICEBOV), and *Reston ebola virus* (REBOV). Fatal infections have been documented for the Zaire and Sudan species, and the mortality rates are based on these data. Only a single nonfatal ICEBOV infection has ever been documented and, despite several documented infections, REBOV has never been known to cause illness in humans.

[b] Mortality rates associated with the most recent outbreak in Durba, DRC, were much higher than previously seen and may have exceeded 80%.

Additional symptoms may include chills, muscle pain, nausea, vomiting, abdominal pain, and/or diarrhea. All patients will show impaired coagulation to some extent, which can manifest itself as conjunctival hemorrhage, bruising, impaired clotting at venipuncture sites, and/or the presence of blood in the urine or feces. Swelling of the lymph nodes, kidneys, and, particularly, the brain can result. Also, there is necrosis of the liver, lymph organs, kidneys, testis, and ovaries. In fatal cases, gross pathological changes include hemorrhagic diatheses into the skin, mucous membranes, visceral organs, and the lumen of the stomach and intestines. Although approximately 50% of individuals develop a maculopapular rash on the trunk and shoulders, massive bleeding is fairly rare and is mainly restricted to the gastrointestinal tract. Severe nausea, vomiting, and prostration, as well as trachypnea, anuria, and decreased body temperature, all indicate impending shock. Death usually occurs between 6 and 9 days after the onset of symptoms (Feldmann *et al.*, 2003; Fisher-Hoch *et al.*, 1985; Murphy *et al.*, 1971; Peters & Zaki, 2002).

Filoviruses are extremely virulent in both human and nonhuman primates. Infection results in visceral organ necrosis, particularly in the liver, spleen, and kidneys, which is due directly to virus-induced cellular damage. Impairment of the microcirculation and the absence of inflammatory infiltration are also characteristic. The initial targets for filovirus replication are macrophages and other cells of the mononuclear phagocytic system, and from there infection spreads to fixed tissue macrophages in the liver, spleen, and other organs (Schnittler and Feldmann, 1998; Zaki and Goldsmith, 1999). Subsequently, progeny virions infect hepatocytes, adrenal cortical cells, fibroblasts, and – late in infection – endothelial cells. Tissue destruction results in the exposure of underlying collagen and the release of tissue factor, which results in the development of disseminated intravascular coagulation (DIC) (Geisbert *et al.*, 2003). Infected macrophages also become activated and, thus, release a number of cytokines and chemokines that upregulate cell surface adhesion and procoagulant molecules (Hensley *et al.*, 2002; Ströher *et al.*, 2001; Villinger *et al.*, 1999). Although these mediators seem to play a role in increasing endothelial permeability, destruction of endothelial cells during infection is also suggested to contribute to the development of hemorrhagic diathesis and shock. In addition, both MARV and EBOV are capable of producing secreted glycoprotein products, which may further contribute to filovirus pathogenesis, although the mechanisms by which this might occur remain unclear (Schnittler and Feldmann, 2003).

2.2. *Arenaviridae*: Lassa, Junin, Machupo, Guanarito, and Sabia

The natural hosts for arenaviruses include several rodent species, in which replication does not result in extensive cell damage and in which a carrier state can, therefore, be established. Transmission of arenaviruses to humans occurs as a result of inhalation of virus in aerosolized urine or feces or ingestion of food contaminated by, or direct contact of mucous membranes or abraded skin with, virus-infected excreta (Johnson *et al.*, 1965, 1966; ter Meulen *et al.*, 1996). Person-to-person transmission is by direct contact with infected blood, tissues, or body fluids, although airborne transmission is suspected in a few cases. Although no transmission has been observed during the incubation period, virus has been detected in semen up to 3 months postonset of symptoms and in urine up to 32 days (Buckley and Casals, 1970) postonset.

Infection in humans is initiated through the nasopharyngeal mucosa, usually following aerosol deposition (Samoilovich *et al.*, 1983). The absence of appreciable cytopathic effects during infection in tissue culture has lead to the suggestion that arenaviruses may exert their pathogenic effects by inducing the secretion of inflammatory mediators from macrophages, which are a primary target cell (Peters *et al.*, 1989). Following early infection of macrophages, virus infection can spread to other cell types, including the epithelial cells of several organs. In particular, infection of the spleen and lymph nodes commonly occurs and may have an influence on the ability of the host immune system to mount an effective response. The development of hemorrhages in some patients appears to be associated with the presence of circulating inhibitors of platelet aggregation and thrombocytopenia (Cummins *et al.*, 1990b). However, unlike filoviruses, the development of DIC does not seem to be a major pathogenic contributor in arenavirus hemorrhagic fevers (Knobloch *et al.*, 1980).

LASV infection is associated with a gradual onset of fever and malaise 5–21 days postinfection (Table 6.3). The severity of fever increases during the course of infection and myalgia and severe prostration may also occur. Gastrointestinal manifestations – such as abdominal pain, nausea, vomiting, diarrhea, and constipation – are common (McCormick *et al.,* 1987). Sore throat also occurs in approximately two-thirds of patients and is typically accompanied by inflammatory or exudative pharyngitis. Symptoms that reflect an increased vascular permeability – such as facial edema or pleural effusion – although uncommon, indicate a very poor prognosis. Similarly, while bleeding diatheses are rare, they also suggest an unfavorable outcome. Fatal cases of LASV result in shock and death. In survivors, symptoms usually last for 2–3 weeks, and there are a number of additional sequelae associated with convalescence (McCormick *et al.*, 1987). Early in convalescence, pericarditis can occur, particularly in male patients. In addition, there is a number of rare, but serious, neurological complications that can arise, including aseptic meningitis, encephalitis and global encephalopathy with seizures (Cummins *et al.*, 1992). Deafness is a very common and often permanent result of LASV infection, occurring in approximately 30% of patients (Cummins *et al.*, 1990a).

JUNV and MACV infections also begin with fever and malaise (Table 6.3). These symptoms are often accompanied by headache, myalgia, and epigastric pain in many cases. After 3–4 days, severe prostration, nausea, vomiting, dizziness, and indications of vascular damage – including conjunctival injection, flushing of the head and upper torso, petechiae, and mild hypotension – may appear (Harrison *et al.*, 1999). There is little evidence of tissue damage with these infections and few dramatic lesions can be observed. Although not prominent, reported lesions include liver or adrenal necroses and interstitial pneumonitis. In severe cases, hemorrhages of the mucous membranes and ecchymoses at injection sites can occur. These manifestations can progress to shock and generally indicate a poor prognosis. In some cases, neurologic complications occur (Harrison *et al.,* 1999). These begin with cerebellar signs, such as intention tremor, dysarthria, and dysphagia, and may then progress to grand mal convulsions and coma, which are almost always fatal. Despite the potential for neurological involvement, virus cannot be detected in either the brain or cerebral spinal fluid of patients. Neutralizing antibodies develop after 10–13 days for JUNV, but often take as long as 30 days to develop following MACV infection, due to the immunosuppressive nature of this virus (de Bracco *et al.,* 1978). In both cases, the development of neutralizing antibodies leads to convalescence, which lasts several weeks and is associated with fatigue, dizziness, and, in some cases, hair loss. Guanarito and Sabia virus infections are clinically very similar to JUNV and MACV infections, except that in Guanarito virus infection thrombocytopenia, bleeding, and neurologic involvement are more prominent (Vainrub and Salas, 1994).

2.3. *Bunyaviridae*: Rift Valley Fever and Crimean-Congo Hemorrhagic Fever

RVFV can be acquired from the bite of an infected mosquito, as well as through contact with infected animal tissues or aerosolized virus from carcasses (Swanepoel and Coetzer, 1994). Epidemiologic evidence also implicates the ingestion of raw milk in RVFV transmission (Jouan *et al.*, 1989), but natural RVF infection is mainly a concern for farmers and others who have close contact with animal tissues or blood. Although there have been no reports to date of person-to-person transmission, laboratory technicians are at considerable risk of infection as a result of inhalation of aerosols created during sample handling (Smithburn *et al.*, 1949; Swanepoel and Coetzer, 1994).

In addition to the possibility of human infection, there are also major agricultural concerns surrounding the possibility of domestic livestock (i.e., sheep, cattle, and goats) becoming infected. Mortality among infected sheep is highest, at around 90% for lambs and 25% for adults, with lower fatalities being observed in cattle and the lowest values in goats (Meegan and Shope, 1981). Infection of pregnant ewes tends to lead to abortion. Infection of domestic large animal species during a biological attack could lead to the establishment of RVFV in new geographic regions, provided one of a wide variety of appropriate mosquito vectors are present in the environment (Swanepoel and Coetzer, 1994). In Canada and the United States, *Aedes* sp., *Anopheles* sp., and *Culex* sp. could function as potential vectors for this virus (Gargan *et al.*, 1988).

Following an incubation period of 2–6 days, patients abruptly develop a fever and may exhibit other influenza-like symptoms (Table 6.3). These symptoms last an additional 2–5 days before convalescence, which may be prolonged, occurs. This process is associated with the development of neutralizing antibodies (Meegan and Shope, 1981). Only a small proportion, estimated to be <5%, of infected individuals go on to develop more serious disease. These include liver necrosis with hemorrhagic phenomena, retinitis with visual impairment, and meningoencephalitis (Meegan and Shope, 1981). The basis for hemostatic derangements observed during RVFV infection remain poorly understood; however, vasculitis and hepatic necrosis are postulated to play a major role in this process (Cosgriff *et al.*, 1989; Peters *et al.*, 1988).

CCHF infection in humans can occur as a result of a bite from an infected *Hyalomma* sp. tick or through direct contact with infected animals or their tissues. As a result agricultural workers, veterinarians, and abattoir workers are at a significant risk (Swanepoel *et al.*, 1987). Person-to-person transmission can also occur and has resulted in a number of nosocomial outbreaks. Transmission to hospital staff typically occurs as a result of contact with infected blood, respiratory secretions, aerosols, or excreta. Following a 1 to 6 day incubation period, there is onset of febrile disease with severe influenza-like symptoms that, after several days, progress to hemorrhagic manifestations, including petechial rash, ecchymoses, bruises, hematemesis, and melena accompanied by thrombocytopenia and leukopenia

(Swanepoel *et al.*, 1987) (Table 6.3). Most CCHF patients show some signs of hepatitis and jaundice, hepatomegaly, and/or elevated serum enzyme levels. Death usually occurs during the second week of illness, and often follows severe hemorrhages, shock, and renal failure. Patients who recover do so without any complications.

3. PATIENT MANAGEMENT

3.1. Clinical Recognition

Due to the rarity of infections that cause hemorrhagic manifestations in regions such as North America, these unusual symptoms may potentially help cases to be identified relatively early. However, it is also possible that the lack of any known risk factors, such as insect bites or travel in the 21 days prior to the onset of symptoms, as well as the nonspecific early manifestations of VHF, their variable clinical presentation, and the lack of familiarity with these disorders among physicians could hinder diagnosis of initial cases. Therefore, identification of an intentional outbreak may not occur until an epidemiologic picture, based on the appearance of large numbers of severely ill patients in a short time span, develops. Following the indication of a biowarfare attack involving a VHF agent, national public health authorities will provide directions to clinical laboratories regarding the processing and transport of samples (e.g., guidelines developed by the Centers for Disease Control and Prevention (CDC), Atlanta, Georgia; "*Canadian Contingency Plan for Viral Hemorrhagic Fevers and Other Related Diseases*"). Specimens must be properly labeled, stored, and transported, and individuals who come into contact with potentially contaminated materials should be identified and monitored for signs of illness. Further details are available on the CDC website at http://www.bt.cdc.gov/Agent/VHF/VHF.asp and in the "*Canadian Contingency Plan for Viral Hemorrhagic Fevers and Other Related Diseases*" (CCDR Supplement, Volume 23S1, 1997). Once a single case of VHF has been identified, the recognition of additional cases can be based on appropriate signs and symptoms in addition to a link to the time and place of exposure.

3.2. Laboratory Diagnosis

As clinical microbiology and public health laboratories are not generally equipped for diagnosis of VHF agents, it is necessary that samples are sent to one of the few designated laboratories capable of performing the required assays. Of the available techniques for diagnosis of VHFs, antigen capture ELISA and RT-PCR are the most useful for making a diagnosis in an acute clinical setting. Serology (IgM capture ELISA and IgG ELISA) is useful for confirmation, but negative serology is not exclusive. Virus isolation should be achieved, although its utility as a diagnostic procedure is restricted by time and biosafety concerns. For nonoutbreak surveillance, immunoperoxidase staining of formalin-fixed biopsies is available for some of the agents (e.g., filoviruses) (Zaki *et al.*, 1999) and has several advantages, including its simplicity, specificity, and the lack of any need for enhanced biocontainment.

Recently, mobile laboratory units have been added to assist case patient management and surveillance efforts during epidemics (e.g., Ebola outbreak in Gulu, Uganda, and Mbomo, The Republic of Congo). In general, these units have been received very well, but experience is rather limited at this point. Mobile units for assistance during intentional release of bioterrorism agents have been established at the National Microbiology Laboratory, Public Health Agency of Canada, which could be deployed to national and international events. However, despite all the achievements in laboratory diagnostics, it should be kept in mind that the diagnosis of VHF will initially have to be based on clinical assessment. For this purpose, contingency plans should be developed that are still missing in many, particularly developing, countries. In addition, many nations encounter difficulties in sample transport, which can cause substantial delays in laboratory response. Once samples are received, laboratory response is fairly reasonable today, and results can be expected within 24–48 hours.

3.3. Treatment

Treatment of VHF infections is mainly supportive in nature and involves a combination of intravenous fluid replacement, the administration of analgesics and standard nursing measures. The maintenance of fluid and electrolyte balance as well as circulatory volume and blood pressure are essential. Additionally, mechanical ventilation, renal dialysis and/or antiseizure therapy may be required, while intramuscular injections, non-steroidal anti-inflammatory and anticoagulant therapies are generally contraindicated. Finally, it is important to note that treatment for other possible etiologic agents (e.g. agents of bacterial sepsis) should not be withheld while a VHF diagnosis is being confirmed.

Ribavirin, a nonimmunosuppressive nucleoside analog, has been found to be somewhat effective in the treatment of bunyavirus and arenavirus infections (Figure 6.2). When administered intravenously within the first 6 days after infection, ribavirin has been shown to decrease LASV mortality from 76% to 9% (Huggins, 1989) and decrease Argentinean hemorrhagic fever mortality from 40% to 12.5% (Enria and Maiztegul, 1994). However, ribavirin does not penetrate into the brain efficiently and, thus, will likely not be effective in countering neurologic symptoms associated with these infections (Huggins, 1989). The main side effect observed with ribavirin therapy is a dose-dependent hemolytic anemia, although a variety of cardiac and pulmonary effects have been associated with combination ribavirin/interferon α treatment in hepatitis C patients. Additional teratogenic and embryolethal effects have been observed in a number of species and, although similar effects have never been reported in humans, ribavirin has been classified as a category X drug and is contraindicated for use during pregnancy. However, due to the enhanced mortality associated with VHF infection during pregnancy, it is likely that the benefits outweigh any potential risk to the fetus and, thus, treatment is still recommended. Ribavirin has never shown any efficacy in the treatment of filovirus or flavivirus infections. Nevertheless, early treatment of a putative VHF case should always include ribavirin. Once the final laboratory diagnosis has been made, treatment should be continued in case of bunyavirus and arenavirus, but stopped in case of filovirus and flavivirus infections (Figure 6.2). In the case of filovirus infections,

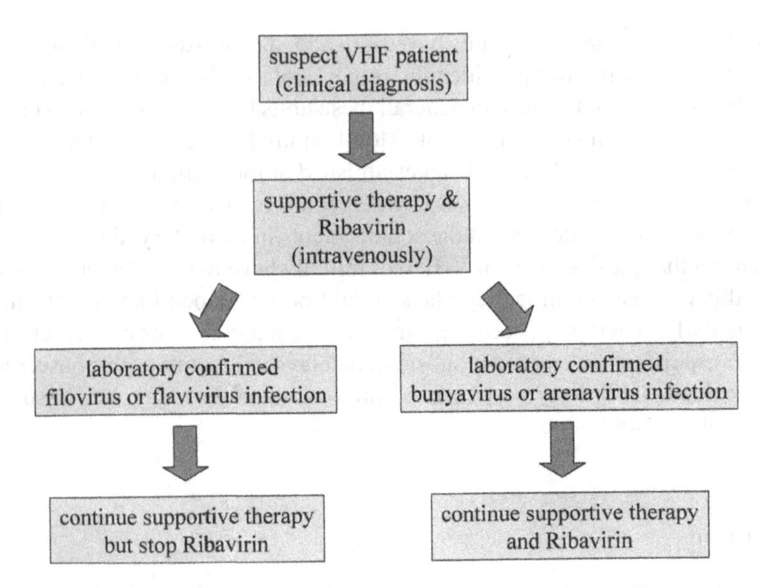

Figure 6.2. Treatment recommendations for viral hemorrhagic fever (VHF) patients.

adenosine analogs have been identified which, through inhibition of the cellular enzyme *S*-adenosylhomocysteine hydrolase, have been shown to significantly reduce replication *in vitro* (Huggins *et al.*, 1999). It was also shown that administration of recombinant nematode anticoagulant protein c2 as late as 24 hours postinfection lead to a 33% survival in a uniformly fatal EBOV-infected macaque model (Geisbert *et al.*, 2003). In addition, the survival time in remaining animals was significantly prolonged, indicating that although this therapy may not be sufficient on its own, it could be a valuable tool in the treatment of filovirus infections, and potentially other hemorrhagic diseases that involve overexpression of procoagulant molecules.

The ability to treat VHF infections using passive immunization depends on the agent in question. Studies have indicated that treatment of CCHFV, as well as LASV and new world arenaviruses, is possible using this method (Enria *et al.*, 1984; Frame *et al.*, 1984; Jahrling *et al.*, 1984; van Eeden *et al.*, 1985). However, the rarity of these infections, as well as the lack of programs to collect and store recovered VHF patient plasma, means that this avenue is unlikely to be part of the early response to a biological event. However, new advances made in the manufacturing of monoclonal antibodies, as well as selecting highly effective human-derived or humanized products, may offer new alternatives in the future.

Following a biological attack, all individuals who were exposed to the VHF, as well as any high-risk or close contacts of these patients, should be placed under medical surveillance. High-risk contacts include those who have had mucosal contact with the patient or received a percutaneous injury that involved exposure to blood, secretions, or excreta from infected patients, whereas close contacts are those who live with or have physical contact with patients with evidence of VHF, as well as those who process lab specimens from or care for these patients prior to the implementation of the appropriate precautions. The exception to this is contacts of RVFV patients or flavivirus-infected patients, since these viruses are not known to be transmitted from person to person. However, contacts responsible for processing lab specimens should still be monitored, since these agents are highly infectious in a laboratory setting. Contacts should be advised to monitor and record their temperature twice daily and report any fever over 101°F (38°C), as well as any suspicious symptoms. Monitoring should continue for 21 days following the suspected exposure or last contact with the patient. If fever develops, treatment with ribavirin should be started immediately, unless another cause for the illness has been diagnosed or the etiologic agent is known to be a filovirus or flavivirus (Figure 6.2). There is some experimental evidence indicating that ribavirin may be effective in delaying, but not preventing, the onset of disease following arenavirus infection when administered postexposure; however, its effectiveness has never been studied in humans (McKee *et al.*, 1988). Regardless, the current CDC guidelines recommend ribavirin treatment for high-risk contacts of LASV patients.

4. VACCINES

The need for vaccines to prevent viral hemorrhagic fevers was recognized long before there was concern over the use of these agents as biological weapons. Early attempts to produce inactivated vaccines for *Ebola virus* sp. were unsuccessful (Feldmann *et al.*, 2003; Geisbert *et al.*, 2002), however, experimental and unlicensed vaccines do exist for JUNV and RVFV (Maiztegui *et al.*, 1998; Pittman *et al.*, 1999). Although it is very unlikely these will ever be fully licensed for human use, they demonstrate that despite the virulence of VHF agents, immunoprophylaxis is a viable option in the prevention of epidemics. Historically, the high level of biological containment required to work with these viruses has been a major block in development of new treatments or vaccines; furthermore, because of the virulence of the wild-type viruses, live attenuated vaccine strains are unlikely to be a viable option for immunization. The development of molecular techniques, enabling the manipulation of RNA genomes (e.g., Neumann *et al.*, 2002; Volchkov *et al.*, 2001), may result in the development of new vaccine strategies. Additionally, the development of effective animal models other than nonhuman primates has been a significant barrier to vaccine testing (e.g., Bray *et al.*, 1998).

Ebola virus has been the focus for a relatively large number of research teams because of the very high mortality, the high public profile of this virus, and the availability of three animal models (nonhuman primates, guinea pig, and mouse). Several vaccine strategies have been successful in protecting rodents from EBOV (reviewed by Hart, 2003); however, almost all were universally unsuccessful in protecting nonhuman primates (Geisbert *et al.*, 2002). The first vaccine to have proven efficacy in nonhuman primates was a DNA prime/adenovirus boost approach (Sullivan *et al.*, 2000); however, the DNA prime/adenovirus boost protocol required months to provide protective immunity making this vaccine unsuitable for use following a bioterrorist attack. However, subsequent studies using only a single dose of the recombinant adenovirus part of the initial vaccine resulted in protection of the nonhuman primates from a high challenge dose (1500 LD_{50}) just 28 days after immunization, indicating this strategy may be useful in the context of a bioterrorist attack (Sullivan *et al.*, 2003).

More recently, a new vaccine strategy using live recombinant, vesicular stomatitis virus (VSV) has been successful in both rodent and nonhuman primate models of

EBOV infection. These vectors have a complete deletion of the wild-type glycoprotein open reading frame that is substituted by the full-length functional glycoprotein of *Zaire ebolavirus* (Garbutt *et al.*, 2004). These recombinant viruses have the tropism of EBOV, but are attenuated *in vivo*. This VSV recombinant vaccine also protected nonhuman primates 28 days after immunization, but was able to protect mice when given 30 minutes after challenge and is effective in mice when administered by the intranasal and oral routes. If these observations can be repeated in the nonhuman primates, there is real potential for rapid mass immunization (Jones *et al.*, 2003). In addition, if the potential of replicating VSV-based vectors as mucosal vaccines is fulfilled, they will be a promising candidate for future vaccine development against other lethal VHF agents.

5. PUBLIC HEALTH MEASURES

5.1. Infection Control

In the absence of effective therapies or vaccines for agents of VHF prevention of infection must rely on patient isolation, careful specimen handling, and appropriate barrier precautions (Figure 6.3). In the majority of cases, these procedures have been suffi-

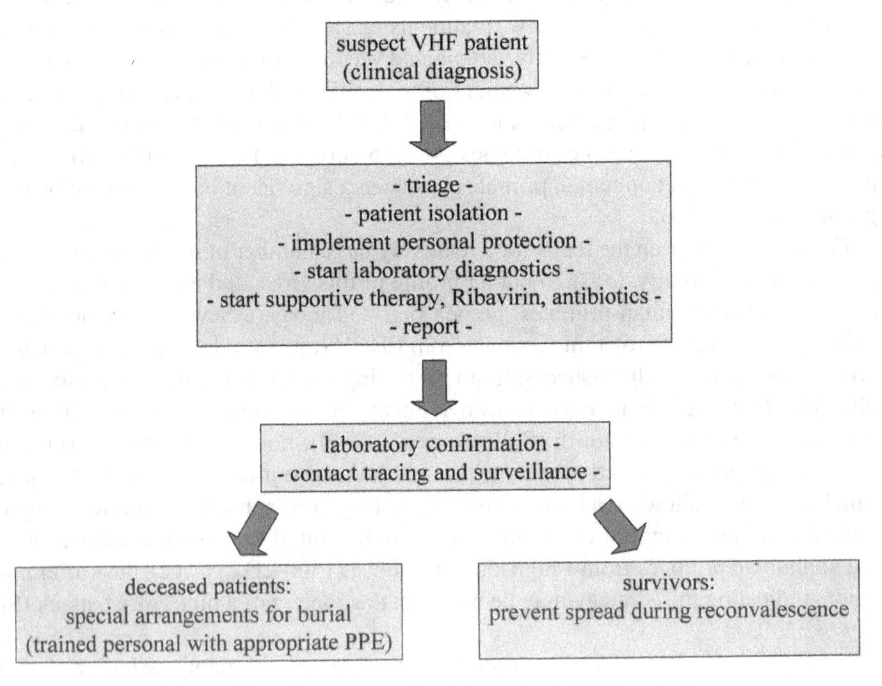

Figure 6.3. Management of viral hemorrhagic (VHF) patients.

cient to prevent further infection of family members, health care workers, and other patients, and should include strict hand hygiene, the use of double gloves, impermeable gowns, face shields, goggles, leg and shoe coverings as well as an N-95 mask or powered air-purifying respirator. Additional precautions should include the use of environmental disinfectant, isolation of patients in negative pressure rooms with restricted access to visitors and, if possible, dedicated medical equipment. Finally, it has been observed that virus may remain in the body fluids of convalescent patients and, thus, VHF patients should be advised to refrain from sexual activity for 3 months following clinical recovery.

While failure to recognize VHF infection in patients and implement appropriate barrier precautions often does not result in high numbers of secondary infections, implementation of infection control procedures will further reduce the possibility of transmission. Staff should be informed of the suspect VHF diagnosis and trained in specimen handling prior to the receipt of samples. Furthermore, appropriate personal protective equipment (PPE) should be worn during all procedures that involve the handling of infectious samples. Respiratory protection should also be used, since some VHFs are highly infectious in the laboratory setting and some may be transmitted by small particle aerosols. As a result, it is also highly recommended that all manipulations be performed under appropriate biosafety guidelines and biocontainment.

The implication of contact with cadavers in disease transmission during several EBOV outbreaks (Centers for Disease Control and Prevention, 2001; Roels *et al.*, 1999) means that, in the event of a biological attack using VHF agents, special arrangements will need to be made to accommodate the burial of deceased patients. In particular, the transport of deceased patients should only be performed by trained individuals using the appropriate PPE and respiratory equipment, as described previously. Postmortem examination of deceased VHF patients should preferentially only be performed in negative pressure rooms by specially trained individuals making use of the appropriate barrier and respiratory precautions. Expeditious burial or cremation of cadavers is also recommended and should involve a minimum of handling.

5.2. Environmental Decontamination

In the event of an undetected aerosol biological attack, it will be at least a week before the first cases of disease become apparent. By this time it has to be assumed that no or little infectious virus remains in the environment, which questions the value of extensive surface decontamination procedures. In the event of other kinds of biological attack, such as those involving virus-containing liquids, decisions regarding suitable decontamination procedures will have to be made together with experts in environmental remediation.

It has been demonstrated for filoviruses that virus particles are stable and remain infectious for several days at room temperature in liquids or dried material. This makes use of appropriate protective equipment, as described earlier, extremely important for all individuals involved in the environmental decontamination process. Whenever possible steam sterilization should be used, as it is the most effective method available, otherwise a 1:100 dilution of household bleach or approved hospital disinfectant, such as those based on phenol or quaternary ammonium compounds, can be used.

6. ONGOING RESEARCH AND PROPOSED AGENDA

A number of significant challenges remain within the hemorrhagic fever virus field with respect to our understanding of the viruses themselves, the disease process, and our ability to prevent and/or manage VHF infections. Issues of major importance include the urgent need to develop rapid diagnostics for VHFs and disseminate safe technology to local laboratories, thus allowing a more expeditious preliminary diagnosis in the event of an outbreak. Rapid, sensitive, and reliable assays carried out at or near the point of care would enhance patient case management and improve care, as well as ensuring that only infected patients are placed into valuable isolation beds. Additionally, the development of similarly rapid, sensitive and reliable assays for the environmental detection of VHF agents will enhance our capacity to both detect and investigate the source of a bioterrorism event (i.e. identifying the dissemination method) before the first cases report to health care facilities. Furthermore, such detection assays may allow us to limit the environmental impact of bioterrorist attacks by ensuring that only areas known to be contaminated with virus are exposed to the decontamination procedures of choice. However, since most of the VHF agents are not particularly stable in the environment, there is a risk of detecting the presence of genome in the absence of live infectious virus. The balance between the best possible detection of the effected areas and the real probability of infection from the detected agent is difficult to assess. Transmission of these viruses is also poorly understood and, in particular, the role of airborne transmission of these agents needs to be clarified, given its relevance to the management of a biological attack. The development and testing of vaccine candidates and therapeutic agents must also be a priority for this field, as is research into the pathogenic mechanisms by which VHFs cause such devastating disease. This kind of basic insight into the pathogenic mechanisms of these viruses has the potential to provide information which could be key to the successful management of infected patients. One major drawback to this research is the biocontainment needed for the animal and tissue culture work with these agents. Building new facilities is one way to respond and political support is mostly guaranteed in crisis situations. But maintaining facilities, long-term funding, and most importantly establishment of a comfort level of well-trained personnel are critical issues that also have to be addressed.

7. CONCLUSIONS

In addition to causing illness and death, a biological attack would aim to cause fear in the general populace and, thus, result in social and economic disruption. Based on their fearsome reputation and dramatization by the popular media, VHF agents would be excellent candidates to serve this purpose. Given the potential for a biological attack to occur, it is of the utmost importance that resources and knowledge are made available to deal effectively with such a situation in a safe and timely manner.

8. ACKNOWLEDGMENTS

The authors gratefully acknowledge the assistance of C.J. Peters and K. Johnson for their valuable discussion and comments, as well as V. Jensen and T. Hoenen for critical reading of the manuscript. Work on viral hemorrhagic fever agents at the Canadian Science Centre for Human and Animal Health is supported by Public Health Agency of Canada, the Canadian Institutes of Health Research (MOP-43921), and the National Institutes of Health (1R21 AI 053560-01). A.G. holds a graduate student award from the Natural Science and Engineering Research Council of Canada (PGSA-254708-2002).

References

Alibek, K., and Handelman, S. (1999). *Biohazard: The Chilling True Story of the Largest Covert Biological Weapons Program in the World, Told from the Inside by the Man Who Ran It.* Random House, New York.

Bazhutin, N., Belanov, E., Spiridonov, V., Voitenko, A., Krivenchuk, N., Krotov, S., Omel' chenko, N., Tereshchenko, A., and Khomichev, V. (1992). The effect of the methods for producing an experimental Marburg virus infection on the characteristics of the course of the disease in green monkeys. *Vopr. Virusol.* 37:153–156.

Belanov, E., Muntianov, V., Kriuk, V., Sokolov, A., Bormotov, N., P'iankov, O., and Sergeev, A. (1996). Survival of Marburg virus infectivity on contaminated surfaces and in aerosols. *Vopr. Virusol.* 41:32–34.

Borio, L., Inglesby, T., Peters, C., Schmaljohn, A., Hughes, J., Jahrling, P., Ksiazek, T., Johnson, K., Meyerhoff, A., Toole, T., Ascher, M., Bartlett, J., Breman, J., Eitzen, E. Jr., Hamburg, M., Hauer, J., Henderson, D., Johnson, R., Kwik, G., Layton, M., Lillibridge, S., Nabel, G., Osterholm, M., Perl, T., Russell, P., and Tonat, K. [Working Group on Civilian Biodefense]. (2002). Hemorrhagic fever viruses as biological weapons: medical and public health management. *J.A.M.A.* 287:2391–2405.

Bray, M., Davis, K., Geisbert, T., Schmaljohn, C., and Huggins, J. (1998). A mouse model for evaluation of prophylaxis and therapy of Ebola hemorrhagic fever. *J. Infect. Dis.* 178:651–661.

Brown, J., Dominik, J., and Morrissey, R. (1981). Respiratory infectivity of a recently isolated Egyptian strain of Rift Valley fever virus. *Infect. Immun.* 33:848–853.

Buckley, S., and Casals, J. (1970). Lassa fever, a new virus disease of man from West Africa. 3. Isolation and characterization of the virus. *Am. J. Trop. Med. Hyg.* 19:680–691.

Center for Nonproliferation Studies. (2000). Chemical and biological weapons: Possession and programs past and present. http://www.cns.miis.edu/research/cbw/possess.htm

Centers for Disease Control and Prevention. (2001). Outbreak of Ebola hemorrhagic fever, Uganda, August 2000 – January 2001. *MMWR Morb. Mort. Wkly. Rep.* 50:73–77.

Centers for Disease Control and Prevention. (1994). Arenavirus infection—Connecticut, 1994. *MMWR Morb. Mort. Wkly. Rep.* 43:635–636.

Centers for Disease Control and Prevention (2000). Biological and chemical terrorism: Strategic plan for preparedness and response. Recommendations of the CDC Strategic Planning Workgroup. *MMWR Recomm. Rep.* 49:1–14.

Colebunders, R., and Borchert, M. (2000). Ebola haemorrhagic fever-a review. *J. Infect.* 40:16–20.

Cosgriff, T., Morrill, J., Jennings, G., Hodgson, L., Slayter, M., Gibbs, P., and Peters, C. (1989). Hemostatic derangement produced by Rift Valley fever virus in rhesus monkeys. *Rev. Infect. Dis.* 11(Suppl 4):S807–S814.

Cummins, D., Bennett, D., Fisher-Hoch, S., Farrar, B., Machin, S., and McCormick, J. (1992). Lassa fever encephalopathy: clinical and laboratory findings. *J. Trop. Med. Hyg.* 9:197–201.

Cummins, D., McCormick, J., Bennett, D., Samba, J., Farrar, B., Machin, S., and Fisher-Hoch, S. (1990a). Acute sensorineural deafness in Lassa fever. *J.A.M.A.* 264:2119.

Cummins, D., Molinas, F., Lerer, G., Maiztegui, J., Faint, R., and Machin, S. (1990b). A plasma inhibitor of platelet aggregation in patients with Argentine hemorrhagic fever. *Am. J. Trop. Med. Hyg.* 42:470–475.

de Bracco, M., Rimoldi, M., Cossio, P., Rabinovich, A., Maiztegui, J., Carballal, G., and Arana, R. (1978). Argentine hemorrhagic fever. Alterations of the complement system and anti-Junin-virus humoral response. *N. Engl. J. Med.* 299:216–221.

Dowell, S., Mukunu, R., Ksiazek, T., Kahn, A., Rollin, P., and Peters, C. (1995). Transmission of Ebola hemorrhagic fever: a study of risk factors in family members, Kikwit, Democratic Republic of the Congo. *J. Infect. Dis.* 179(Suppl 1): S87–S91.

Enria, D., Briggiler, A., Fernandez, N., Levis, S., and Maiztegui, J. (1984). Importance of dose of neutralising antibodies in treatment of Argentine haemorrhagic fever with immune plasma. *Lancet* 2:255–256.

Enria, D., and Maiztegui, J. (1994). Antiviral treatment of Argentine hemorrhagic fever. *Antiviral Res.* 23:23–31.

Feldmann, H., Jones, S., Klenk, H.-D., and Schnittler, H. (2003). Ebola virus: from discovery to vaccine. *Nat. Rev. Immunol.* 3:677–685.

Fisher-Hoch, S., Platt, G., Neild, G., Southee, T., Baskerville, A., Raymond, R., Lloyd, G., and Simpson, D. (1985). Pathophysiology of shock and hemorrhage in a fulminating viral infection (Ebola). *J. Infect. Dis.* 152:887–894.

Frame, J., Verbrugge, G., Gill, R., and Pinneo, L. (1984). The use of Lassa fever convalescent plasma in Nigeria. *Trans. R. Soc. Trop. Med. Hyg.* 78:319–324.

Garbutt, M., Liebscher, R., Wahl-Jensen, V., Jones, S., Moeller, P., Wagner, R., Volchkov, V., Klenk, H.D., Feldmann, H., and Stroeher, U. (2004). Properties of replication-competent vesicular stomatitis virus vectors expressing glycoproteins of filoviruses and arenaviruses. *J. Virol.* 78:5458–5465.

Gargan, T. 2nd, Clark, G., Dohm, D., Turell, M., and Bailey, C. (1988). Vector potential of selected North American mosquito species for Rift Valley fever virus. *Am. J. Trop. Med. Hyg.* 38:440–446.

Geisbert, T., Hensley, L., Jahrling, P., Larsen T., Geisbert, J., Paragas, J., Young, H., Fredeking, T., Rote, W., and Vlasuk, G. (2003). Treatment of Ebola virus infection with a recombinant inhibitor of factor VIIa/tissue factor: a study in rhesus monkeys. *Lancet* 362:1953–1958.

Geisbert, T., Pushko, P., Anderson, K., Smith, J., Davis, K., and Jahrling, P. (2002). Evaluation in nonhuman primates of vaccines against Ebola virus. *Emerg. Infect. Dis.* 8:503–507.

Global Proliferation of Weapons of Mass Destruction (1996). Hearings Before the Permanent Subcommittee on Investigations of the Committee on Governmental Affairs, United States Senate, 104th Cong, 1st and 2nd Sess.

Harrison, L., Halsey, N., McKee, K., Peters, C., Barrera Oro, J., Briggiler, A., Feuillade, M., Maiztegui, J. (1999). Clinical case definitions for Argentine hemorrhagic fever . *Clin. Infect. Dis.* 28:1091–1094 .

Hart, M. (2003). Vaccine research efforts for filoviruses. *Int. J. Parasitol.* 33:583–595.

Hensley, L., Young, H., Jahrling, P., and Geisbert, T. (2002). Proinflammatory response during Ebola virus infection of primate models: possible involvement of the tumor necrosis factor receptor superfamily. *Immunol. Lett.* 80:169–179.

Huggins, J. (1989). Prospects for treatment of viral hemorrhagic fevers with ribavirin, a broad-spectrum antiviral drug. *Rev. Infect. Dis.* 11(Suppl 4):S750–S761.

Huggins, J., Zhang, Z., and Bray, M. (1999). Antiviral drug therapy of filovirus infections: S-adenosylhomocysteine hydrolase inhibitors inhibit Ebola virus in vitro and in a lethal mouse model. *J. Infect. Dis.*179(Suppl 1):S240–S247.

Jaax, N., Davis, K., Geisbert, T., Vogel, P., Jaax, G., Topper, M., and Jahrling, P. (1996). Lethal experimental infection of rhesus monkeys with Ebola-Zaire (Mayinga) virus by the oral and conjunctival route of exposure. *Arch. Pathol. Lab. Med.* 120:140–155.

Jahrling, P., Peters, C., and Stephen, E. (1984). Enhanced treatment of Lassa fever by immune plasma combined with ribavirin in cynomolgus monkeys. *J. Infect. Dis.* 149:420–427.

Johnson, K., Kuns, M., Mackenzie, R., Webb, P., and Yunker, C. (1966). Isolation of Machupo virus from wild rodent *Calomys callosus*. *Am. J. Trop. Med. Hyg.* 15:103–106.

Johnson, K., Mackenzie, R., Webb, P., and Kuns, M. (1965). Chronic infection of rodents by Machupo virus. *Science*. 150:1618–1619.

Jones, S., Geisbert, T., Ströher, U., Geisbert, J., Bray, M., Jahrling, P., Geisbert, T., and Feldmann, H. (2003). Replicating vectors for vaccine development. Symposium on Viral hemorrhagic Fevers, Vaccine Research Center, NIAID, NIH, DHHS, Bethesda, Md.

Jouan, A., Coulibaly, I., Adam, F., Philippe, B., Riou, O., Leguenno, B., Christie, R., Ould Merzoug, N., Ksiazek, T., and Digoutte, J. (1989). Analytical study of a Rift Valley fever epidemic. *Res. Virol.* 140:175–186.

Kaplan, D. (2000). Aum Shinrikyo. In: Tucker, J. (ed.), *Toxic Terror: Assessing Terrorist Use of Chemical and Biological Weapons*. MIT Press, Cambridge, mass., pp. 207–226.

Kenyon, R., McKee, K. Jr., Zack, P., Rippy, M., Vogel, A., York, C., Meegan, J., Crabbs, C., and Peters, C. (1992). Aerosol infection of rhesus macaques with Junin virus. *Intervirology* 33:23–31.

Knobloch, J., McCormick, J., Webb, P., Dietrich, M., Schumacher, H., and Dennis, E. (1980). Clinical observations in 42 patients with Lassa fever. *Tropenmed. Parasitol.* 31:389–398.

LeDuc, J. (1989). Epidemiology of hemorrhagic fever viruses. *Rev. Infect. Dis.* 11(Suppl 4): S730–S735.

Maiztegui, J., McKee, K. Jr., Barrera Oro J., Harrison, L., Gibbs, P., Feuillade, M., Enria, D., Briggiler, A., Levis, S., Ambrosio, A., Halsey, N., and Peters, C. (1998). Protective efficacy of a live attenuated vaccine against Argentine hemorrhagic fever. AHF Study Group. *J. Infect. Dis.* 177: 277–283.

Martini, G. (1969). Marburg agent disease in man. *Trans. R. Soc. Trop. Med. Hyg.* 63:295–302.

McCormick, J., King, I., Webb, P., Johnson, K., O'Sullivan, R., Smith, E., Trippel, S., and Tong, T. (1987). A case-control study of the clinical diagnosis and course of Lassa fever. *J. Infect. Dis.* 155:445–455.

McKee, K. Jr., Huggins, J., Trahan, C., and Mahlandt, B. (1988). Ribavirin prophylaxis and therapy for experimental argentine hemorrhagic fever. *Antimicrob. Agents Chemother.* 32:1304–1309.

Meegan J., and Shope, R. (1981). Emerging concepts on Rift Valley fever. *Perspect. Virol.* 11: 267–387.

Miller, J., Engelberg, S., and Broad, W. (2002). *Germs: Biological Weapons and America's Secret War*. GK Hall, Waterville, Me.

Murphy, F., Simpson, D., Whitfield, S., Zlotnik, I., and Carter, G. (1971). Marburg virus infection in monkeys. Ultrastructural studies. *Lab. Invest.* 24:279–291.

Neumann, G., Feldmann, H., Watanabe, S., Lukashevich, I., and Kawaoka, Y. (2002). Reverse genetics demonstrates that proteolytic processing of the Ebola virus glycoprotein is not essential for replication in cell culture. *J. Virol.* 76:406–410.

Peters, C., Jones, D., Trotter, R., Donaldson, J., White, J., Stephen, E., and Slone, T. Jr. (1988). Experimental Rift Valley fever in rhesus macaques. *Arch. Virol.* 99:31–44.

Peters, C., Liu, C., Anderson, G. Jr, Morrill, J., and Jahrling, P. (1989). Pathogenesis of viral hemorrhagic fevers: Rift Valley fever and Lassa fever contrasted. *Rev. Infect. Dis.* 11(Suppl 4): S743–S749.

Peters, C., and Zaki, S. (2002). Role of the endothelium in viral hemorrhagic fevers. *Crit. Care Med.* 30(Suppl 5):S268–S273.

Pittman, P., Liu, C., Cannon, T., Makuch, R., Mangiafico, J., Gibbs, P., and Peters, C. (1999). Immunogenicity of an inactivated Rift Valley fever vaccine in humans: a 12-year experience. *Vaccine* 18: 181–189.

Rodriguez, L., De Roo, A., Guimard, Y., Trappier, S., Sanchez, A., Bressler, D., Williams, A., Rowe, A., Bertolli, J., Khan, A., Ksiazek, T., Peters, C., and Nichol, S. (1999). Persistence and genetic stability of Ebola virus during the outbreak in Kikwit, Democratic Republic of the Congo, 1995. *J. Infect. Dis.* 179(Suppl 1):S170–S176.

Roels, T., Bloom, A., Buffington, J., Muhungu, G., MacKenzie, W., Khan, A., Ndambi, R., Noah, D., Rolka, H., Peters, C., and Ksiazek, T. (1999). Ebola hemorrhagic fever, Kikwit, Democratic Republic of the Congo, 1995: risk factors for patients without a reported exposure. *J. Infect. Dis.* 179(Suppl 1):S92–S97.

Samoilovich, S., Carballal, G., and Weissenbacher, M. (1983). Protection against a pathogenic strain of Junin virus by mucosal infection with an attenuated strain. *Am. J. Trop. Med. Hyg.* 32:825–828.

Schnittler, H., and Feldmann, H. (2003). Viral hemorrhagic fever – a vascular disease. *Thromb. Haemost.* 89:967–972.

Schnittler, H., and Feldmann, H. (1998). Marburg and Ebola hemorrhagic fevers: does the primary course of infection depend on the accessibility of organ-specific macrophages? *Clin. Infect. Dis.* 27: 404–406.

Simpson, D. (1969). Marburg agent disease. *Trans. R. Soc. Trop. Med. Hyg.* 63:303–309.

Slenczka, W. (1999). The Marburg virus outbreak of 1967 and subsequent episodes. *Curr. Top. Microbiol. Immunol.* 235:49–75.

Smithburn, K., Mahaffy, A., Haddow, A., Kitchen, S., and Smith, J. (1949). Rift Valley fever: accidental infection among laboratory workers. *J. Immunol.* 62:213–227.

Stephenson, E., Larson, E., Dominik, J. (1984). Effect of environmental factors on aerosol-induced Lassa virus infection. *J. Med. Virol.* 14:295–303.

Ströher, U., West, E., Bugany, H., Klenk, H.-D., Schnittler, H., and Feldmann, H. (2001). Infection and activation of monocytes by Marburg and Ebola viruses. *J. Virol.* 75:11025–11033.

Sullivan, N., Geisbert, T., Geisbert, J., Xu, L., Yang, Z., Roederer, M., Koup, R., Jahrling, P., and Nabel, G. (2003). Accelerated vaccination for Ebola virus haemorrhagic fever in non-human primates. *Nature* 424:681–684.

Sullivan, N., Sanchez, A., Rollin, P., Yang, Z., and Nabel, G. (2000). Development of a preventive vaccine for Ebola virus infection in primates. *Nature*. 408:605–609.

Swanapoel, R., and Coetzer, J. (1994). Rift Valley Fever. In: *Infectious Diseases of Livestock With Special Reference to Southern Africa*. New York, Oxford University Press.

Swanepoel, R., Shepherd, A., Leman, P., Shepherd, S., McGillivray, G., Erasmus, M., Searle, L., and Gill, D. (1987). Epidemiologic and clinical features of Crimean-Congo hemorrhagic fever in southern Africa. *Am. J. Trop. Med. Hyg.* 36:120–32.

ter Meulen, J., Lukashevich, I., Sidibe, K., Inapogui, A., Marx, M., Dorlemann, A., Yansane, M., Koulemou, K., Chang-Claude, J., and Schmitz, H. (1996). Hunting of peridomestic rodents and consumption of their meat as possible risk factors for rodent-to-human transmission of Lassa virus in the Republic of Guinea. *Am. J. Trop. Med. Hyg.* 55:661–666.

Vainrub B, and Salas R. (1994). Latin American hemorrhagic fever. *Infect. Dis. Clin. North. Am.* 8: 47–59.

van Eeden, P., van Eeden, S., Joubert, J., King, J., van de Wal, B., and Michell, W. (1985). A nosocomial outbreak of Crimean-Congo haemorrhagic fever at Tygerberg Hospital. Part II. Management of patients. *S. Afr. Med. J.* 68:718–721.

Villinger, F., Rollin, P., Brar, S., Chikkala, N., Winter, J., Sundstrom, J., Zaki, S., Swanepoel, R., Ansari, A., and Peters, C. (1999). Markedly elevated levels of interferon (IFN)-gamma, IFN-alpha, interleukin (IL)-2, IL-10, and tumor necrosis factor-alpha associated with fatal Ebola virus infection. *J. Infect. Dis.* 179(Suppl 1):S188–S191.

Volchkov, V., Volchkova, V., Muhlberger, E., Kolesnikova, L., Weik, M., Dolnik, O., and Klenk, H.-D. (2001). Recovery of infectious Ebola virus from complementary DNA: RNA editing of the GP gene and viral cytotoxicity. *Science* 291:1965–1969.

World Health Organization. (1978). Ebola hemorrhagic fever in Zaire, 1976. *Bull. World Health Org.* 56:271–293.

Zaki, S., and Goldsmith, C. (1999). Pathologic features of filovirus infection in humans. *Curr. Top. Microbiol. Immunol.* 235:97–115.

Zaki, S., Shieh, W., Greer, P., Goldsmith, C., Ferebee, T., Katshitshi, J., Tshioko, F., Bwaka, M., Swanepoel, R., Calain, P., Khan, A., Lloyd, E., Rollin, P., Ksiazek, T., and Peters, C. (1999). A novel immunohistochemical assay for the detection of Ebola virus in skin: implications for diagnosis, spread, and surveillance of Ebola hemorrhagic fever. *J. Infect. Dis.* 179(Suppl 1):S36–S47.

7

Botulism as a Potential Agent of Bioterrorism

Thomas P. Bleck, MD, FCCM

1. INTRODUCTION

The clostridial neurotoxins are among the most potently lethal substances in the world, with median lethal doses (LD_{50}) for humans in the nanogram/kilogram range. The toxins are closely related proteins, produced as a single polypeptide chain and subsequently altered to produce a heavy chain and a light chain connected by disulfide bonds. The genetic information encoding the seven botulinum toxins is encoded on the bacterial chromosome; in contrast, tetanospasmin, the tetanus neurotoxin, is encoded on a plasmid. Both botulinum and tetanus toxins are relatively simple to produce. The botulinum toxins affect the neuromuscular junction and muscarinic peripheral autonomic synapses; their major manifestations are weakness and autonomic dysfunction. In stark relief, the predominant effect of tetanospasmin is in the central nervous system, where it produces failure of inhibition leading to hypertonia and spasms.

Tetanospasmin is not a useful candidate for weaponization because of widespread immunity due to vaccination, although at least one group has attempted this (Williams and Wallace, 1989). The botulinum toxins are a more interesting target for weapons development, since the public health strategy for this toxin has been to prevent exposure.

Part of the potential value of these toxins to an aggressor lies in the mortality they could produce. In the developed world, however, the expense (and diversion of resources) involved in trying to support a large number of patients with ventilatory failure would itself constitute a potential target. The cost of care per botulism patient in Canada and the United States was estimated to be $340,000 in 1989 (Todd, 1989).

Clostridium botulinum produces most cases of botulism, with a few other clostridial strains accounting for the remainder. Botulinum toxins are designated types A through G based on antigenic differences (Hatheway, 1989). Types A, B, E, and (uncommonly) F

Bioterrorism and Infectious Agents
Edited by Fong and Alibek, Springer Science+Business Media, Inc., New York, 2005

produce human disease, whereas types C and D are almost exclusively confined to animals (Oguma. *et al.*, 1990). Type G has not yet been associated with natural illness. The clinical forms of botulism include *foodborne botulism, infant botulism, wound botulism*, and *botulism of undetermined etiology*. In the past decade, botulinum A toxin has achieved prominence as a therapeutic modality in conditions resulting from excessive muscle activity, such as torticollis. The type B toxin is also produced commercially for this purpose. Since muscle activity also contributes to some types of skin lines and some types of headache, the range of conditions treated with botulinum toxin has expanded considerably in recent years. Botulinum toxin has also been developed as a weapon, which could be used to contaminate food or beverage supplies, or be aerosolized.

2. HISTORY OF BOTULISM

The term "botulism" derives from the Latin word *botulus* or sausage. Outbreaks of poisoning related to sausages and other prepared foods first occurred in Europe in the 19th century. Justinus Kerner, a district health officer in southern Germany, first recognized the connection between sausage ingestion and paralytic illness in 230 patients in 1820 and made sausage poisoning a reportable disease (Kerner, 1928). Around the same time, physicians in Russia recognized a disease with similar symptoms, termed *fish poisoning* (Young, 1976). Van Ermengen described *C. botulinum* in 1897 and demonstrated that the organism elaborated a toxin that could induce weakness in animals (van Ermengen, 1897). van Ermengen's toxin was subsequently shown to be the type A toxin; the type B toxin was discovered in 1904 (Landman, 1904). Wound botulism was described in 1943 (Davis et al, 1951) and infant botulism in 1976 (Midura and Arnon, 1976). The occurrence of sporadic cases without an apparent etiology, predominantly related to gastrointestinal colonization, was first reported in 1986 (Chia *et al.*, 1986).

Rumors of the intentional use of this toxin date to the turn of the 20th century (Centers for Disease Control and Prevention, 1999). However, its first known development as a biological weapon took place in Manchuria during the 1930s under the auspices of Unit 731, the infamous Japanese biological warfare research unit (Williams and Wallace, 1989). During World War II, the United States also developed methods for the large-scale production of botulinum toxin. Scientists who worked in this program did not discuss the subject of their work by name, but rather referred to it as agent X (Middlebrook and Franz, 1997). The United States also produced large quantities of botulinum toxoid for use in vaccination (Arnon *et al.*, 2001). Because the Allied leadership feared that their troops would be attacked with botulinum toxin during the Normandy invasion, Allied troops involved in the Normandy invasion may have been immunized with this toxoid. There remains debate about this possibility in the historical record (Middlebrook and Franz, 1997; Williams and Wallace, 1989). The Germans were likely dissuaded from using biological weapons in this war because of concern that the Allies would retaliate with their own large arsenal of such weapons. Other countries have also experimented with techniques to weaponize botulinum toxin, including the former Soviet Union (Arnon *et al.*, 2001).

Iraq conducted the largest known military botulinum toxin programs, during which it produced 19,000 liters of concentrated toxin, much of which was loaded into missiles and artillery shells (Zilinskas, 1997). The Iraqi military deployed several other biological weapons systems prior to the first Gulf War. The seed cultures for the Iraqi biological weapons programs were purchased legally from an American microbiological supply house. To prepare for potential exposure to botulinum toxin, approximately 8,000 American service personnel were vaccinated in 1991 (Middlebrook and Franz, 1997).

The Japanese cult Aum Shinrikyō, best known for their attacks on civilians using the nerve agent sarin (Kortepeter *et al.*, 2001) disseminated botulinum toxin as a weapon on at least three occasions in the early 1990s (Lifton, 2000). Although their attempts were not successful (perhaps because of the reticence of some cult members to carry out their orders), the ease with which this group was able to produce botulinum toxin demonstrates the potential for small groups to use this substance as an agent of bioterrorism. At least some of the *C. botulinum* cultures Aum Shinrikyō grew came from spores obtained from local soil, highlighting the ubiquitous nature of the organism, and our inability to eliminate this threat by controlling the commercial supply of microorganisms.

3. BOTULINUM TOXIN AS A WEAPON

Natural poisoning with botulinum toxin is normally results from ingestion of preformed toxin produced in improperly prepared or stored food; the exceptions include infant botulism, gastrointestinal botulism in adults, and wound botulism. A terrorist attack with this toxin could involve intoxication via ingestion or an aerosol. Botulinum toxin could be introduced into foods or beverages at many points in their manufacturing cycles. The clinical manifestations of the victims would not be distinguishable from a natural botulism outbreak, although epidemiologic investigation might reveal the common food ingested was not one typically associated with botulism, or that different foods in the same geographic area were contaminated. The introduction of toxin into milk trucks or other large beverage transports would produce more sporadic cases. In such a circumstance, clinicians would be likely to overlook an attack early in its course. Automated systems for the collection and analysis of epidemiologic data would probably be required to accomplish this goal (M'ikantha *et al.*, 2003).

The gastrointestinal symptoms of nausea, vomiting, diarrhea, and abdominal cramping heralding natural gastrointestinal botulism are not manifestations of botulinum toxin, which by itself would cause constipation. Smith (1977) has argued that these effects are a consequence of other bacterially derived substances. Consequently, an outbreak of botulism without these other symptoms would raise the suspicion of a bioterrorist attack. Toxin would be detectable in serum, and in stool as well, if the gastrointestinal tract was the route of entry. Inhalation of toxin would probably not produce measurable levels in stool, but it might be detectable in nasal secretions.

The classic presentation of botulism is the rapid evolution of bilateral cranial neuropathies and symmetric descending weakness in an afebrile patient. The Centers for Disease Control and Prevention (CDC) recommends attention to five cardinal features: (1) fever is absent (unless a complicating infection occurs); (2) the neurologic manifestations are symmetric; (3) the patient remains responsive; (4) the heart rate is normal or slow in the absence of hypotension; and (5) sensory deficits do not occur (except for blurred vision) (Smith, 1977). The first two features are important for the exclusion of poliomyelitis and other viral infections causing acute flaccid paralysis. Rare exceptions have been reported to most of these generalizations, but in an attack the vast majority of cases would be expected to follow the usual presentation and progression.

Symptoms of foodborne botulism usually develop 12–36 hours after the toxin has been ingested. The patient initially complains of nausea and a dry mouth, and diarrhea may occur at this stage (which probably reflects a different manifestation of the bacterial infection than those due to botulinum toxin; see previous data). Cranial nerve dysfunction usually starts with the pupils, reflecting parasympathetic involvement (blurred vision due to pupillary dilation) or diplopia from impairment of cranial nerves III, IV, or VI (Terranova *et al.*, 1979). The pupillary reactions may remain abnormal for months after recovery of motor functions (Friedman *et al.*, 1990). Nystagmus is noted occasionally, usually in patients with type A disease. Dysfunction of the lower cranial nerves presents as dysphagia, dysarthria, and tongue weakness. The weakness then spreads to the upper extremities, the trunk, and eventually the lower extremities. Respiratory dysfunction results either from upper airway obstruction (the weakened glottis tending to close during attempted inspiration) or from weakness of the diaphragm and parasternal intercostals muscles. Patients who require mechanical ventilation, respectively, need average periods of 58 days (type A) and 26 days (type B) for weaning (Hughes *et al.*, 1981). Recovery may not begin for as long as 100 days (Colerbatch *et al.*, 1989). Possible autonomic problems include altered gastrointestinal function, alterations in heart rate, loss of cardiac accelerator responsive to hypotension or postural change, hypothermia, and urinary retention (Vita *et al.*, 1987).

Hughes *et al.* (1981) summarized several published reports of botulism to detect differences in the clinical findings produced by different toxin types (see Table 7.1). Type A is more commonly associated with dysarthria, blurred vision, dyspnea, diarrhea, sore throat, dizziness, ptosis, ophthalmoplegia, facial paresis, and upper extremity weakness; types B and E produce more autonomic dysfunction. None of these differences is diagnostic of the toxin type, however. The pupils are either dilated or unreactive in less than 50% of patients; although these are very useful signs when present, their absence does not diminish the likelihood of botulism.

Patients with infant botulism present with constipation, often followed by feeding difficulties, hypotonia, increased drooling, and weak crying (Cornblath *et al.*, 1983). Upper airway obstruction may on occasion be the initial sign (Oken *et al.*, 1992) and is the major indication for intubation (Schreiner *et al.*, 1991). The condition progresses in severe cases to include cranial neuropathies and respiratory weakness, with ventilatory failure in about half of the diagnosed patients. The condition progresses for 1–2 weeks and stabilizes for another 2–3 weeks before recovery starts (Vita *et al.*, 1987) and relapses may occur (Glauser *et al.*, 1990).

Wound botulism lacks the typical prodromal gastrointestinal disorder of the foodborne form, but is otherwise similar in presentation. Fever, if present at all, reflects the wound infection rather than botulism. The wound may itself in rare instances appear to be healing well even while neurologic manifestations are occurring. *C. botulinum* infection may occasionally produce abscesses (Elston *et al.*, 1991), and botulism has occurred as a consequence of sinusitis with this organism following inhalation of cocaine (Kudrow *et al.*, 1988). The incubation period varies from 4 to 14 days.

The signs and symptoms exhibited by victims of inhalational botulism are the same as those seen with ingestion, except for the lack of prodromal gastrointestinal disturbances

Table 7.1
Symptoms and Signs in Patients with the Common Types of Human Botulism[a]

	Type A (%)	Type B (%)	Type E (%)
Neurologic symptoms			
Dysphagia	96	97	82
Dry mouth	83	100	93
Diplopia	90	92	39
Dysarthria	100	69	50
Upper extremity weakness	86	64	NA
Lower extremity weakness	76	64	NA
Blurred vision	100	42	91
Dyspnea	91	34	88
Paresthesiae	20	12	NA
Gastrointestinal symptoms			
Constipation	73	73	52
Nausea	73	57	84
Vomiting	70	50	96
Abdominal cramps	33	46	NA
Diarrhea	35	8	39
Miscellaneous symptoms			
Fatigue	92	69	84
Sore throat	75	39	38
Dizziness	86	30	63
Neurologic findings			
Ptosis	96	55	46
Diminished gag reflex	81	54	NA
Ophthalmoparesis	87	46	NA
Facial paresis	84	48	NA
Tongue weakness	91	31	66
Pupils fixed or dilated	33	56	75
Nystagmus	44	4	NA
Upper extremity weakness	91	62	NA
Lower extremity weakness	82	59	NA
Ataxia	24	13	NA
DTRs diminished or absent	54	29	NA
DTRs hyperactive	12	0	NA
Initial mental status			
Alert	88	93	27
Lethargic	4	4	73
Obtunded	8	4	0

DTRs, deep tendon reflexes; NA, not available.
[a] Complied using data from Hughes (1991), Tacket and Rogawski (1989), and Weber *et al.* (1993).

as discussed previously. The latency from exposure to clinically apparent disease following inhalation is between 12 hours and 3 days, with maximal disease by about 5 days (Arnon *et al.*, 2001).

Dysphagia and other symptoms of neuromuscular impairment have been reported in rare cases after the therapeutic use of botulinum A toxin (Comella *et al.*, 1992).

4. DIAGNOSIS

If botulinum toxin was used as a biological weapon, diagnosis would depend on recognizing the route of exposure. Contaminated food or beverages would produce an epidemic resembling that of a natural foodborne outbreak, except for the absence of the initial irritative gastrointestinal symptoms. The diagnosis of inhalational botulism would depend on the recognition of cases whose common exposure was presence in a particular geographic area at a particular time, rather than having ingested the same food. The amount of inhaled toxin producing disease would probably not produce measurable toxin in blood or other patient samples, except perhaps for nasopharyngeal secretions. More detailed information about weaponized botulinum toxin is available at http://www.usamriid.army.mil/education/bluebook.html and https://ccc.apgea.army.mil/sarea/products/textbook/Web_Version/ index.htm (the latter site requires registration).

An appropriate history for the type of botulism suspected is the most important diagnostic test in naturally occurring botulism. If others are already affected, the condition is easily recognized. However, since the toxin may not be evenly distributed in foodstuffs, the absence of other patients does not eliminate the diagnosis. In an attack with botulinum toxin, one might anticipate large numbers of patients presenting with intoxication without the gastrointestinal prodrome of the naturally acquired disease.

Botulism has a limited differential diagnosis. Myasthenia gravis and the Eaton-Lambert myasthenic syndrome share some of the characteristics of botulism, but these conditions are rarely fulminant and usually lack the autonomic dysfunctions typical of botulism. An edrophonium test may be considered, but improving strength after administration of this agent is not pathognomonic of myasthenia gravis and has been reported in botulism (Edell *et al.*, 1983). Tick paralysis is excluded by a careful physical examination, since the Dermacentor tick will still be attached, and the weakness resolves rapidly after the tick is removed. Classic acute inflammatory polyneuropathy (AIPN; Guillain-Barré syndrome) frequently begins with sensory complaints, rapidly progresses to areflexia, rarely begins with cranial nerve dysfunction, and almost never alters pupillary reactivity. Patients with botulism are not areflexic until the affected muscle group is completely paralyzed. The Miller Fisher variant of AIPN presents with oculomotor dysfunction and may produce other cranial neuropathies, but includes prominent ataxia; this ataxia is absent in botulism. Patients with poliomyelitis and other viral causes of acute flaccid paralysis are typically febrile on presentation, and their weakness is almost always asymmetric. Magnesium intoxication can mimic botulism (Cherington, 1990). Botulism may be suspected in cases of diphtheria, organophosphate poisoning, or brainstem infarction (Dunbar, 1990).

Laboratory evaluation includes anaerobic cultures and toxin assays of serum, stool, and potentially implicated foodstuff if available. Confirmation and toxin typing are successful in almost 75% of cases (Dowell *et al.*, 1977). Early cases of natural botulism are more likely to be diagnosed by the toxin assay, whereas those studied later in the disease are more likely to have a positive culture than a positive toxin assay (Woodruff *et al.*, 1992). In an attack with botulinum toxin, no clostridial bacteria would be anticipated. The mouse bioassay remains the most sensitive test for botulinum toxin (Notermans and

Nagel, 1989). Serum from patients with AIPN can produce paralysis when injected into mice, however, so this test is not completely diagnostic of botulism (Notermans *et al.*, 1972). The toxin can also be detected by gel hydrolysis or ELISA. Patients may excrete toxin for up to 1 month after the onset of illness, and stool cultures may remain positive for a similar period.

The increasing concern about bioterrorism in recent years has resulted in several new rapid detection methods, some of which are potentially applicable to both environmental and patient samples (Ahn-Yoon, 2004; Liu *et al.*, 2003; Peruski *et al.*, 2002).

Neurophysiologic studies reveal nerve conduction velocities to be normal (or unchanged in the case of a preexisting neuropathy); the amplitude of compound muscle action potentials is reduced in 85% of cases, although not all motor units demonstrate this (Cherington, 1982). Repetitive nerve stimulation at high rates (20 Hz or greater, as opposed to the 4 Hz rate used in the diagnosis of myasthenia gravis) may reveal a small increment in the motor response (see Figure 7.1). This test is quite uncomfortable and should not be performed unless botulism or the Lambert-Eaton myasthenic syndrome is a serious consideration. Botulism can be distinguished clinically and electrophysiologically from the Lambert-Eaton syndrome (Gutmann and Pratt, 1976). In infants, the increment after high-frequency stimulation may be very dramatic. Single-fiber electromyographic studies may be useful in questionable cases, but in a biological attack this would probably not be needed. The therapeutic use of botulinum A toxin for dystonic disorders can produce electrophysiologic evidence of toxin dissemination to distant sites (Buchman *et al.*, 1993)

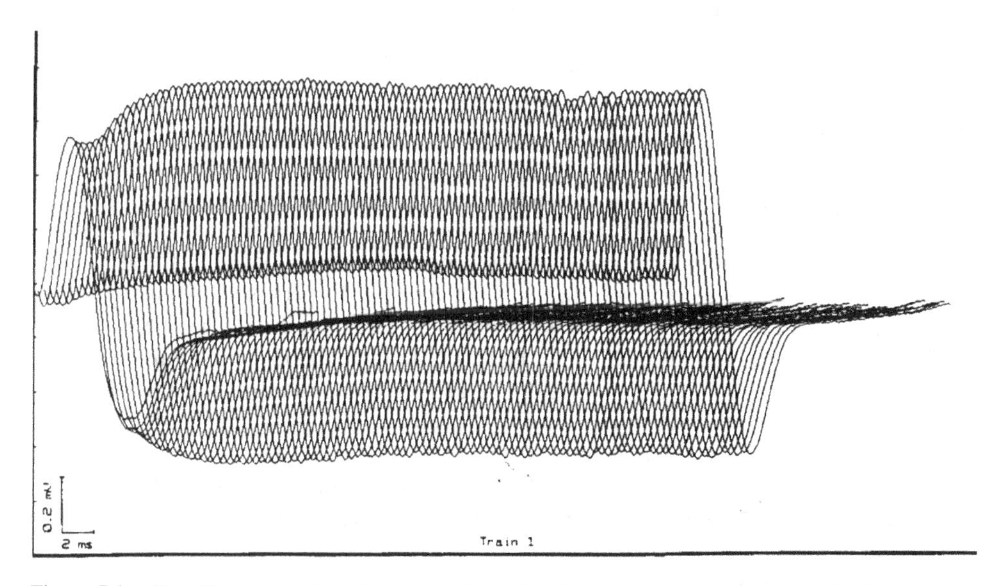

Figure 7.1. Repetitive nerve stimulation in botulism. Note the increment in response amplitude during the initial stimulations. (Courtesy of Vern Juel, M.D., Department of Neurology and Laboratory of Electromyography, University of Virginia.)

5. TREATMENT

The importance of supportive therapy is underlined by the progressive improvement in botulism mortality rates with advances in critical care, especially mechanical ventilation. The decision to intubate a patient with suspected botulism should be based on (1) bedside assessment of upper airway competency and (2) changes in vital capacity (in general, an appropriately performed vital capacity measurement below 12 mL/kg frequently indicates intubation). One should not wait for the $PaCO_2$ to rise or the oxygen saturation to fall before intubation commences. In a mass casualty situation, however, repeated measurements of vital capacity would not be feasible. Simpler screening methods would be necessary, such as asking the victim to take the deepest possible breath and count as high as possible in a loud voice.

In contrast to tetanus, autonomic dysfunction in botulism is rarely life-threatening, and patients who receive appropriate airway and ventilator management are expected to recover unless complications supervene. Patients intubated with high-volume, low-pressure endotracheal tubes should not automatically undergo tracheostomy, regardless of the duration of intubation, unless required for mechanical reasons (Barrett, 1991). The detailed critical care management of botulism patients is beyond the scope of this test; Tacket and Rogawski (1989) have presented a useful approach.

If contaminated material may still reside in the gastrointestinal tract, purgatives may be useful, unless ileus has occurred.

In a mass casualty setting, supplies of antitoxin would quickly be exhausted. They should probably be reserved initially for patients who are weakening slowly in the hope of avoiding the need for intubation and mechanical ventilation. If supplies are adequate, however, everyone with symptoms suggesting impending respiratory failure should be treated. Antitoxin therapy is usually carried out with a trivalent (types A, B, and E) equine serum; in the United States, it is obtained through state health departments, which then contact the CDC (770-488-7100). Its use is supported by inferential studies in the absence of clinical trials. Hypersensitivity rates to this equine serum vary between 9–20% (Black and Gunn, 1980). Skin testing is performed prior to administering the antitoxin; the package includes a regimen for desensitization. The standard antitoxin dose is one vial intravenously and one vial intramuscularly; although the package insert recommends repeating the dose in 4 hours in severe or progressive cases; this does not appear to be necessary (Centers for Disease Control and Prevention, 1998). The Department of Defense has a pentavalent antitoxin, but it is not available for public use. Human botulinum immune globulin is available for infant botulism (Infant Botulism Treatment and Prevention Program, California State Department of Health Services); http://www.dhs.cahwnet.gov/ps/dcdc/html/ibtindex.htm, 510-540-2646) (Frankovich and Arnon, 1991).

Agents that may improve acetylcholine release at the neuromuscular junction have been tried in botulism without success. Guanidine has received the greatest attention (Kaplan *et al.*, 1979), but other drugs are under study.

Although the bulk of the improvement in muscle strength occurs in the first 3 months of recovery, patients may continue to improve in strength and endurance for a year after onset (Wilcox *et al.*, 1990). Recovery may also be followed by persistent psychological problems requiring mental health intervention (Cohen *et al.*, 1988). This would be even more likely following a bioterror release of botulinum toxin.

Predicting the consequences of environmental dissemination is problematic, since data about the stability of the toxin in water or sunlight are lacking. One CDC expert estimated that an aerosol release of toxin could intoxicate 10% of people within 500 meters (Arnon *et al.*, 2001). The decay rate of toxin in the atmosphere is estimated at 1–4% per minute. Models of aerosol exposures suggest that substantial inactivation may take up to 2 days (Arnon *et al.*, 2001), which would be accelerated in extremes of temperature and humidity. Aerosol exposure to the toxin would not be predicted to result in recoverable toxin in stool, but it could still be detected in serum and potentially in respiratory secretions.

The only report of human inhalational botulism followed the accidental exposure of three German laboratory workers who performed necropsies on monkeys subjected to a type A botulinum aerosol (Holzer, 1962). They inhaled a small amount of toxin wafted into the air from the fur of the animals. For the first 2 days after their exposure, they showed no symptoms, but on the 3rd day began to have difficulty swallowing and coryza. On the 4th day, they had generalized weakness and difficulties with extraocular movements, as well as modest pupillary dilation and slight rotary nystagmus, along with dysarthria and gait disturbance. They were given antitoxin on days 4 and 5 after exposure. Their symptoms stopped progressing after antitoxin administration, and they were discharged from the hospital within 2 weeks.

Primate experiments designed to study treatments for inhaled toxin indicate that doses of 5–10 LD_{50} cause death in 2–4 days in animals neither immunized nor treated with antitoxin (Franz *et al.*, 1993). Signs of intoxication developed about 12–18 hours before death, beginning with diffuse weakness, and followed by ptosis and neck weakness.

6. MANAGEMENT

In the event of a bioterrorist attack, several logistical issues would arise. From the limited information available, a large-scale attack (either foodborne or aerosol) would probably not begin to produce symptomatic victims for over 1 day. In a large common-source foodborne outbreak, the initial neurologic symptoms developed over a range of 24–108 hours after exposure (Terranova *et al.*, 1978). Therefore, recognition of the attack, and the potential magnitude of the problem, would initially be difficult. Vaccination to limit disease spread, although an important strategy in a contagious disease such as smallpox, would not be useful after botulinum exposure since (1) as an intoxication, there is no risk of person-to-person transmission and (2) the vaccination series takes about 12 weeks. Immunization is therefore only useful for those in whom exposure is anticipated. The current investigational pentavalent vaccine (types A–E) is reserved for laboratory workers and the military.

The cornerstones of management are airway protection, mechanical ventilation, antitoxin administration, and supportive therapy. Individual cases are usually observed for problems with airway protection (such as difficulty handling secretions), and are monitored for ventilatory failure by measuring vital capacity, negative inspiratory pressure, and respiratory rate. Endotracheal intubation is performed when either problem threatens the patient.

The U.S. Strategic National Stockpile push-packs contain laryngoscopes, endotracheal tubes, and Ambu bags, along with a small number of ventilators. More detailed information is available at http://www.bt.cdc.gov/stockpile/index.asp. However, this equipment would not arrive until 24 hours after being requested by the state governor, and state health departments are responsible for on-scene management at all times. Because the number of patients requiring ventilation may easily exceed the number of available ventilators, it may become necessary to recruit healthy civilians to perform bag ventilation on these patients. This process was very successful in saving lives in Scandinavia during the poliomyelitis epidemics of the 1950s (Wackers, 1994).

References

Ahn-Yoon, S., DeCory, T.R., and Durst, R.A. (2004). Ganglioside-liposome immunoassay for the detection of botulinum toxin. *Anal. Bioanal. Chem.* 378:68–75.

Arnon, S.S., Schechter, R., Inglesby, T.V., Henderson, D.A., Bartlett, S.G., Ascher, M.S., Eitzen, E., Hauer, J., Layton, M., Lillibridge, S., Osterholm, M.T., O'Toole, T., Parker, G., Perl, T.M., Russell, P.K., Swerdlow, D.L., and Tonat, K, for the Working Group on Civilian Biodefence. (2001). Botulinum toxin as a biological weapon: medical and public health management. *J.A.M.A.* 285:1059–1070.

Barrett, D.H. (1991). Endemic food-borne botulism: clinical experience, 1973–1986. *Alaska. Med.* 33:101–108.

Black, R.E., and Gunn, R.A. (1980). Hypersensitivity reactions associated with botulinal antitoxin. *Am. J. Med.* 69:567–70.

Buchman, A.S., Comella, C.L., Stebbins, G.T., Tanner, C.M., and Goetz, C.G. (1993). Quantitative electromyographic analysis of changes in muscle activity following botulinum toxin therapy for cervical dystonia. *Clin. Neuropharmacol.* 16:205–10.

Centers for Disease Control and Prevention. (1998). Botulism in the United States 1899–1996. In: Handbook for epidemiologists, clinicians, and laboratory workers. Centers for Disease Control and Prevention, Atlanta, Ga.

Centers for Disease Control and Prevention. (1999). The history of bioterrorism. [Video]. Available at http://www.bt.cdc.gov/training/historyofbt/index.asp.

Cherington, M. (1990). Botulism. *Semin. Neurol.* 10:27–31.

Cherington, M. (1982). Electrophysiologic methods as an aid in diagnosis of botulism. A review. *Muscle. Nerve.* 6:528–529.

Chia, J.K., Clark, J.B., Ryan, C.A., and Pollack, M.(1986). Botulism in an adult associated with food-borne intestinal infection with *Clostridium botulinum*. *N. Engl. J. Med.* 315:239–241.

Cohen, F.L., Hardin, S.B., Nehring, W., Keough, M.A., Laurenti, S., McNabb, J., Platis, C., and Weber, C. (1998). Physical and psychosocial health status 3 years after catastrophic illness - botulism. *Issues Mental Health Nurs.* 9:387–398.

Colerbatch, J.G., Wolff, A.H., Gilbert, R.J., Mathias, C.J., Smith, S.E., Hirsch, N., and Wiles, C.M. (1989). Slow recovery from severe foodborne botulism. *Lancet.* 2:1216–1217.

Comella, C.L., Tanner, C.M., DeFoor-Hill, L., and Smith, C. (1992). Dysphagia after botulinum toxin injections for spasmodic torticollis: clinical and radiologic findings. *Neurology* 42(7):1307–1310.

Cornblath, D.R., Sladky, J.T., and Sumner A.J. (1983). Clinical electrophysiology of infantile botulism. *Muscle. Nerve.* 6:448–452.

Davis, J.B., Mattman, L.H., and Wiley, M. (1951). *Clostridium botulinum* in a fatal wound infection. *J.A.M.A.* 146:646–648.

Dowell, V.R. Jr., McCroskey, L.M., Hatheway, C.L., Lombard, G.L., Hughes, J.M., and Merson, M.H. (1977). Coproexamination for botulinal toxin and *Clostridium botulinum*. A new procedure for laboratory diagnosis of botulism. *J.A.M.A.* 238:1829–1832.

Dunbar, E.M. (1990). Botulism. *J. Infect.* 20:1–3.

Edell, T.A., Sullivan, C.P., Jr., Osborn, K.M., Gambin, J.P., and Brenman, R.D. (1983). Wound botulism associated with a positive tensilon test. *West. J. Med.* 139:218–219.

Elston, H.R., Wang, M., and Loo, L.K. (1991). Arm abscesses caused by *Clostridium botulinum*. *J. Clin. Microbiol.* 29:2678–2379.

Frankovich, T.L., and Arnon, S.S. (1991). Clinical trial of botulism immune globulin for infant botulism. *West. J. Med.* 154:103.

Franz, D.R., Pitt, L.M., Clayton, M.A., Hanes, M.A., and Rose, K.J. (1993). Efficacy of prophylactic and therapeutic administration of antitoxin for inhalation botulism. In: Das Gupta, B.R. (ed.), *Botulinum and Tetanus Neurotoxins: Neurotransmission and Biomedical Aspects*. Plenum, New York, pp. 473–476.

Friedman, D.I., Fortanasce, V.N., and Sadun, A.A. (1990). Tonic pupils as a result of botulism. *Am. J. Ophthalmol.* 109:236–237.

Glauser, T.A., Maquire, H.C., and Sladky, J.T. (1990). Relapse of infant botulism. *Ann. Neurol.* 28:187–189.

Gutmann, L., and Pratt, L. (1976). Pathophysiologic aspects of human botulism. *Arch. Neurol.* 33:175–179.

Hatheway, C.L. (1989). Bacterial sources of clostridial neurotoxins. In: Simpson, L.L. (ed.), *Botulinum Neurotoxin and Tetanus Toxin*. Academic Press, San Diego, pp. 4–25.

Holzer, V.E. (1962). Botulismus durch inhalation. *Med. Klin.* 57:1735–1738.

Hughes, J.M., Blumenthal, J.R., Merson, M.H., Lombard, G.L., Dowell, V.R. Jr., and Gangarosa, E.J. (1981). Clinical features of types A and B foodborne botulism. *Ann. Intern. Med.* 95:442–445.

Kaplan, J.E., Davis, L.E., Narayan, V., Koster, J., and Katzenstein, D. (1979). Botulism, type A, and treatment with guanidine. *Ann. Neurol.* 6:69–71.

Kerner, J. (1928). Neue Beobachtungen über die in Würtemburg so haüfig vorfallen Vergiftung durch den Genuss gerauchter Würst. Tubingen, 1820. Quoted in: Damon, S.R. (ed.), *Food Infections and Food Intoxications*. Williams & Wilkins, Baltimore, p. 67.

Kortepeter, M.G., Cieslak, T.J., and Eitzen, E.M. (2001). Bioterrorism. *J. Environ. Health* 63:21–24.

Kudrow, D.B., Henry, D.A., Haake, D.A., Marshall, G., and Mathiesen, G.E. (1988). Botulism associated with *Clostridium botulinum* sinusitis after intranasal cocaine abuse. *Ann. Intern. Med.* 109:984–985.

Landman, G. (1984). Ueber die Ursache der Darmstadter Bohnen Vergiftung. *Hyg. Rundsch.* 14:449–452.

Lifton, R.J. (2000). *Destroying the World to Save It*. Henry Holt and Company, New York, pp. 39, 186–188.

Liu, W., Montana, V., Chapman, E.R., Mohideen, U., and Parpura, V. (2003). Botulinum toxin type B micromechanosensor. *Proc. Natl. Acad. Sci. U.S.A.* 100:13621–13625.

Middlebrook, J.L., and Franz, D.R. (1997). Botulinum toxins. In: Sidell, F.R., Takafuji, E.T., and Franz, D.R. (eds), *Medical Aspects of Chemical and Biological Warfare*. Office of the Surgeon General, Washington, D.C. Available online at: https://ccc.apgea.army.mil.

Midura, T.F., and Arnon, S.S. (1976). Infant botulism: identification of *Clostridium botulinum* and its toxin in faeces. *Lancet.* 2:934–936.

M'ikantha, N.M., Southwell, B., and Lautenbach, E. (2003). Automated laboratory reporting of infectious diseases in a climate of bioterrorism. *Emerg. Infect. Dis.* 9:1053–1057.

Notermans, S., and Nagel, J. (1989). Assays for botulinum and tetanus toxins. In: Simpson, L.L. (ed.), *Botulinum Neurotoxin and Tetanus Toxin*. Academic Press, San Diego, pp. 319–331.

Notermans, S.H.W., Wokke, J.H.J., and van. den. Berg., L.H. (1982). Botulism and Guillain-Barré syndrome. *Lancet*. 340:303.
Oguma, K., Yokota, K., Hayashi, S., Takeshi, K., Kumagai, M., Itosh, N., and Chiba S. (1990). Infant botulism due to *Clostridium botulinum* type C toxin. *Lancet*. 336:1449–1450.
Oken, A, Barnes, S., Rock, P., and Maxwell, L. (1992). Upper airway obstruction and infant botulism. *Anesth. Analg*. 75:136–138.
Peruski, A.H., Johnson, L.H. 3rd, and Peruski, L.F. Jr. (2002). Rapid and sensitive detection of biological warfare agents using time-resolved fluorescence assays. *J. Immunol. Methods*. 263:35–41.
Schreiner, M.S., Field, E., and Ruddy, R. (1991). Infant botulism: a review of 12 years experience at the Children's Hospital of Philadelphia. *Pediatrics*. 87:159–165.
Smith, L.D.S. (1977). *Botulism: The Organism, Its Toxins, The Disease*. CC Thomas, Springfield, 236 pp.
Tacket, C.O., and Rogawski, M.A. (1989). Botulism. In: Simpson, L.L. (ed.), *Botulism Neurotoxin and Tetanus Toxin*, Academic Press, San Diego, pp. 351–378.
Terranova, W., Breman, J.G., Locey, R.P., and Speck, S. (1978). Botulism type B: epidemiologic aspects of an extensive outbreak. *Am. J. Epidemiol*. 108:150–156.
Terranova, W., Palumbo, J.N., and Berman, J.G. (1979). Ocular findings in botulism type B. *J.A.M.A*. 241:475–477.
Todd, E.C.D. (1989). Costs of acute bacterial foodborne disease in Canada and the United States. *Int. J. Food Microbiol*. 9:313–326.
van Ermengen, E. (1897). Ueber einen neuen anaëroben Bacillus und seine Beziehungen zum Botulismus. *Zeitschrift für Hygiene und Infektionskrankheiten* 26:1–256.
Vita, G., Girlanda. P., Puglisi, R.M., Marbello, L., and Messina, C. (1987). Cardiovascular-reflex testing and single-fiber electromyography in botulism. A longitudinal study. *Arch. Neurol*. 44:202–206.
Wackers, G.L. (1994). Modern anaesthesiological principles for bulbar polio: manual IPPR in the 1952 polio-epidemic in Copenhagen. *Acta. Anaesthesiol. Scand*. 38:420–431.
Wilcox, P.G., Morrison, N.J., and Pardy, R.L. (1990), Recovery of the ventilatory and upper airway muscles and exercise performance after type A botulism. *Chest*. 98:620–626.
Williams, P., and Wallace, D. (1989). *Unit 731: Japan's Secret Biological Warfare in World War II*. The Free Press, New York, pp. 27–28.
Woodruff, B.A., Griffin, P.M., McCroskey, L.M., Smart, J.F., Wainwright, R.B., Bryant, R.G., Hutwagner, L.C., and Hatheway, C.L. (1992). Clinical and laboratory comparison of botulism from toxin type A, B, and E in the United States, 1975–1988. *J. Infect. Dis*. 166:1281–1286.
Young, J.H. (1976). Botulism and the ripe olive scare of 1891–1920. *Bull. History Med*. 50:372–391.
Zilinskas, R.A. (1997). Iraq's biological weapons. The past as future? *J.A.M.A*. 278:418–424.

8

Ricin: A Possible, Noninfectious Biological Weapon

Maor Maman, MD, and Yoav Yehezkelli, MD

1. INTRODUCTION

On September 7, 1978, 49-year-old Bulgarian exile named Georgi Markov was hit by an umbrella's tip while waiting in a bus station in London. The next day, he was admitted to a hospital in a severe condition, which rapidly deteriorated, terminating in his death 4 days later. A tiny pellet was removed from his thigh in autopsy. Based on the clinical course and on the pellet dimensions, it was concluded that Markov was assassinated using the poison ricin (Crompton and Gall, 1980; Franz and Jaax, 1997).

This unique event raised the interest in ricin as a biological weapon. Being known for centuries as a phytotoxin produced from Castor beans, ricin was developed in modern times as a weapon in between the two world wars and later on. Ricin, suspected of being produced as a biological weapon, was found in London and in Paris as recently as 2003 (Koppel *et al.*, 2003; Mayor, 2003). Its plant origin makes ricin available and easy to produce – thus attractive to terrorist groups and perhaps to rogue countries. In this chapter, we will review various aspects of ricin as a biological weapon.

2. HISTORY

The castor beans are known for their high toxicity for centuries. In ancient times, farmers knew to keep their livestock away from the castor plant or else they would risk loosing them. The seeds have been also used in folk medicine against a wide variety of diseases (Franz and Jaax, 1997). The castor bean plant *Ricinus communis* originated from Asia and Africa, but today it can be found also in Europe and America (Olsnes and Pihl,

Bioterrorism and Infectious Agents
Edited by Fong and Alibek, Springer Science+Business Media, Inc., New York, 2005

1976). It is assumed that the kikaion mentioned in the Bible is a variant of the plant (*Old Testament*, 1982). The castor beans are commonly used as ornamental beans, prayer beads, bracelets, or necklaces. They have an outer shell that must be broken to cause toxicity (Hostetler, 2003). Castor oil – an extract of the castor bean plant – has been used for a number of purposes in ancient Egypt.

Stilmark, in the late 19th century, was the first to obtain evidence that the toxicity of the castor bean refers to a toxic protein he named ricin. In his extensive research, he observed that the toxin causes agglutination of erythrocytes and precipitation of serum proteins (Olsnes and Pihl, 1976).

The ricin toxin was first developed as weaponry by the United States, Canada, and the UK during and in between the two world wars. The U.S. army named it: "compound W." Hundreds of kilograms were produced and armed into bombs, which were never used (Franz and Jaax, 1997). Ricin's production and research for offensive use is prohibited according to the Biological and Toxin Weapons Convention from 1972. Ricin, along with saxitoxin, are the only toxins in which their development, production, and stockpiling are also prohibited according to the Chemical Weapons Convention from 1993. Nevertheless, ricin has become a favorite tool of radical groups and individuals, and is related to several incidents over the last decades. The most famous one is the assassination of Georgi Markov using a gun disguised as an umbrella. Markov's story will be detailed later.

The ricin toxin was part of Iraq's weapon of mass destruction program between 1985 and 1991. About 10 liters of concentrated ricin solution was produced, and few artillery shells were filled with it for field testing (Zilinskas, 1997).

Between 1995 and 1997, several individuals were arrested in the United States for possessing ricin they had produced with homemade equipment. In addition, an extremist group in Minnesota was discovered and arrested for planning to kill a U.S. marshal using a mix of ricin with the solvent dimethyl sulfoxide (DMSO) (Franz and Jaax, 1997; Maman *et al.*, 2003).

The latest incident was just at the beginning of 2003 when the British police uncovered a domestic laboratory that had already managed to produce a small quantity of ricin designated for use as weaponry (Mayor, 2003).

2.1. The Story of a "Death Umbrella"

Georgi Markov was a 49-year-old Bulgarian novelist and playwriter. He had to leave Bulgaria for London in the 1970s after he had put on a controversial play. In London, he had published and broadcasted anticommunist views. On September 7, 1978, while waiting in a bus station, he felt a painful blow to his right thigh. When he turned, he saw a man holding an umbrella. The next day, he was admitted to the hospital with a high temperature, vomiting, and difficulty in speaking. He appeared toxic, and had a 6-cm diameter region of inflammation and induration in his thigh. Three blood cultures were negative. His white blood cell count was 10,600/μL. The day after, he suffered from a septic shock-like syndrome with vascular collapse, and was sweating and dizzy. His white blood cell count rose to 26,300/μL. Afterward, he had stopped passing urine and the vomiting became bloody. Four days after the attack, his electrocardiogram showed complete conduction block. A few

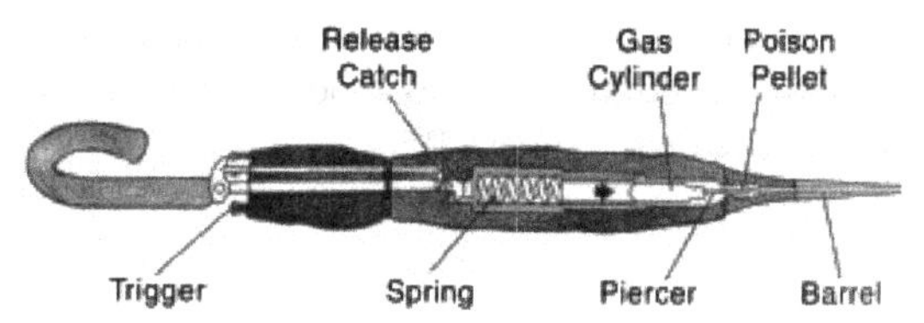

Figure 8.1. A schematic drawing of the umbrella used in Markov's assassination. [From Eitzen and Takafuji (1997), with permission from the publishers.]

hours later, he died. Autopsy revealed pulmonary edema, fatty change of the liver, hemorrhagic necrosis of the small intestines, and interstitial hemorrhage in the testicles, pancreas, and inguinal lymph nodes (Crompton and Gall, 1980).

Vladimir Kostov was another Bulgarian exile in Paris. On August 26, 1978, just 2 weeks before Markov's assassination, he felt a similar blow to his back while he was on the Metro. He also heard a sound like an air pistol shot. He had fever for which he was hospitalized for 12 days, and he recovered completely. X-ray showed a foreign body in his back. It was a tiny pellet. An identical one was removed from Markov's thigh (Crompton and Gall, 1980).

The metallic pellet was 1.5 mm in diameter, and it had two holes drilled through it. The holes were sealed with wax intended to melt at body temperature (Christopher *et al.*, 1997; Crompton and Gall, 1980). Kostov wore heavy clothing so the pellet did not penetrate deep enough in his body for the wax coat to melt. It was estimated that the holes could have contained 500 μg of material inside. Although no substance was ever found in those two pellets, several agents – such as diphtheria toxin, clostridial toxin, and endotoxin – were considered to be possible causes. The circumstances suggested that it was probably ricin that was used

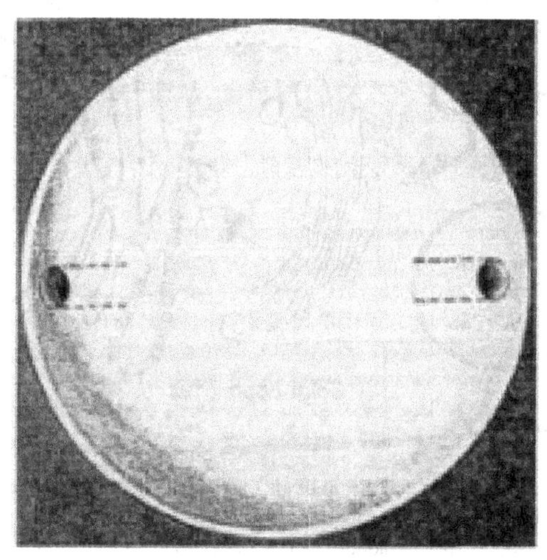

Figure 8.2. The tiny, cross-drilled and ricin-containing pellet that was removed from Markov's thigh after the assassination. [From Eitzen and Takafuji (1997), with permission from the publishers.]

in the attacks (Crompton and Gall, 1980). These two assasinations were said to have been carried out by the Bulgarian government. It was reported that the ricin was produced and sent to Bulgaria by the Soviet Union (Alibek, 1999).

3. THE TOXIN

The fibrous portion that remains after the extraction of oil from the castor bean contains the protein ricin – the most toxic substance in the plant kingdom (Ellenhorn, 1997).

Every year, a total amount of 1 million tons of castor beans are processed worldwide in the production of castor oil. One gram of ricin can be extracted from the waste mash, which remains after processing 1 kg of castor beans (Kortepeter *et al.*, 2001).

Ricin is a 2.5 Å heterodimeric protein consisting of two glycoprotein chains of approximately equal molecular mass (~30 kDa) that are connected through a disulfide bridge. One chain (B) is a lectin that binds to a glycoprotein in the cell membrane, facili-

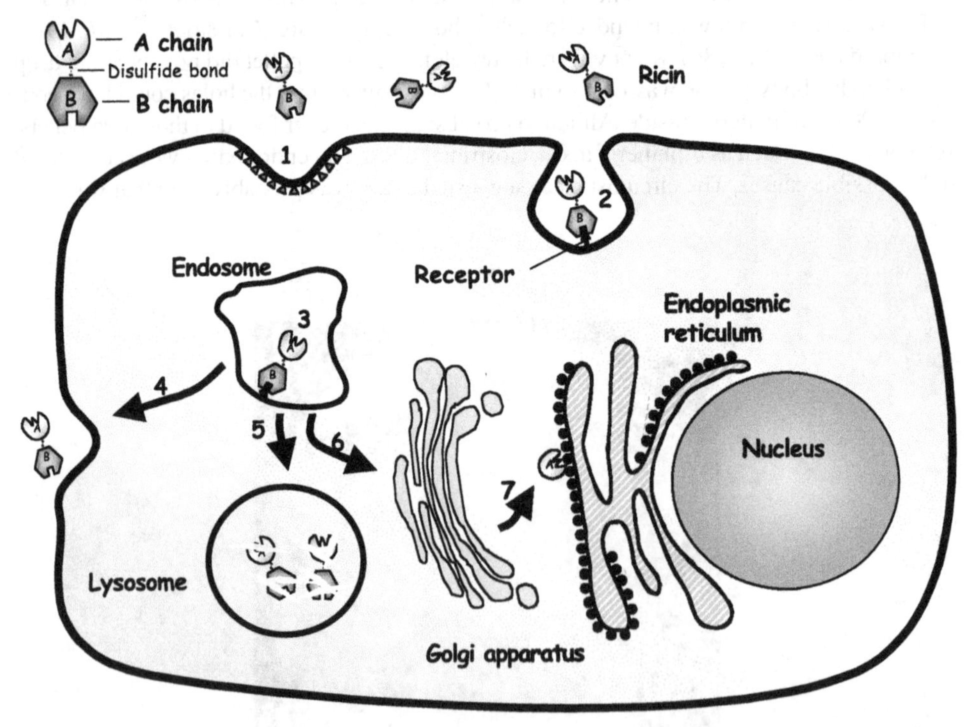

Figure 8.3. The pathway of ricin in the cell. Ricin binds to receptors with terminal galactose at either (*1*) coated pits or (*2*) smooth pits. (*3*) The ricin–receptor complex is endocytosed. (*4*) The complex can be recycled out of the cell or (*5*) transported to lysosomes for degradation. (*6*) Ricin A chain is transported to the *trans*-Golgi network. (*7*) Through the *trans*-Golgi apparatus, the A chain is transported to the endoplasmic reticulum, where it exerts its toxic effects on ribosomes.

tating endocytosis of the toxin to the cytosol. The other chain (A) has RNA *N*-glycosidase activity, removing a specific adenine base from 28S rRNA of the 60S ribosomal subunit, causing inactivation of the ribosome, preventing polypeptide elongation, and leading to cell death (Sandvig and Van Deurs, 2000).

Ricin is synthesized in the plant seeds as a single polypeptide, which is later cleaved into the A and B chains (Frigerio and Roberts, 1998; Kaku, 1998). The A chain is composed of 267 amino acids and the B chain is composed of 262 amino acids (Lord *et al.*, 1994).

Ricin belongs to the group of type 2 ribosome inactivating proteins (RIPs). RIPs can be found in a wide variety of plants. Two types of RIPs are known. Type 1 RIPs are present in wheat and barley. They are not cytotoxic because they lack the means of entering the cell to inactivate ribosomes (Lord *et al.*, 1994).

Type 2 RIPs, which have a galactose-binding lectin that assists cell entrance, is a group of extremely potent toxins that also includes the toxin abrin, isolated from the seeds of *Abrus precatorius* (or rosary pea) (Kaku, 1998; Lord *et al.*, 1994).

Ricin's ability to bind both glycoproteins and glycolipids with terminal galactose, located on the cell surface, enables it to enter the cell exploiting all endocytosis mechanisms operating in the cell (Sandvig and van Deurs, 2002). On entering the cell inside an endosome, the toxin–receptor complex can be either delivered toward lysosomes where it would be destroyed or it can be recycled back to the cell surface. Apparently, only a small fraction of about 5% ricin will finally enter cytosol (Olsnes and Kozlov, 2001). It is not clear yet exactly when the disulfide bond is dissociated to yield the two separate chains A and B; yet, after it occurs, the A chain is taken up by the Golgi apparatus and transported retrogradely to the rough endoplasmic reticulum, where it exerts its toxic effect. In contrast, diphtheria and shiga toxins – which have a similar structure of two functionally different parts – enter the cytosol directly from the endosome (Olsnes and Kozlov, 2001).

The ricin A chain is a very efficient substance. Only one molecule is sufficient to block the cell's protein synthesis as a result of destroying about 2,000 ribosomes per minute. Thus, one molecule can kill the affected cell (Ellenhorn, 1997; Sandvig and van Deurs, 2002).

Ricin can be prepared and remain stable in various forms: liquid, crystalline, and even aerosol. Detoxification can be achieved by heating at 80°C for 10 minutes or 50°C for an hour at pH 7.8 or by chlorine [99.8% inactivation by 100 mg/L free available chlorine (FAC) for 20 minutes]. The toxin will remain stable in low chlorine concentrations (10 mg/L FAC) and in iodine at up to 16 mg/L (Kortepeter *et al.*, 2001).

3.1. Toxicity

The toxicity of ricin is dependent on the route of exposure and the amount of the toxin that was administrated. Ricin is several hundreds times less toxic by ingestion than by parenteral administration, probably because of the enzymatic degradation in the digestive tract and poor absorption (Ellenhorn, 1997; Franz and Jaax, 1997; Maman *et al.*, 2003).

Table 8.1
Toxicity of Ricin by Route of Exposure

		Ingestion	Inhalation	Intravenous injection	Subcoutaneous injection
Mice	LD_{50}	20 mg/kg[a]	3–5 μg/kg[a]	5 μg/kg[a]	24 μg/kg[a]
	Time to death (hrs)	85[a]	36–72[b]	90[a]	100[a]
Human	LD_{50}	30 μg/kg[c]	3 μg/kg[c]	3 μg/kg[c]	500 μg (based on Markov's assassination)
	Time to death	6–8 days	36–72 hrs[a, c]	36–72 hrs	3–6 days

[a] From Franz and Jaax (1997).
[b] Kortepeter *et al.* (2001).
[c] Mirarchi and Allswede (2003).

Ingestion of eight castor beans is said to be enough to kill a person (Ellenhorn, 1997). Fatal dose is considered as 5–6 castor beans for a child and 20 beans for an adult. Data concerning human toxicity is limited. It is based on reports describing animal experiments and accidental exposures of humans (Table 8.1).

4. RICIN AS A POTENTIAL BIOWEAPON

The Center for Disease Control and Prevention (CDC) listed ricin in category B of moderate threat biowarfare agents. Pathogens that pose a potential risk if used as a biological weapon are divided into three categories:

- **Category A agents** can be easily disseminated or transmitted from person-to-person, can cause high mortality, and have the potential for major public health impact. This category includes agents like smallpox, anthrax, plague, botulinum toxin, and Ebola hemorrhagic fever.
- **Category B agents** are moderately easy to disseminate, can cause moderate morbidity and low mortality, and include brucellosis, Q fever, glanders, ricin *Staphylococcus enterotoxin B*, and other pathogens that are food- or waterborne like Salmonella species and Shigella dysenteriae.
- **Category C agents** could be engineered for mass dissemination in the future because of availability, ease of production and dissemination, and potential for high morbidity and mortality and major health impact. This category includes yellow fever, multidrug-resistant tuberculosis, hantaviruses, and others (CDC, 2000).

Ricin is not an efficient weapon for mass destruction. Dispersion of ricin over a wide area is possible though logistically impractical. It is estimated that if eight metric tons of the toxin were scattered over a 100 km^2 area, about 50% of the population would die. In contrast, the same death toll could be "achieved" by using only kilogram quantities of anthrax spores (Kortepeter and Parker, 1999). Castor beans' availability and the relative ease and inexpensive extraction in large quantities are factors that favor the use of ricin as a biological weapon in deliberate (terrorist) poisoning and assassinations. It can be injected into a target, used to contaminate food and water supplies, or can be dispersed as an aerosol (Kortepeter *et al.*, 2001).

5. CLINICAL PRESENTATION

Castor beans are highly allergic and may cause anaphylaxis (Ellenhorn, 1997). The clinical presentation depends on the route of exposure.

1. **Gastrointestinal:** Rauber and Heard (1985) described more than 750 cases of intoxication following castor bean ingestion, 14 of them resulted in a fatality. This and other reports of serious cases of intoxication describes typical clinical manifestations (Challoner and McCarron, 1990; Rauber and Heard, 1985; Wedin *et al.*, 1986). Allergic reaction may occur immediately after exposure (Ellenhorn, 1997). During the first few days following consumption of seeds, patients remain asymptomatic except for gradual loss of appetite and nausea. The most common initial symptoms result from gastrointestinal irritation and include burning in the alimentary tract, vomiting, diarrhea, and colicky abdominal pain. In severe poisoning, the symptoms progress to gastrointestinal bleeding with necrosis of the liver, spleen, and kidneys; dehydration; and even vascular collapse and shock. Children are more prone than adults to dehydration because of fluid losses through vomiting and diarrhea (Ellenhorn, 1997; Maman *et al.*, 2003; Rauber and Heard, 1985).
2. **Parenteral administration:** Fostad *et al.* (1984) reported a clinical trial carried out on 54 cancer patients who were given intravenous injection of low-dose ricin (4.5–23 μg/m^2 of body surface area). Side effects were insignificant up to 18–20 μg/m^2, but 4–6 hours after administration of higher dose, flu-like symptoms with fatigue, muscular pain, nausea, and vomiting were observed and lasted for 1–2 days (Fodstad *et al.*, 1984). Injection of a higher dose of ricin would cause vascular endothelial injury manifested by perivascular edema (Kortepeter *et al.*, 2001).
3. **Inhalation:** The impact of inhaled ricin in rodents is characterized by necrosis of the upper and lower respiratory epithelium (Kortepeter *et al.*, 2001). Symptoms emerge within 8 hours after inhalation and include fever, dyspnea, cough, respiratory distress, and nausea. Pulmonary edema, cyanosis, hypotension, respirataory failure, and circulatory collapse may be developing subsequently.
4. **Intramuscular injection:** At low doses, flu-like symptoms, myalgias, nausea, vomiting, and localized pain and swelling at the injection site are expected. Severe intoxication causes also muscle and regional lymph node necrosis, as well as visceral organ involvment (gastrointestinal hemorrhage, diffuse hepatic, splenic, and renal necrosis) described in Markov's assassination.
5. **Dermal exposure:** Dermal exposure is usually insignificant because of poor absorption through the skin unless enhanced with a strong solvent like DMSO.

5.1. Prognosis

Death from ricin intoxication could occur within 36–48 hours, independent of the route of exposure. If death has not occurred within 3–5 days, the patient usually survives. The study of Rauber and Heard (1985) described a mortality rate of 1.9% (14 of 751 patients) after castor bean ingestion. This rate is lower than previously known. Mortality rate from inhalation of ricin in humans is unknown, but potentially could be substantially higher.

5.2. Diagnosis

The diagnosis of ricin poisoning can be made on a clinical and epidemiological basis. Ricin intoxication should be suspected if a cluster of cases with acute lung injury has occurred. A covert dispersion of aerosolized ricin is expected to be diagnosed, post factum, only after clinical symptoms occur (Kortepeter *et al.*, 2001). Confirmation of ricin exposure is possible by ELISA analysis of nasal or throat swabs. The CDC can detect ricin in environmental specimens using a time-resolved fluorescence immunoassay.

- **Laboratory findings:** Patients may present with neutrophilic leukocytosis, arterial hypoxemia, bilateral infiltrates on chest radiograph, and protein-rich bronchial aspirate (Maman *et al.*, 2003; Mirarchi and Allswede, 2003). Identification of the toxin in body fluids is very difficult. ELISA may be useful (Shyu *et al.*, 2002). Being immunogenic, ricin is expected to induce an antibody response. Thus, antibody level should be measured in the sera of patients 2 weeks after exposure (Franz and Jaax, 1997). A polymerase chain reaction is done only at CDC to detect the DNA of the gene that produces the ricin toxin in environmental samples.
- **Differential diagnosis:** Differential diagnosis includes community-acquired pneumonia, staphylococcal enterotoxin B, inhalational anthrax, oxides of nitrogen and phosgene, Q fever and tularemia. If antibiotic treatment did not halt progress of symptoms, an infectious agent should be ruled out – thus suggesting ricin intoxication (Kortepeter *et al.*, 2001). Table 8.2 compares symptoms and clinical signs, and X-ray and laboratory findings of three major differential diagnoses: community-acquired pneumonia, inhalational anthrax, and inhalational ricin intoxication.

6. TREATMENT

There are no antidotes for ricin. Prevention of exposure is always best. Aerosol exposure can be effectively prevented using a protective mask. Treatment is basically supportive and depends on the route of exposure and clinical manifestations. Symptomatic patients

Table 8.2
Comparison of Signs and Symptoms, X-ray, and Laboratory Findings of Three Major Differential Diagnoses

	Inhalational ricin intoxication	Community-acquired pneumonia	Inhalational anthrax
Signs and symptoms	Fever Dyspnea Cough Respiratory distress Nausea Cyanosis Hypotension Respiratory failure	Cough Sputum production Dyspnea Tachypnea Cyanosis Bronchial breath Pleuritic chest pain Fever	Flu-like symptoms (fever and chills, muscle aches and headaches) Nonproductive cough Nausea and vomiting Dyspnea Chest discomfort Headache Tachycardia Respiratory failure Shock Meningitis may develop
X-ray	Bilateral infiltrates Pulmonary edema	Pulmonary infiltrates	Mediastinal widening Pleural effusion
Laboratory findings	Neutrophilic leukocytosis	Leukocytosis	Neutrophilic leukocytosis Elevated transaminases *Bacillus anthracis* growth in blood cultures

require hospitalization for supportive care, fluids, and electrolyte replacement. Ricin is water-soluble, and little is extracted in the urine. Therefore, there is no indication for hemodialysis or induced diuresis (Ellenhorn, 1997).

1. **Ingestion:** Effort should first be taken to prevent ricin absorption. This should include gastric emptying/lavage, syrup of ipecac, cathartics, and some even recommend the use of activated charcoal (Ellenhorn, 1997; Maman *et al.*, 2003). If the patient chewed some beans and presented asymptomatic, he should remain under observation for 4–6 hours after ingestion (Ellenhorn, 1997).
2. **Inhalation:** Respiratory support is given as needed. The patient may require treatment of pulmonary edema and the use of positive end-expiratory pressure (Kortepeter *et al.*, 2001).
3. **Dermal exposure:** Supportive treatment.

7. PREVENTION AND VACCINE

Prevention: Early detection of the toxin is essential to promote an adequate response. Several locations in the United States, like government and U.S. Postal Service offices, use sensors for detection of chemical and biological agents, such as ricin. If ricin is suspected to be present, a sample is referred for confirmation in a reference laboratory (Centers for Disease Control and Prevention, 2003). When entering a contaminated area, one should wear adequate personal protective equipment, especially a protective mask as protection against inhalation. Health care workers should use standard precautions when treating a patient with ricin intoxication.

Vaccine: No vaccine is currently available. Experimental vaccines are under development and have shown effectiveness in animals (Kende *et al.*, 2002; Yan *et al.*, 1996). Two vaccines have been used to confer complete protection against inhalation exposure: toxoid-formalin inactivated toxin and deglycosylated ricin toxin A unit. Nonetheless, inactivation of the toxin by formalin is not complete, and the preparations still have residual toxic effect (Smallshaw *et al.*, 2002). Smallshaw *et al.* (2002) developed novel mutant and non-toxic ricin toxin A unit that protected mice exposed to 10 times mice LD_{50} of ricin.

The latest report in September 2004 (Hampton, 2004) indicated that US Army scientists have developed a new experimental vaccine that is safe and protective in mice. Further animal studies are planned in the future in order to achieve a vaccine safe to humans.

8. MEDICAL USE OF RICIN

Ricin, being constructed of two parts, one of which has a receptor-binding capability, is used to develop novel therapeutics, such as immunotoxins (Oeltmann and Frankel, 1991; Weissmann-Brenner *et al., 2002*). The receptor-binding B moiety can be replaced by either antibodies to targets, such as cancer-associated antigens, or by other binding proteins like hormones or growth factors. The new-bound protein should have the same affinity to be guided to the endoplasmic reticulum through the *trans*-Golgi network to reach the ribosomes (Kreitman, 1999, 2001).

9. CONCLUSION

Ricin is a potent and easy to extract plant toxin. Its characteristics make it a potentially dangerous biological weapon. We understand now, better than ever, the pathogenesis of ricin poisoning. But treatment of ricin poisoning is still mainly supportive. More research is needed to develop specific and effective modalities of treatment.

The threat of bioterrorism is no longer as remote as it was in the past. The medical community should be familiar with the clinical presentation and treatment of ricin poisoning. Knowledge will allow better recognition and response to an attack.

ACKNOWLEDGMENTS

We would like to express our sincere appreciation to Mrs. Sarah Efrati for her tremendous effort and assistance in information collection, and for being available around the clock.

References

Alibek, K. (1999). *Biohazard.* New York, Random House, Inc.

CDC. (2000). Biological and chemical terrorism: strategic plan for preparedness and response recommendations of the CDC Strategic Planning Workgroup. *MMWR* 49(RR-4):1–14.

Centers for Disease Control and Prevention. (2003). Laboratory testing for ricin in environmental samples—fact sheet. Available at: http://www. bt.cdc. gov/agent/ricin/labtesting.asp.

Challoner, K.R., and McCarron, M.M. (1990). Castor bean intoxication. *Ann. Emerg. Med.* 19:1177–1183.

Christopher, G.W., Cieslak, T.J., Palvin, J.A., and Eitzen, E.M. (1997). Biological warfare. A historical perspective. *J.A.M.A.* 278:412–417.

Crompton, R., and Gall, D. (1980). Georgi Markov—death in a pellet. *Medico Legal J.* 48(2):51–62.

Eitzen, E.M., and Takafuji, E.T. (1997). Historical overview of biological warfare. In: Sidell, F.R., Takafuji, E.T., and Franz, D.R. (eds.), *Textbook of Military Medicine: Medical Aspects of Chemical and Biological Warfare.* TMM Publications, Washington, D.C.

Ellenhorn, M.J. (1997). Plants – mycotoxins – mushrooms. In: *Ellenhorn's Medical Toxicology: Diagnosis and Treatment of Human Poisoning.* Baltimore, Williams & Wilkins, pp. 1847– 1849.

Fodstad, O., Kvalheim, G., Godal, A., Lotsberg, J., Aamdal, S., Host, H., and Pihl, A. (1984). Phase I study of plant protein ricin. *Cancer Res.* 44:862–865.

Franz, D.R., and Jaax, N.K. (1997). Ricin toxin. In: Sidell, F.R., Takafuji, E.T., and Franz, D.R. (eds.), *Textbook of Military Medicine: Medical Aspects of Chemical and Biological Warfare.* TMM Publications, Washington, D.C.

Frigerio, L., and Roberts, L.M. (1998). The enemy within: ricin and plant cells. *J. Exp. Botany* 49(326):1473–1480.

Hampton, T. (2004). Ricin vaccine developed. *JAMA* 292:1419–a.

Hostetler, M.A. (2003). Toxicity, plants – castor bean and jequirity bean. Available at: http://www.emedicine.com/ped/topic331.htm.

Kaku, H. (1998). Ricin related protein family. Available at: http://www.glycoforum.gr.jp/science/word/lectin/LEA08E.html.

Kende, M., Yan, C., Hewetson, J., Frick, M.A., Rill, W.L., and Tammariello, R. (2002). Oral immunization of mice with ricin toxoid vaccine encapsulated in polymeric microspheres against aerosol challenge. *Vaccine* 20(11–12):1681–1691.

Koppel, A., MacVicar, S., Schuster, H., and Moniquet, C. (2003). Deadly ricin at Paris rail station. Available at: http://www.cnn.com/2003/WORLD/europe/03/20/france.ricin/index.html.

Kortepeter, M.G., Christopher, G., Cieslak, T., Culpepper, R., Darling, R., Palvin, J., Rowe, J., McKee, K., and Eitzen, E. (2001). *USAMRIID's Medical Management of Biological Casualties Handbook*, 4th ed. U.S. Army Medical Research Institute of Infectious Diseases, Fort Detrick, Frederick, MD, pp. 130–137.

Kortepeter, M.G., and Parker, G.W. (1999). Potential biological weapons threats. *Emerg. Infect. Dis.* 5:523–527.

Kreitman, R.J. (1999). Immunotoxins in cancer therapy. *Curr. Opin. Immunol.* 11:570–578.

Kreitman, R.J. (2001). Toxin-labeled monoclonal antibodies. *Curr. Pharm. Biotechnol.* 2:313–325.

Lord, J.M., Roberts, L.M., and Robertus J.D. (1994). Ricin: structure, mode of action, and some current applications. *FASEB J.* 8:201–208.

Maman, M., Sagi, R., Moseri, R., and Yehezkelli, Y. (2003). Ricin—a potent plant toxin. In: Shoenfeld, Y., and Shemer, J. (eds.), *Terror and Medicine.* Pabst Science Publishers, Berlin.

Mayor, S. (2003). UK doctors warned after a ricin poison was found in police raid. *Br. Med. J.* 326:126.

Mirarchi, F.L., and Allswede, M. (2003). CBRNE – ricin. Available at: http://www.emedicine.com/emerg/topic889.htm.

Oeltmann, T.N., and Frankel, A.E. (1991). Advances in immunotoxins. *FASEB J.* 5(10):2334–2337.

Old Testament. (1982). Jonah 4:6–9. Copyright by Y. Orestein, "Yavne" Publishing House, Ltd., Tel-Aviv.

Olsnes, S., and Kozlov, J.V. (2001). Ricin. *Toxicon* 39:1723–1728.

Olsnes, S., and Pihl, A. (1976). Abrin, ricin and their associated agglutinins. In Cuatrecasas, P. (eds.), *Receptors and Recognition. Series B. The Specificity and Action of Animal, Bacterial and Plant Toxins.* Chapman and Hall, London, pp. 129–173.

Rauber, A., and Heard, J. (1985). Castor bean toxicity re-examined: a new prospective. *Vet. Hum. Toxicol.* 27:498–502.

Sandvig, K., and Van Deurs, B. (2000). Entry of ricin and Shiga toxin into cells: molecular mechanisms and medical perspectives. *EMBO J.* 19:5943–5950.

Sandvig, K., and Van Deurs, B. (2002). Transport of protein toxins into cells: pathways used by ricin, cholera toxin and shiga toxin. *FEBS Lett.* 529(1):49–53.

Shyu, H.F., Chaio, D.J., Liu, H.W., and Tang, S.S. (2002). Monoclonal antibody-based enzyme immunoassay for detection ricin. *Hybrid Hybridomics* 21:69–73.

Smallshaw, J.E., Firan, A., Fulmer, J.R., Ruback, S.L, Ghetie, V., and Vitetta, E.S. (2002). A novel recombinant vaccine which protects mice against ricin intoxication. *Vaccine* 20:3422–3427.

Wedin, G.P., Neal, J.S., Everson, G.W., and Krenzelok, E.P. (1986). Castor bean poisoning. *Am. J. Emerg. Med.* 4(3):259–261.

Weissmann-Brenner, A., Brenner, B., Kats, L., and Hourvitz, A. (2002). Ricin – from a Bulgarian umbrella to an optional treatment of cancer. *Harefuah* 141:153–156. [Hebrew].

Yan, C., Rill, W.L., Malli, R., Hewetson, J., Naseem, H., Tammariello, R., and Kende, M. (1996). Intranasal stimulation of long lasting immunity against aerosol ricin challenge with ricin toxoid vaccine encapsulated in polymer microspheres. *Vaccine* 14:1031–1038.

Zilinskas, R.A. (1997). Iraq's biological weapons: the past as future? *J.A.M.A.* 278:418–424.

9

Bioterrorism Alert for Health Care Workers

Theodore J. Cieslak, MD, George W. Christopher, MD, and Edward M. Eitzen, Jr., MD, MPH

1. INTRODUCTION

When the agent used in a biological attack is known, response to such an attack is considerably simplified. The first eight chapters of this text deal with agent-specific concerns and strategies for dealing with infections due to the intentional release of these agents. A larger problem arises when the identity of an agent is not known. In fact, in some cases, an attack may be threatened or suspected, but it may remain unclear as to whether such an attack has actually occurred. Moreover, it may be unclear whether casualties are due to a biological agent, a chemical agent, or even a naturally occurring infectious disease process or toxic exposure. Recent experience with West Nile Virus (Fine and Layton, 2001), Severe Acute Respiratory Syndrome (SARS) (Lampton, 2003), and monkeypox highlight this dilemma. In each of these cases, the possibility of bioterrorism was raised, and rightly so, although each outbreak ultimately proved to have a natural origin. This chapter provides a framework for dealing with outbreaks of unknown origin and etiology. Furthermore, it addresses several related concerns and topics not covered elsewhere in this text.

When dealing with the unknown, it is often helpful to use an algorithmic approach. This would be especially true in a medical mass casualty (MASCAL) incident, where a considerable precedent exists for the use of standardized approaches as embodied in the Advanced Trauma Life Support (ATLS) model sponsored by the American College of Surgeons (1989). It would also be especially true under austere conditions, such as might be seen on the battlefield. In fact, as will be seen, much of the dogma and technology, and many of the strategies advocated for use in responding to a bioterrorist attack, derive from military biowarfare defense doctrine and research. Considerable parallels exist between military biowarfare concerns and civilian bioterrorism concerns. Although major medical centers and research institutions may possess sophisticated diagnostic and response capabilities, the busy clinician practicing in a small town may find their situation much more akin to that of the military on the battlefield. That is, they may be required to make quick therapeutic decisions based on incomplete information and with little immediate support.

Bioterrorism and Infectious Agents
Edited by Fong and Alibek, Springer Science+Business Media, Inc., New York, 2005

In the setting of a biological (or chemical) attack, similar to the setting of a MASCAL trauma event, such decisions may have life-and-death implications. In such situations, an algorithmic approach becomes invaluable.

We advocate a 10-step approach to the management of casualties that might result from biological (or chemical) terrorism. The derivation of this approach has been reported elsewhere (Cieslak *et al.*, 2000; Cieslak and Henretig, 2001, 2003a). We expand upon it here.

2. STEP 1. MAINTAIN A HEALTHY "INDEX OF SUSPICION" (OR, "HOW TO RECOGNIZE ILLNESS DUE TO BIOLOGICAL WEAPONS")

In the case of chemical terrorism, the sinister nature of an attack might be obvious. Here, victims would likely succumb in close temporal and geographic proximity to a dispersal device. Complicating discovery of the sinister nature of a biological attack, however, is the fact that biological agents possess inherent incubation periods, a characteristic not shared with conventional, chemical, and nuclear weapons. These incubation periods, typically days to even weeks in length, allow for the wide dispersion of victims in time and space. Moreover, they make it likely that the "first responder" to a biological attack would not likely be the traditional first responder (fire, police, and paramedical personnel), but rather primary care physicians, emergency room personnel, and public health officials. In such circumstances, the maintenance of a healthy "index of suspicion" is imperative.

In certain cases, maintenance of suspicion is made easier by the fact that diseases caused by biological agents may present with characteristic "hallmark" clinical findings. Moreover, in many cases, such findings present a very limited differential diagnosis. Smallpox is characterized by a very unique exanthem. The classic finding in inhalational anthrax is a widened mediastinum; in botulism, it is a descending, symmetric, flaccid paralysis. Although a single patient with flaccid paralysis should call to mind the possibility of disorders such as the Guillan-Barré syndrome and myasthenia gravis, the near-simultaneous presentation of multiple patients with flaccid paralysis should point one to a diagnosis of botulism. Similarly, plague victims often develop hemoptysis in the later stages of illness. Such a finding is otherwise uncommon among previously healthy individuals, but can be caused by tuberculosis, Staphylococcal and Klebsiella pneumonia, and carcinoma, among others. Multiple patients with hemoptysis, however, should prompt a diagnosis of plague. Yet, by the time each of these characteristic findings develops, treatment is likely to be ineffective. Therefore, therapy is best instituted during the incubation or prodromal phases of these diseases if it is likely to be of benefit. Because of this, and because many potential biological warfare diseases – such as tularemia, brucellosis, melioidosis, Q-fever, staphylococcal enterotoxin intoxication, and Venezuelan equine encephalitis are likely to present simply as undifferentiated febrile illnesses – prompt diagnosis is a possibility only with the maintenance of a high index of suspicion.

Epidemiologic clues can assist the clinician in suspecting that a disease outbreak may be sinister in origin (Pavlin, 1999). The presence of large numbers of victims clustered

Table 9.1
Epidemiologic Clues to a Bioterrorist Attack

Epidemiologic Clues to a Bioterrorist Attack
Presence of an unusually large epidemic
High infection rate
Disease limited to a discrete population
Unexpected severity of disease
Evidence of an unusual route of exposure
Disease in an atypical geographic locale
Disease occurring outside normal transmission seasons
Disease occurring in the absence of usual vector
Simultaneous outbreaks of multiple diseases
Simultaneous occurrence of human and zoonotic disease
Unusual organism strains
Unusual antimicrobial sensitivity patterns
Disparity in attack rates among persons indoors and outdoors
Terrorist claims
Intelligence reports
Discovery of unusual munitions

in time and space, or limited to a discrete population, should raise the clinician's suspicion. Similarly, cases of unexpected death or unexpectedly severe illness merit concern. Diseases unusual in a given locale, in a given age group, or during a certain season likewise warrant further investigation. Simultaneous outbreaks of a disease in noncontiguous areas should cause one to consider an intentional release, as should outbreaks of multiple diseases in the same area. Conversely, even a single case of rare disorders – such as anthrax or certain viral hemorrhagic fevers (Ebola, Marburg, Lassa, and many others) – would be suspicious, and a single case of smallpox would almost certainly be the result of an intentional release. The presence, in the community or the environment, of dying animals (or the simultaneous occurrence of zoonotic outbreaks among humans and animals) might provide evidence of an unnatural aerosol release. Evidence of a disparate attack rate between those known to be indoors and outdoors at a given time should also be sought and evaluated. Finally, intelligence reports, terrorist claims, and the finding of aerosol spray devices would obviously lend credence to the theory that a disease outbreak was of sinister origin. The epidemiologic clues to a biological attack are summarized in Table 9.1.

3. STEP 2. PROTECT THYSELF FIRST

Providers are of little use if they themselves become casualties. Before approaching the victims of a potential terrorist attack, then, clinicians should be familiar with basic steps that might be taken to protect themselves. These steps generally fall into one of three categories: physical protection, chemical protection, and immunologic protection. In a given set of circumstances, clinicians might be advised to avail themselves of one or more of these forms of protection.

3.1. Physical Protection

The military's answer to the question of physical protection on the battlefield often involves "gas" masks and charcoal-impregnated chemical protective overgarments. Some have advocated the issuance of similar ensembles to civilians, and, in fact, the Israeli government has issued masks to civilians (principally with chemical agents in mind). Military-style protective clothing and masks, even if offered, however, would likely be unavailable at the precise moment of a release of agent. Moreover, the unannounced release by terrorists of odorless, colorless, virtually undetectable biological agents would afford no opportunity to don such gear, even if it were available. Furthermore, the misuse of protective equipment in the past has led to fatalities, including infants left to suffocate in "protective" ensembles (Hiss and Arensburg, 1992). A simple surgical mask will protect against inhalation of infectious aerosols of virtually any of the biological agents typically mentioned in a terrorism context. The lone exception would be smallpox, where a high-efficiency particulate air filter mask would be ideal. With the exception of smallpox, pneumonic plague, and, to a lesser degree, certain viral hemorrhagic fevers, the agents in the Centers for Disease Control and Prevention's (CDC's) categories A and B (Table 9.2) are not contagious via the respiratory route. Respiratory protection is thus necessary when operating in an area of primary release, but would not be required in most patient-care settings.

3.2. Chemical Protection

In early 2003, the U.S. Food and Drug Administration (FDA) gave its final approval for the use of pyridostigmine bromide as a *pre*-exposure means of prophylaxis against intoxication with the chemical agent soman (one of the organophosphate-based "nerve" agents). It is conceivable, given credible and specific intelligence, that similar strategies might be used

Table 9.2
Critical Agents for Health Preparedness

Category A	Category B	Category C
Variola virus	*Coxiella burnetii*	Other biological agents that may emerge as future threats to public health, such as:
Bacillus anthracis	*Brucellae*	Nipah virus
Yersinia pestis	*Burkholderia mallei*	Hantaviruses
Botulinum toxin	*Burkholderia pseudomallei*	Yellow Fever virus
Francisella tularensis	Alphaviruses	Drug-resistant tuberculosis
Filoviruses and Arenaviruses	Certain toxins (ricin, SEB, Trichothecenes)	Tick-borne encephalitis
	Food safety threat agents (*Salmonellae, E. coli* O157:H7)	
	Water safety threat agents (*Vibrio cholera*, etc.)	

Category A – agents with high public health impact requiring intensive public health preparedness and intervention; Category B – agents with a somewhat lesser need for public health preparedness. Adapted from Kortepeter and Cieslak (2003).

against biological weapons. For example, if a terrorist group were known to be operating in a given locale, and were known to possess a specific weaponized agent, then conceivably, public health authorities might contemplate the widespread distribution of a prophylactic antibiotic. Obviously, the opportunities to use such a strategy are likely to remain few.

3.3. Immunologic Protection (Including "Pros and Cons of Mass Vaccination")

For the near future, active immunization is likely to provide one of the most practical methods for providing pre-exposure prophylaxis against biological attack. This is not to say that immunization against a specific agent is necessarily appropriate. Rather, the decision to offer a specific vaccine to the general population is a complex one, which must take into account a careful risk-benefit calculation. Factors that would influence a decision by public health officials to recommend immunization include intelligence (How likely and/or plausible is an attack? How imminent is the threat? How specific is the threat?), vaccine safety, vaccine availability, disease consequences (Is the threat from a lethal agent? Or merely from an incapacitant?), and the feasibility of postexposure prophylaxis and/or therapy. Recently, public health and policy planners have given some consideration to the widespread civilian distribution of two vaccines: anthrax and smallpox.

Anthrax. Anthrax vaccine–adsorbed (AVA; Bioport, Lansing, Michigan) was licensed by the U.S. FDA in 1970 and consists of a purified preparation of protective antigen (a potent immunogen, protective antigen is critical for entry of lethal and edema factors into mammalian cells; it is nonpathogenic when given alone). In a large controlled trial, AVA was shown effective at preventing cutaneous anthrax among textile workers (Brachman *et al.*, 1962). Based on an increasing amount of animal data, there is every reason to believe that this vaccine is quite effective at preventing inhalational anthrax as well. Moreover, at least 18 studies now attest to the safety of AVA (Cieslak *et al.,* in press). Nonetheless, logistical and other considerations make large-scale civilian employment of AVA impractical at present. The vaccine is licensed as a six-dose series, given at 0, 2, and 4 weeks, and again at 6, 12, and 18 months. Yearly boosters are recommended for those at ongoing risk. Further complicating any potential anthrax immunization strategy is the fact that AVA is approved by the U.S. FDA only for those persons 18–65 years of age. Although a large-scale pre-exposure offering of AVA to the public might thus be problematic, some authors allow that three doses of the vaccine, given in conjunction with antibiotics, may enhance protection and/or enable the clinician to shorten a postexposure antibiotic course (Centers for Disease Control and Prevention, 2002). In this regard, a three-dose series of AVA (given at time 0 and at 2 and 4 weeks after the initial dose), combined with 30 days of antibiotics, may be an acceptable alternative to longer (60–100 days) antibiotic courses alone in the treatment of, or postexposure prophylaxis against, inhalational anthrax. It should be noted that AVA is not licensed by the U.S. FDA for postexposure use.

Smallpox. Widespread civilian immunization against smallpox is equally controversial and problematic, although U.S. President George Bush, on December 13, 2002, announced a plan to vaccinate selected American health care workers and military personnel. Although universal civilian vaccination was not recommended, the possibility of such a strategy in the future was allowed for. Moreover, plans were made to provide vaccination to those members of the general public who specifically requested it.

The wisdom of widespread civilian vaccination is difficult to assess. Most medical decisions involve a (sometimes subconscious) risk-benefit analysis on the part of the responsible clinician. In the case of smallpox vaccination, the risks are well known, and they are significant (Centers for Disease Control and Prevention, 2003a, 2003b). The vaccine currently used in the United States is Dryvax® (Wyeth Laboratories, Marietta, Pa.), a preparation derived from the harvested lymph of calves inoculated with a strain of Vaccinia, an orthopoxvirus closely related to Variola. Production of Dryvax® ceased in 1981 and lots currently in use are more than 20 years old. New generation vaccines are in production, however, and should be available very soon. Many health authorities have chosen to wait until these new products are available before readdressing the question of civilian vaccination. These new vaccines are produced in cell culture rather than calf lymph. It is unlikely that this will significantly diminish the risk of adverse reactions, however, as the new vaccines will use the same live strain of vaccinia virus. The vast majority of adverse reactions to current vaccinia-containing vaccines derive from the live nature of the virus, rather than the method of preparation.

Although the risks of vaccination are well known, the benefits are less clear. On the one hand, the global eradication of smallpox ranks as one of the greatest public health accomplishments of our time. Furthermore, this eradication was made possible entirely by live vaccinia-containing vaccines. In an era when naturally occurring smallpox posed a valid threat, there was little question as to the wisdom of vaccination. On the other hand, the likelihood of contracting smallpox today via a terrorist attack is unknown. In this regard, the risk-benefit calculation is not based on medical, but rather on intelligence, considerations, to which few, if any, are privy.

In summary, a prerelease mass vaccination program for the general population is potentially the most effective approach to dealing with the terror threat posed by smallpox. By conferring individual and herd immunity, and by obviating the mayhem of implementing postrelease vaccine and quarantine programs, such a program possesses distinct advantages over other response plans. However, such an approach is currently hampered not only by the unknown risk of a smallpox release, but also by vaccine supply, safety, and logistics issues (Fauci, 2002; Amorosa and Isaacs, 2003). The increasing number of persons at risk for severe vaccine reactions due to compromised immunity or dermatologic conditions (such as eczema) raises concern about the safety of a pre-exposure vaccination program in the era of epidemic HIV, organ transplantation, and immunosuppressive therapy (Bozzette *et al.*, 2003; Centers for Disease Control and Prevention, 2003b; Grabenstein and Winkenwerder, 2003; Kemper *et al.*, 2002). In addition, the incidence of rare but severe smallpox vaccine complications in otherwise healthy recipients could result in a number of deaths and cases of severe disease that would be unacceptable were the risk of a smallpox release low. Risk analysis favors prerelease mass vaccination of the general population only if the probability of a large-scale or multifocal attack is high. However, prerelease mass vaccination of health care workers is favored at lower probabilities of attack, because of an increased probability of exposure while caring for patients, and the value of keeping health care workers healthy and functioning in the setting of an epidemic (Bozzette *et al.*, 2003).

Effective measures to enhance the safety of a pre-exposure mass vaccination program were implemented by the U.S. Department of Defense in conjunction with its December 2002 smallpox vaccination program kickoff. Service members deploying to locations at risk for biological attack, and members of designated "smallpox epidemiologic and clinical response teams" were selected for vaccination. The program emphasized staff training, recipient education, the use of rigorous standardized screening tools for vaccine contraindications, and local vaccine site care. More than 450,000 service members were vaccinated, with the incidence of most adverse events occurring at frequencies below those historically noted. The success of this program suggests that mass vaccination can be accomplished with greater safety than previously thought possible (Grabenstein and Winkenwerder, 2003).

Since Dryvax® stocks are controlled by the CDC, vaccine for use in the civilian population is currently released only under conditions set forth by that agency (Centers for Disease Control and Prevention, 2003b). The current CDC smallpox response strategy is based on pre-exposure vaccination of carefully screened members of first response teams, epidemiologic response teams, and clinical response teams at designated facilities. Safety concerns have resulted in a program to treat certain severe complications with vaccinia immune globulin under an investigational new drug protocol, as well as the establishment of a Smallpox Vaccine Injury Compensation Program (Health Resources and Services Administration, 2003), which provides compensation to persons developing severe adverse reactions. More than 38,000 civilian health care and public health workers have been vaccinated, with no deaths and no cases of eczema vaccinatum, fetal vaccinia, or progressive vaccina. Although the emergence of myopericarditis as a complication of vaccination (Centers for Disease Control and Prevention, 2003c, 2003d; Halsell *et al.*, 2003) lead to a revision of prevaccine screening (candidates with multiple cardiac risk factors are now excluded), rates of other adverse reactions were low. Vaccinia immune globulin has been used once, to treat ocular vaccinia in a contact of a vaccinee (Centers for Disease Control and Prevention, 2003d).

The plan then calls for a "ring vaccination" policy following a smallpox release: identification and isolation of cases, with vaccination and active surveillance of contacts. Mass vaccination would be reserved for instances when the number cases or the location of cases renders the ring strategy inefficient, or if the risk of additional smallpox releases is high (Centers for Disease Control and Prevention, 2003e). Although ring vaccination was successful in containing smallpox in the setting of herd immunity, mathematical models predict that this strategy may not initially contain large or multifocal epidemics (Kaplan *et al.*, 2002). Furthermore, there is disagreement regarding the predicted benefit of postrelease mass vaccination, due to the lack of herd immunity, a highly mobile population, a relatively long incubation, and difficulties implementing quarantine and mass vaccination promptly (Mack, 2003; Mortimer, 2003). This brings us back to the dilemma posed by pre-exposure mass vaccination in the context of uncertain risk, and the realization that vaccination is only one component of a multifaceted response, to include farsighted planning and logistical preparation, risk communication, surveillance, isolation, quarantine, and humane treatment of patients.

Other Agents. Few authorities have recommended mass immunization against potential agents of bioterrorism other than anthrax and smallpox, and the implementation of any such strategy would be quite problematic at present. A vaccine against plague, previously licensed in the United States, is currently out of production. It required a three-dose primary series followed by annual boosters. Moreover, it was licensed only for persons 18–61 years of age. Finally, although reasonably effective against bubonic plague, it likely afforded little protection against pneumonic plague, the form of disease likely to be associated with a terrorist attack. A vaccine against yellow fever is widely available, but this hemorrhagic fever is not regarded as a significant weaponization threat by most policymakers and health officials. Additionally, a vaccine against Q-fever (Q-Vax) is licensed in Australia. Given the self-limited nature of Q-fever, however, it is unlikely that widespread use of this vaccine would ever be contemplated outside of military circles. Research efforts are underway in an attempt to improve upon anthrax, smallpox, and plague vaccines. Similarly, vaccines effective against tularemia, brucellosis, botulism, the equine encephalitides, staphylococcal enterotoxins, ricin, and several viral hemorrhagic fevers – as well as other potential agents of bioterrorism – are in various stages of development (Cieslak *et al.*, 2004).

4. STEP 3. SAVE THE PATIENT'S LIFE ("THE PRIMARY ASSESSMENT")

Once reasonable steps have been taken to protect him- or herself, the clinician is now ready to approach the MASCAL scenario and begin assessing patients (the "primary survey," in keeping with ATLS guidelines (American College of Surgeons, 1989). This initial assessment must be brief and limited to discovering and treating those conditions that present an immediate threat to life or limb. Victims of biological (or chemical) terrorism may also have conventional injuries; attention should thus be focused at this point on maintaining a patent airway and providing for adequate breathing and circulation. The need for decontamination and for the administration of antidotes for rapid-acting chemical agents (nerve agents and cyanide) should be determined at this time. An "ABCDE" algorithm aids the clinician in recalling the steps to be taken during the primary assessment. "A" stands for airway, which should be assessed for the possibility of conventional injury, but also because exposure to certain chemical agents (such as mustard, lewisite, or phosgene) can cause upper airway pathology. "B" denotes breathing; many agents of biological (and chemical) terrorism may present with respiratory difficulty. Examples include anthrax, plague, tularemia, botulism, Q-fever, and staphylococcal enterotoxin or ricin exposure, as well as cyanide, nerve agents, and phosgene. "C" denotes circulation, which may be compromised due to conventional or traumatic injuries sustained during a MASCAL event, but may also be directly affected by plague, the viral hemorrhagic fevers, and cyanide exposure. "D" refers to disability, usually taken to denote neuromuscular disability. It is important to note that botulism and nerve agent exposures are likely to present as a neuromuscular syndrome. Finally, "E" refers to "exposure." In a typical MASCAL, this reminds the provider to remove the victim's clothing to perform a more thorough "secondary assessment." It is here that one considers the need for decontamination and disinfection.

5. STEP 4. DISINFECT OR DECONTAMINATE AS APPROPRIATE

Once patients have been stabilized, decontamination can be accomplished where appropriate. It should be pointed out, however, that decontamination is rarely necessary after a biological attack (the same cannot always be said following a chemical attack). This is due to the inherent incubation periods of biological agents. Because most victims will not become symptomatic until several days after exposure, they will likely have bathed and changed clothing (often several times) before presenting for medical care, thus effectively accomplishing “self-decontamination.” Exceptions might include personnel near ground zero in an observed attack or persons encountering a substance in a threatening letter, where common sense might dictate topical disinfection. Even in these situations, bathing with soap and water and conventional laundry measures would likely be adequate. Moreover, it should be kept in mind that situations such as the case of the threatening letter represent crime scenes, wherein medical interests must be weighed against those of law enforcement. In such cases, hasty and ill-considered attempts at decontamination can destroy vital evidence. Furthermore, millions of dollars have been wasted and significant psychological stress has been caused by unnecessary decontamination attempts in the past (Cole, 2000). Some of these attempts have involved forced disrobing and showering in public streets, under the watchful eye of media cameras. To avoid such problems, we advocate the following measured response (Kortepeter and Cieslak, 2003).

In the case of the announced threat or hoax, the existence of a definitive crime scene and the need to preserve evidence and maintain a chain-of-custody when handling that evidence are important considerations. Although human and environmental health protection would take precedence over law enforcement concerns, threat and hoax scenarios require the early involvement of law enforcement personnel and a respect for the need to maintain an uncompromised crime scene. Decontamination or disinfection would not likely be necessary.

In the case of a telephoned threat, and in the absence of a “device,” local law enforcement and public health authorities should be alerted. An envelope containing nothing other than a written threat poses little risk and should be handled in the same manner as a telephoned threat. Because the envelope constitutes evidence in a crime, however, further handling should be left to law enforcement professionals. In these cases, no decontamination is necessary pending results of the legal and public health investigation. If a threat is subsequently deemed credible, public health authorities should contact potentially exposed individuals, obtain appropriate information, and consider instituting prophylaxis or therapy.

When a package is found to contain powder, liquid, or other physical material, response should be individualized. In most cases, the package should be left in place, the room should be vacated, additional untrained persons should be prohibited from handling the material or approaching the scene, and law enforcement and public health officials should again be contacted. Persons coming in physical contact with contents should remove clothing as soon as practical and should seal it in a plastic bag. Victims should then wash with soap and water (Centers for Disease Control and Prevention, 1999) and, in most cases, may be sent home after adequate instructions are provided and contact information obtained. Specific antibiotic prophylaxis would not typically be necessary prior to the preliminary identification of package contents by a competent laboratory, although such decisions to provide or withhold postexposure prophylaxis are best made following consultation with public health authorities. Floors, walls, and furniture would not require decontamination before laboratory analysis is completed. Nonporous contaminated personal items such as eyeglasses and jewelry may be washed with soap and water or wiped clean with 0.5% hypochlorite (household bleach diluted 10-fold) if a foreign substance has contacted the items.

In the event that a device or other evidence of a credible aerosol threat is discovered, the room (and potentially the building) should be vacated. Law enforcement and public health personnel should be notified immediately and further handling of the device left to personnel with highly specialized training (in the United States, this might include the Federal Bureau of Investigation’s Hazardous Materials Response Unit). Contact information should be obtained from potential victims and detailed instructions provided. Clothing removal, soap and water showering, and decontamination of personal effects should be accomplished as described. Decisions regarding institution of empiric postexposure prophylaxis pending determination of the nature of the threat and identification of the involved biological agents should again be left to local and state public health authorities.

In providing a rational and measured response to each situation, public health and law enforcement personnel can assist in minimizing the disruption and cost associated with biological threats and hoaxes. Large-scale decontamination, costly HAZMAT unit involvement, broad institution of therapeutic interventions, and widespread panic can hopefully be avoided by such a measured and logical response.

6. STEP 5. ESTABLISH A DIAGNOSIS ("THE SECONDARY ASSESSMENT")

Now that decontamination has been considered, the clinician may perform a more thorough and targeted assessment aimed at establishing a diagnosis (the "secondary survey"). The completeness and accuracy with which one establishes a diagnosis will vary depending on the circumstances the clinician finds him- or herself in. At a tertiary care center, the clinician may well have access to infectious disease consultants, microbiology professionals, and sophisticated diagnostic assays. Under such circumstances, it may be possible to arrive at a microbiologic diagnosis fairly promptly. On the other hand, it is equally conceivable that the primary care provider, practicing in more austere circumstances, may need to intervene promptly based on limited information and without immediate access to subspecialty consultation. Even in such cases, however, reasonable care can be instituted based simply on a "syndromic diagnosis." An "AMPLE" (A = allergies, arthropod exposures; M = medications; P = past illnesses and immunizations; L = last meal; E = environment) history may aid in establishing this diagnosis. In this regard, most victims of a biological or chemical attack will likely present with a predominance of respiratory, neuromuscular, or dermatologic findings. Victims of bioterrorism might also present, as previously noted, with little more than an undifferentiated febrile illness. By categorizing victims in this manner, logical empiric therapy decisions can be facilitated (Cieslak *et al.*, 2000; Cieslak and Henretig, 2003a; Henretig *et al.*, 2002).

7. STEP 6. PROVIDE PROMPT THERAPY

Once a diagnosis is established, therapy must be provided promptly. In the cases of anthrax and plague, in particular, survival is inversely related to the delay in providing proper therapy. A delay of more than 24 hours in the treatment of either disease leads to a uniformly grim prognosis. When the identity of a bioterrorist agent is known, the provision of proper therapy is straightforward. Our recommendations for such therapy are provided in Table 9.3. When a clinician, in relatively austere circumstances, however, is faced with

Table 9.3
Recommended Therapy of (and Prophylaxis Against) Diseases Caused by "Category A" Biothreat Agents

Condition	Adults	Children
Anthrax, inhalational, Therapy[e] (patients who are clinically stable after 14 days can be switched to a single oral agent [ciprofloxacin or doxycycline] to complete a 60-day course[d])	Ciprofloxacin[c] 400 mg IV q12h OR Doxycycline 100 mg IV q12h AND Clindamycin[a] 900 mg IV q8h AND Penicillin G[b] 4 mU IV q4h	Ciprofloxacin[c] 10–15 mg/kg IV q12h OR Doxycycline 2.2 mg/kg IV q12h AND Clindamycin[a] 10–15 mg/kg IV q8h AND Penicillin G[b] 400–600 kU/kg/day IV ÷ q4h
Anthrax, inhalational, postexposure prophylaxis (60-day course[d])	Ciprofloxacin 500 mg PO q12h OR Doxycycline 100 mg PO q12h	Ciprofloxacin 10–15 mg/kg PO q12h OR Doxycycline 2.2 mg/kg PO q12h
Anthrax, cutaneous in setting of terrorism, therapy[f]	Ciprofloxacin 500 mg PO q12h OR Doxycycline 100 mg PO q12h	Ciprofloxacin 10–15 mg/kg PO q12h OR Doxycycline 2.2 mg/kg PO q12h
Plague, therapy	Gentamicin 5 mg/kg IV qd OR Doxycycline 100 mg IV q12h OR Ciprofloxacin 400 mg IV q12h	Gentamicin 2.5 mg/kg IV q8h OR Doxycycline 2.2 mg/kg IV q12h OR Ciprofloxacin 15 mg/kg IV q12h
Plague, prophylaxis	Doxycycline 100 mg PO q12h OR Ciprofloxacin 500 mg PO q12h	Doxycycline 2.2 mg/kg PO q12h OR Ciprofloxacin 20 mg/kg PO q12h
Tularemia, therapy, and prophylaxis	Same as for plague	Same as for plague
Smallpox, therapy	Supportive care	Supportive care
Smallpox, prophylaxis	Vaccination may be effective if given within the first several days after exposure	Vaccination may be effective if given within the first several days after exposure
Botulism, therapy	Supportive care; antitoxin may halt the progression of symptoms, but is unlikely to reverse them	Supportive care; antitoxin may halt the progression of symptoms, but is unlikely to reverse them
Viral hemorrhagic fevers, therapy	Supportive care; ribavirin may be beneficial in select cases	Supportive care; ribavirin may be beneficial in select cases

[a] Rifampin or clarithromycin may be acceptable alternatives to clindamycin as drugs that target bacterial protein synthesis. If ciprofloxacin or another quinolone is used, doxycycline may be used as a second agent, as it also targets protein synthesis.

[b] Ampicillin, imipenem, meropenem, or chloramphenicol may be acceptable alternatives to penicillin as drugs with good central nervous system penetration.

[c] Levoflxacin or ofloxacin may be acceptable alternatives to ciprofloxacin.

[d] Assuming the organism is sensitive, children may be switched to oral amoxicillin (80 mg/kg/day ÷ q8h) to complete a 60-day course. We recommend that the first 14 days of therapy or postexposure prophylaxis, however, include ciprofloxacin and/or doxycycline regardless of age. A three-dose series of anthrax vaccine–adsorbed may permit shortening of the antibiotic course to 30 days.

[e] In a mass casualty setting, where resources are severely constrained, oral therapy may need to be substituted for the preferred parenteral option.

[f] Ten days of therapy may be adequate for endemic cutaneous disease. We recommend a full 60-day course in the setting of terrorism, however, because of the possibility of a concomitant inhalational exposure.

multiple victims and the nature of the illness is not known, empiric therapy must be instituted. Guidelines for the provision of empiric therapy in such situations have been published (Cieslak *et al.*, 2000), and we advocate that doxycycline or ciprofloxacin be administered empirically to patients with significant pulmonary symptoms when exposure to a bioterrorist attack is considered a strong possibility.

8. STEP 7. INSTITUTE PROPER INFECTION CONTROL MEASURES

The clinician must practice proper infection control procedures to ensure that contagious diseases are not propagated among patients. The vast majority of agents commonly regarded as biological weapons threats are not contagious. Among these are the causative agents of anthrax, botulism, tularemia, brucellosis, Q-fever, the alphaviral equine encephalitides, glanders, melioidosis, and many others. "Standard Precautions" alone should suffice when caring for victims of such diseases (Garner and The Hospital Infection Control Practices Advisory Committee, 1996). More stringent "transmission-based precautions" are applied to patients with certain infectious diseases. Three subcategories of transmission-based precautions exist: "droplet precautions" should be used in managing the pneumonic plague victim. Ordinary surgical masks are a component of proper droplet precautions and are adequate protection against plague; "contact precautions" should be used in managing certain viral hemorrhagic fever patients; and "airborne precautions," ideally including a HEPA filter mask, should be used when managing smallpox victims.

Table 9.4

Conventional Infectious Diseases and Diseases Potentially Resulting from an Act of Bioterrorism: Required Hospital Infection Control Precautions

Standard precautions	Contact precautions	Droplet precautions	Airborne precautions
All patients	MRSA, VRE	Meningococcal disease	Pulmonary tuberculosis
	Enteric infections	Resistant pneumococci	Measles
	Skin infections	Pertussis	Varicella
	Lice	Group A Streptococci	
	Scabies	Mycoplasma	
	Clostridium difficile disease	Adenovirus	
	RSV, parainfluenza	Influenza	
Anthrax	Certain VHFs	Pneumonic plague	Smallpox
Botulism	Ebola		
Tularemia	Marburg		
Brucellosis	Lassa fever		
Q-fever			
Glanders			
Melioidosis			
Ricin intoxication			
SEB Intoxication			
T-2 Intoxication			
VEE, EEE, WEE			

A summary of hospital infection control precautions as they apply to victims of biological terrorism is presented in Table 9.4.

9. STEP 8. ALERT THE PROPER AUTHORITIES ("WHICH AGENCY SHOULD ONE NOTIFY FOR SUSPICIOUS CASES?")

As soon as it is suspected that a case of disease might be the result of terrorism, the proper authorities must be alerted so that the appropriate warnings may be issued and outbreak control measures implemented. Typically, such notification would be made through local and/or regional health department channels. In the United States, a few larger cities have their own health departments. In most other areas, the county represents the lowest echelon health jurisdiction. In some rural areas, practitioners would access the state health department directly. The situation in Canada is analogous; practitioners would contact their local or provincial health authorities. Once alerted, local and regional health authorities are well versed on the mechanisms for requesting additional support from health officials at higher jurisdictions. Each practitioner should have a point of contact with such agencies and should be familiar with mechanisms for contacting them before a crisis arises. A list of useful points of contact is provided in Table 9.5.

If an outbreak proves to be the result of terrorism, or if the scope of the outbreak strains resources available at the local level, a regional or national response becomes imperative. Under such circumstances, an extensive array of supporting assets and capabilities may be called on. In the U.S. system, the "Incident Command System" (ICS) provides a standardized approach to command and control of an incident scene (Emergency

Table 9.5
Bioterrorism: Points of Contact and Training Resources

Local Law Enforcement Authorities[a]	
Local or County Health Department[a]	
State Health Department[a]	
CDC Emergency Response Hotline	770-488-7100
CDC Bioterrorism Preparedness and Response Program	404-639-0385
CDC Emergency Preparedness Resources	http://www.bt.cdc.gov
Strategic National Stockpile	Access through State Health Department
FBI (general point of contact)	202-324-3000
FBI (suspicious package info)	http://www.fbi.gov/pressrel/pressrel01/mail3.pdf
Health Canada (suspicious package info)	http://www.hc-sc.gc.ca/english/epr/packages.html
USAMRIID General Information	http://www.usamriid.army.mil
USAMRICD Training Materials	http://ccc.apgea.army.mil
U.S. Army Medical NBC Defense Information	http://www.nbc-med.org
Johns Hopkins Center for Civilian Biodefense	http://www.hopkins-biodefense.org
Infectious Diseases Society of America	http://www.idsociety.org/bt/toc.htm

CDC, Centers for Disease Control and Prevention; FBI, Federal Bureau of Investigation; USAMRIID, U.S. Army Medical Research Institute of Infectious Diseases.

[a] Clinicians and response planners are encouraged to post this list in an accessible location. Specific local and state points of contact should be included.

Table 9.6
The Laboratory Response Network

Level A Labs: These labs, found in many hospitals and local public health facilities, have the ability to "rule out" specific bioterrorism threat agents, to handle specimens safely, and to forward specimens on to higher echelon laboratories within the network.
Level B Labs: These labs – found in larger hospitals, medical centers, and health departments – use BSL-2/3 practices (see Table 9.7) and have the ability to "rule in" and confirm specific bioterrorism threat agents. In addition, they can perform antimicrobial sensitivity testing.
Level C Labs: These labs, typically found in State Health Departments, use BSL-3 practices, and can conduct nucleic acid amplification and molecular typing studies. They serve as "back-up" to Level B labs.
Level D Labs: These labs, at the CDC and USAMRIID, can use BSL-4 practices, and serve as the final authority in the work-up of bioterrorism specimens. These labs provide specialized reagents to lower level labs and have the ability to bank specimens, perform serotyping, and detect genetic recombinants and chimeras.

BSL, biosafety level; CDC, Centers for Disease Control and Prevention; USAMRIID, U.S. Army Medical Research Institute of Infectious Diseases.

Management Institute, 1998). Local officials use this system when responding to natural disasters, as well as to man-made incidents. Its use would be equally appropriate in the response to a biological attack. Under the ICS, a designated official, often the fire chief or the chief of police, serves as local incident commander. In any incident, when local resources or capabilities are exceeded, the local incident commander may request assistance from the state through the State Coordinating Officer (SCO). The SCO works with the governor and other state officials to make state-level assets available. State Health Departments and Public Health Laboratories, as well as State Police capabilities, are among these assets. Most state public health laboratories possess "Level C" capabilities, and can provide sophisticated confirmatory diagnosis and typing of biological agents (Gilchrist, 2000; Morse *et al.*, 2003). [An overview of public health laboratory capabilities is provided in Table 9.6. The biosafety level (U.S. Department of Health and Human Services, 1999) precautions they employ are outlined in Table 9.7.] Moreover, State Police can provide law enforcement assistance and forensic laboratory analysis. Finally, state governors can access military assets directly through National Guard units under their direct control. These units can provide law enforcement, public works assistance, mobile "field" hospital bed capacity, and other support. Specifically, many state governors now have, at their disposal, military "Weapons of Mass Destruction—Civil Support" teams, which can offer expert advice and provide liaison to more robust military assets at the federal level.

When state capabilities are overwhelmed or insufficient, the SCO can contact his or her counterpart at the federal level. This Federal Coordinating Officer assists in activating the Federal Response Plan (FRP) (Federal Emergency Management Agency, 2003), which guides the delivery of federal assets and provides for a coordinated multiagency federal response.

Federal response and support to the states, according to the FRP, is organized into 12 emergency support functions (ESFs). ESF 8 provides for health and medical services. Although a specific agency takes primary responsibility for each of the 12 ESFs, 26 different federal agencies (and the American Red Cross) can, by law, be tasked to provide assistance. Federal disaster medical support is primarily the responsibility of the Department of

Table 9.7
Biosafety Levels

Biosafety Level 1: Involves practices used by a microbiology lab that deals only with well-characterized organisms that do not typically produce disease in humans. Work is conducted on open benchtops using standard microbiologic practices. A high school biology lab might use BSL-1 practices.
Biosafety Level 2: Involves practices used by labs that deal with most human pathogens of moderate potential hazard. Lab coats and gloves are typically worn, access to the lab is restricted to trained personnel, and safety cabinets are often used. A clinical hospital laboratory would typically utilize BSL-2 practices.
Biosafety Level 3: Involves practices used by labs that work with agents with the potential to cause serious and lethal disease via the inhalational route of exposure. Work is generally conducted in safety cabinets, workers are often immunized against the agents in question, and respiratory protection is worn. Clothing (such as "scrub suits") is exchanged upon exiting the lab. Labs are negatively pressurized. A State Health Department lab would typically use BSL-3 practices.
Biosafety Level 4: Also involves practices used by labs working with highly hazardous human pathogens infectious via the inhalational route. BSL-4 organisms differ from those requiring BSL-3 precautions in that no vaccine or antibiotic therapy is available. Personnel may only enter the lab through a series of changing and shower rooms. Equipment and supplies enter via a double-door autoclave. Strict and sophisticated engineering controls are used and personnel wear sealed positive pressure "space suits" with supplied air. Labs are negatively pressurized. Labs at CDC, USAMRIID, the Canadian Science Center for Human and Animal Health, and a few other research facilities, are equipped with BSL-4 controls.

BSL, biosafety level; CDC, Centers for Disease Control and Prevention; USAMRIID, U.S. Army Medical Research Institute of Infectious Diseases.

Health and Human Services, although the Office of Emergency Response, part of the new Department of Homeland Security, oversees the National Disaster Medical System (NDMS) (Knouss, 2001). The NDMS includes numerous Disaster Medical Assistance Teams, consisting of trained medical volunteers who can arrive at a disaster site within 8–16 hours. Another important aspect of the NDMS involves excess hospital bed capacity at numerous Department of Veterans Affairs, military, and civilian hospitals throughout the nation.

Finally, several other federal agencies may play an important role in the response to disasters, in general, and biological attacks in particular. The CDC and the U.S. Army Medical Research Institute of Infectious Diseases (USAMRIID) can provide "Level D" reference laboratories, which support the Level C labs at the state level and are capable of dealing with virtually all potential biological threat agents (Centers for Disease Control and Prevention, 2000). The Canadian Science Center for Human and Animal Health in Winnipeg provides a similar level of expertise. Expert consultation and assistance with epidemiological investigations is also available from the CDC, and threat evaluation and medical consultation is similarly available through USAMRIID. Additionally, the military can provide expert advice and assistance to civilian authorities through the Chemical/Biological Rapid Response Team (CBRRT), which can arrive at a disaster site within a few hours of notification. Another potentially useful military asset is the Chemical/Biological Incident Response Force, a Marine Corps unit capable of providing reconnaissance, decontamination, and field treatment. Like the CBRRT, this unit can potentially be available within a few hours of notification. Military support, when requested, would be subordinate to civilian authorities, and would be provided and tailored by the Joint Task Force for Civil Support. This task force, a component of U.S. Northern Command, is specifically designed to provide command and control for all military assets involved in disaster response missions and contingencies within the United States. Finally, the CDC has developed the Strategic National Stockpile, whereby critical drugs and vaccines necessary to combat a large disaster or terrorist attack are stockpiled at several locations throughout the country, available for rapid deployment to an affected area (Esbitt, 2003). Release of stockpile components is currently controlled by the Department of Homeland Security. An analogous asset, the National Emergency Services Stockpile System, provides Canada with similar capabilities.

10. STEP 9. CONDUCT AN EPIDEMIOLOGIC INVESTIGATION (AND MANAGE THE MEDICAL AND PSYCHOLOGICAL AFTERMATH OF A BIOTERROR ATTACK)

The clinician must be prepared to assist in an epidemiological investigation, which will be necessary in the case of a suspected terrorist attack. Although health department personnel will be invaluable in the course of such an investigation, the clinician should, nonetheless, have a working knowledge of basic epidemiology and the steps necessary in conducting an epidemiological investigation. These steps, the so-called "epidemiological sequence," are published elsewhere (Centers for Disease Control and Prevention, 1998) and summarized in Table 9.8. Although the well-prepared clinician may impact positively on the health and well-being of individual patients, it is only through the rapid conduct of a competent epidemiologic investigation that large numbers of exposed persons are likely to be reached, and positively impacted, prior to the widespread outbreak of disease.

In addition to implementing specific medical countermeasures against biological agent exposures, and instigating an epidemiologic investigation, the clinician must be prepared to address the psychological effects of a known, suspected, or feared exposure (Holloway *et al.,* 1997). An announced or threatened bioterroism attack can provoke fear, uncertainty, and anxiety in the population, resulting in overwhelming numbers of patients seeking medical evaluation for unexplained symptoms, and demanding antidotes for feared exposure. Such a scenario could also follow a covert release when the resulting epidemic is characterized as the consequence of a bioterror attack. Symptoms due to anxiety and autonomic arousal, and side effects of postexposure antibiotic prophylaxis may suggest

Table 9.8
The Epidemiologic Sequence

1. Make an Observation
2. Count cases
3. Relate cases to population
4. Make comparisons
5. Develop the hypothesis
6. Test the hypothesis
7. Make scientific inferences
8. Conduct studies
9. Intervene and evaluate

prodromal disease due to biological agent exposure, and pose challenges in differential diagnosis. This "behavioral contagion" is best prevented by risk communication from health and government authorities that includes a realistic assessment of the risk of exposure, information about the resulting disease, and what to do and whom to contact for suspected exposure. Risk communication must be timely, accurate, consistent, and well coordinated. As the epidemic subsides and public knowledge increases, public anxiety will decrease to realistic levels. This cycle of uncertainty, panic, response, and resolution occurred during the October 2001 anthrax bioterror event (Rundell, 2003). The CDC has taken a proactive approach, featuring the development of internet-accessible, agent-specific information packages for local public health authorities and the general public (Centers for Disease Control and Prevention, 2003e).

Effective risk communication is predicated upon the pre-existence of well-conceived risk communication plans and tactics. Similarly, plans must be made to rapidly deploy personnel from local centers for the initial evaluation and administration of postexposure prophylaxis (ideally decentralized to residential areas). Finally, plans must be made to proactively develop patient and contact tracing and vaccine screening tools, to access stockpiled vaccines and medications, and to identify and prepare local facilities and healthcare teams for the care of mass casualties. The CDC smallpox response plan (Centers for Disease Control and Prevention, 2003e) provides a template for such a coordinated, multifaceted approach. The benefits of farsighted planning and coordination were demonstrated by the efficient mass prophylaxis of more than 10,000 individuals in New York City during the anthrax bioterror event of 2001 (Blank *et al.*, 2003).

11. STEP 10. MAINTAIN A LEVEL OF PROFICIENCY

Once response plans have been developed, they must be exercised. Local exercises designed to test incident command and control, communications, logistics, laboratory coordination, and clinical capabilities are thus a final, and necessary, preparation. Such exercises may involve only the leadership of an organization and focus on planning and decision-making (the "command post exercise"), they may involve notional "play" around a "table-top exercise," or they may involve actual "hands-on" training and evaluation in a "disaster drill" or "field-training exercise." In fact, the Joint Commission on the Accreditation of Healthcare Organizations (JCAHO) requires hospitals to conduct a hazard vulnerability analysis, develop an emergency management plan, and evaluate this plan annually (Joint Commission on the Accreditation of Healthcare Organizations, 2003). Moreover, JCAHO specifically mandates that hospitals provide facilities (and training in the use of such facilities) for radioactive, biological, and chemical isolation and decontamination.

Many resources are now available to assist the clinician and public health professional in planning for, and maintaining proficiency regarding, the management of real or threatened terror attacks. Moreover, electronic resources of a similar nature have been developed (U.S. Army Medical Research Institute of Infectious Diseases, 2000; U.S. Army Medical Research Institute of Infectious Diseases and U.S. Food and Drug Administration, 2000) and multiple websites provide a wealth of training materials and information on-line (Ferguson *et al.*, 2003) (Table 9.5). Finally, as discussed under step 8, numerous governmental, military, and civilian organizations now stand ready to provide assistance and consultation to the clinician faced with planning for, and treating, the victims of a potential terrorist attack. It is assistance that, if sought for planning purposes, will hopefully never be needed for patient management.

ACKNOWLEDGMENTS

The opinions and assertions contained herein are the private views of the authors and are not to be construed as official or as necessarily reflecting the views of the U.S. Department of Defense, the U.S. Department of Health and Human Services, or their component services, agencies, and institutions.

References

Amorosa, V.K., and Isaacs, S.N. (2003). Separate worlds set to collide: smallpox, vaccinia virus vaccination, and human immunodeficiency virus and acquired immunodeficiency syndrome. *Clin. Infect. Dis.* 37:426–432.

Blank, S., Moskin, L.C., and Zucker, J.R. (2003). An ounce of prevention is a ton of work: mass antibiotic prophylaxis for anthrax, New York City, 2001. *Emerg. Infect. Dis.* 9:615–622.

Bozzette, S.A., Boer, R., Bhatnagar, V., Brower, L.J., Keeler, E.B., Morton, S.C., and Stoto, M.A. (2003). A model for smallpox-vaccination policy. *N. Engl. J. Med.* 348:416–425.

Brachman, P.S., Gold, H., Plotkin, S.A., Fekerty, F.R., Werrin, M., and Ingraham, N.R. (1962). Field evaluation of a human anthrax vaccine. *Am. J. Public Health* 52:632– 645.

Centers for Disease Control and Prevention. (1998). Investigating an outbreak. In: *Principles of Epidemiology: Self Study Course SS3030*, 2nd ed. Centers for Disease Control and Prevention, Atlanta, pp. 347–424.

Centers for Disease Control and Prevention. (1999). Bioterrorism alleging use of anthrax and interim guidelines for management- United States, 1998. *MMWR. Morb. Mortal. Wkly. Rep.* 48:69–74.

Centers for Disease Control and Prevention. (2000). Biological and chemical terrorism: strategic plan for preparedness and response. *MMWR. Morb. Mortal. Wkly. Rep.* 49(RR-04):1–14.

Centers for Disease Control and Prevention. (2002). Notice to readers: use of anthrax vaccine in response to terrorism: supplemental recommendations of the advisory committee on immunization practices. *MMWR. Morb. Mortal. Wkly. Rep.* 51:1024–1026.

Centers for Disease Control and Prevention. (2003a). Smallpox vaccination and adverse reactions: guidance for clinicians. *MMWR. Morb. Mortal. Wkly. Rep.* 52(Dispatch):1–29.

Centers for Disease Control and Prevention. (2003b). Recommendations for using smallpox vaccine in a pre-event vaccination program. *MMWR. Morb. Mortal. Wkly. Rep.* 52(RR-07):1–16.

Centers for Disease Control and Prevention. (2003c). Update: cardiac and other adverse events following civilian smallpox vaccination—United States. *MMWR. Morb. Mortal. Wkly. Rep.* 52:639–642.

Centers for Disease Control and Prevention. (2003d). Update: Adverse events following civilan smallpox vaccination—United States, 2003. *MMWR. Morb. Mortal. Wkly. Rep.* 52:819–820.

Centers for Disease Control and Prevention. (2003e). Smallpox response plan and guidelines (version 3.0). Accessed August 19, 2003. Available at: http://www.bt.cdc.gov/agent/smallpox/response-plan/index.asp.

Centers for Disease Control and Prevention. (2003f). Public health emergency preparedness and response. Accessed 31 August 2003. Available at: http://www.cdc.bt.gov.

Cieslak T.J., Christopher, G.W., Kortepeter. M.G., Rowe, J.R., Pavlin, J.A., Culpepper, R.C., and Eitzen, E.M. (2000). Immunization against potential biological warfare agents. *Clin. Infect. Dis.* 30:843–850.

Cieslak, T.J., and Henretig, F.M. (2001). Medical consequences of biological warfare: the ten commandments of management. *Milit. Med.* 166(Suppl 2):11–12.

Cieslak, T.J., and Henretig, F.M. (2003a). Bioterrorism. *Pediatr. Ann.* 32:154–165.

Cieslak, T.J., and Henretig, F.M. (2003b). Biological and chemical terrorism. In: Berman, R.E., Kliegman, R.M., and Jenson, H.B. (eds.), *Nelson Textbook of Pediatrics*, 17th ed. Saunders, Philadelphia, pp. 2378–2385.

Cieslak, T.J., Kortepeter, M.G., and Eitzen, EM. (2004). Vaccines against agents of bioterrorism. In: Levine, M.M., Kaper, J.B., Rappouli, R., Liu, M., and Good, MF. (eds.), *New Generation Vaccines*, 3rd ed. Marcel Dekker, New York.

Cieslak, T.J., Rowe, J.R., Kortepeter, M.G., Madsen, J.M., Newmark, J., Christopher, G.W., Culpepper, R.C., and Eitzen, E.M. (2000). A field-expedient algorithmic approach to the clinical management of chemical and biological casualties. *Milit. Med.* 165:659–662.

Cole, L.A. (2000). Bioterrorism threats: learning from inappropriate responses. *J. Public Health Manag. Pract.* 6:8–18.

Committee on Trauma, American College of Surgeons. (1989). Initial assessment and management. In: *Advanced Trauma Life Support Student Manual*. American College of Surgeons, Chicago, pp. 9–30.

Emergency Management Institute. (1998). *Incident Command System Independent Study Guide* - (IS-195). Federal Emergency Management Agency, Washington, D.C.

Esbitt, D. (2003). The Strategic National Stockpile: roles and responsibilities of health care professionals for receiving the stockpile assets. *Disaster Manag. Response*. 1:68–70.

Fauci, A.S. (2002). Smallpox vaccination policy-the need for dialogue. *N. Engl. J. Med.* 346:1319–1320.

Federal Emergency Management Agency. (2003). *The Federal Response Plan* (9230.1PL, interim). U.S. Government Printing Office, Washington, D.C.

Ferguson, N.E., Steele, L., Crawford, C.Y., Huebner, N.L., Fonseka, J.C., Bonander, J.C., and Kuehnert, M.J. (2003). Bioterrorism web site resources for infectious disease clinicians and epidemiologists. *Clin. Infect. Dis.* 36:1458–1473.

Fine, A., and Layton, M. (2001). Lessons from the West Nile viral encephalitis outbreak in New York City, 1999: implications for bioterrorism preparedness. *Clin. Infect. Dis.* 32:277–282.

Garner, J.S., and The Hospital Infection Control Practices Advisory Committee (1996). Guideline for infection precautions in hospitals. *Infect. Control. Hosp. Epidemiol.* 17:53–80.

Gilchrist, M.J.R. (2000). A national laboratory network for bioterrorism: evolution from a prototype network of laboratories performing routine surveillance. *Milit. Med.* 165(Suppl 2):28–31.

Grabenstein, J.D., and Winkenwerder, W., Jr. (2003). US military smallpox vaccination experience. *J.A.M.A.* 289;3278–3282.

Halsell, J.S., Riddle, J.R., Atwood, J.E., Gardner, P., Shope, R., Poland, G.A., Gray, G.C., Ostroff, S., Eckart, R.E., Hospenthal, D.R., Gibson, R.L., Grabenstein, J.D., Arness, M.K., Tornberg, D.N., and Department of Defense Smallpox Vaccination Clinical Evaluation Team. (2003). Myopericarditis following smallpox vaccination among vaccinia-naive US military personnel. *J.A.M.A.* 289:3283–3289.

Health Resources and Services Administration, Department of Health and Human Services. (2003). Smallpox vaccine injury compensation program: smallpox (vaccinia) vaccine injury table. *Fed. Reg.* 68:51492–51499.

Henretig, F.M., Cieslak, T.J., Kortepeter, M.G., and Fleisher, G.R. (2002). Medical management of the suspected victim of bioterrorism: an algorithmic approach to the undifferentiated patient. *Emerg. Med. Clin. North Am.* 20:351–364.

Hiss, J., and Arensburg, B. (1992). Suffocation from misuse of gas masks during the Gulf War. *Br. Med. J.* 304:92.

Holloway, H.L., Norwood, A.E., Fullerton, C.S., Engel, C.C., and Ursano, R.J. (1997). The threat of biological weapons: prophylaxis and mitigation of psychological and social consequences. *J.A.M.A.* 278:425–427.

Joint Commission on Accreditation of Healthcare Organizations. (2003). *Hospital Accreditation Standards.* JCAHO, Oakbrook Terrace, Illinois, pp. 221–224.

Kaplan, E.H., Craft, D.L., and Wein, L.M. (2002). Emergency response to a smallpox attack: the case for mass vaccination. *Proc. Natl. Acad. Sci. U.S.A.* 99:10935–10940.

Kemper, A.R., Davis, M.M., and Freed, G.L. (2002). Expected adverse events in a mass smallpox vaccination campaign. *Eff. Clin. Pract.* 5:84–90.

Knouss, R.F. (2001). National disaster medical system. *Public Health Rep.* 116(Suppl 2):49–52.

Kortepeter, M.G., and Cieslak, T.J. (2003). Bioterrorism: plague, anthrax, and smallpox. In: Baddour, L., and Gorbach, S.L. (eds.), *Therapy of Infectious Diseases.* W.B. Saunders, Philadelphia, pp. 723–740.

Lampton, L.M. (2003). SARS, biological terrorism, and mother nature. *J. Miss. State Med. Assoc.* 44:151–152.

Mack, T. (2003). A different view of smallpox and vaccination. *N. Engl.J. Med.* 348:460–463.

Morse, S.A., Kellogg, R.B., Perry, S., Meyer, R.F., Bray, D., Nichelson, D., and Miller, J.M. (2003). Detecting biothreat agents: the laboratory response network. *ASM News.* 69:433–437.

Mortimer, P.P. (2003). Can postexposure vaccination against smallpox succeed? *Clin. Infect. Dis.* 36:622–629.

Pavlin, J.A. (1999). Epidemiology of bioterrorism. *Emerg. Infect. Dis.* 5:528–530.

Rundell, J.R. (2003). A consultation-laison psychiatry approach to disaster/terrorism victim assessment and management. In: Ursano, R.J., Fullerton, A.E., and Norwood, C.S. (Eds.), *Terrorism and Disaster: Individual and Community Mental Health Interventions.* Cambridge University Press, New York, pp. 107–120.

U.S. Army Medical Research Institute of Infectious Diseases. (2000). Medical management of biological warfare casualties CD-ROM: U.S. Army Medical Research Institute of Infectious Diseases, Ft. Detrick, MD.

U.S. Army Medical Research Institute of Infectious Diseases, U.S. Food and Drug Administration. (2000). Biological warfare and terrorism: medical issues and response [satellite television broadcast]. U.S. Army Medical Research Institute of Infectious Diseases, U.S. Food and Drug Administration, Gaithersburg, MD.

U.S. Department of Health and Human Services. (1999). *Biosafety in Microbiological and Biomedical Laboratories*, 4th ed. U.S. Government Printing Office, Washington, D.C.

10

The Economics of Planning and Preparing for Bioterrorism

Martin I. Meltzer

1. INTRODUCTION

The end of the cold war between the "super powers," and the subsequent reduction in the probability of a large-scale nuclear war, brought into being a world with new political realities. Governments around the world now have to face the possibility of terrorist attacks on the civilian population, in which biological, chemical, and nuclear agents could be used. Prior to the collapse of the Soviet Union in the early 1990s, planning and preparing responses to large-scale catastrophic infectious disease events among civilian populations, such as a bioterrorist attack, was not an activity to which governments typically devoted many resources.

In response to the threat of a bioterrorist attack, governments have begun to stockpile drugs and vaccines, and have started to examine an array of potential response plans. For economists, such planning and preparations raise two central questions: How many resources, and what type of resources, should be devoted to planning and preparing a response to a bioterrorist attack or any other catastrophic infectious disease event? This chapter will be devoted to providing some guiding principles that will help policy makers answer both of these questions.

2. HOW MANY RESOURCES?: BASIC CONCEPT

Assuming that the probability of a large-scale bioterrorist attack is greater than zero, the public and policy makers will probably agree that it is logical to spend resources to plan and prepare a set of responses for such an event. Obviously, neither group wants to spend

Bioterrorism and Infectious Agents
Edited by Fong and Alibek, Springer Science+Business Media, Inc., New York, 2005

"too little" and be woefully underprepared. On the other hand, spending "too many" resources on planning and preparing would be a waste. Thus, there is a need to define the optimal amount of resources that logically should be spent on planning and preparing for any identified threat, where the term "optimal" essentially means "not too much nor too little."

To calculate the optimal amounts of resources for planning and preparing, consider the problem similar to calculating an optimal insurance premium. To calculate the optimal insurance premium that society should pay each year for protection against a bioterrorist attack, we need to know two pieces of information:

1. Estimates of the size of the potential losses (e.g., $ value) without any planning and preparing
2. Probability of such losses occurring

Using these two pieces of information, the optimal annual insurance premium is then calculated using this simple formula:

Equation 1:
Optimal annual amount to be spent on planning and preparing = $ value of potential losses × annual probability of that loss actually occurring.

3. REFINING THE BASIC CONCEPT: BEING MORE REALISTIC

There is one problem with the basic equation, given above, defining the optimal annual amount to be spent on planning and preparing for a bioterrorist event. In the equation, it is assumed that the planning and preparation will be 100% effective in preventing the estimated dollar loss – that is, the equation represents the optimal annual expenditure when there is "perfect defense." This is, of course, unrealistic. Potential bioterrorists have a distinct strategic advantage over a society trying to defend itself against an array of probable threats. Terrorist organizations potentially interested in attacking the United States (or any other country) with biological agents have the advantage of concentrating their resources on a few select pathogens and a few selected targets. Furthermore, an actual bioterrorist attack most likely needs a relatively small number of terrorists to carry out the attack. A small attacking force – combined with the fact that often a small amount of biological agent is all that is needed to mount a potentially large attack – provides the terrorists with a degree of flexibility. The ability to pick the date and time of an attack, and the option to delay that attack (sometimes for years), provides additional flexibility. Thus, up until an attack is actually launched (or just before it), the terrorists can rapidly change targets and plans. A society facing a possible attack does not have such flexibility. Society must make plans and preparations against a number of potential attack scenarios, and the large population size of many countries reduces the possibilities of providing a perfectly flexible, 100% effective pre-attack defense.

Thus, it must be anticipated that an attack will actually occur, resulting in some casualties. Part of the annual amount spent on planning and preparing for such an attack (the insurance premium) must therefore include preparation to limit the number and severity of health outcomes (e.g., provide postexposure prophylaxis, provide adequate medical treatment). The optimal insurance amount must therefore be adjusted to account for the fact that casualties will occur. The adjusted formula is as follows:

Equation 2:
Optimal annual amount to be spent on planning and preparing = $ value of potential losses actually averted × annual probability of that loss actually occurring.

Where $ value of potential losses actually averted is a function of the type of intervention(s) being evaluated, the cost of that intervention, an estimate of the casualties averted, and the value of those casualties averted, The estimate of the casualties averted is, in turn, a function of the estimate of casualties without the intervention less the estimated casualties after the intervention. Table 10.1 provides an outline of the various elements in the above equation, along with an outline of the remainder of this chapter.

3.1. Cost of Deploying a Planned Intervention

The cost associated with actually deploying a planned intervention can impact the optimal amount to be spent on planning and preparing. Some of the money spent on planning and preparing could be spent on stockpiling supplies likely to be needed if an event actually occurs, as well as training personnel to improve the response (a form of stockpiling services and human capital). However, during an actual event, most likely there will be a need for materials or services that have not been stockpiled, and these will have to be paid for at the time of use. Unless part of the planning and preparation includes the formation of a reserve fund to pay for such expenses (i.e., essentially prepay such expenses), cost incurred at deployment should be accounted for when calculating the optimal insurance premium. Essentially, cost incurred at the time of deployment of a response will reduce the optimal amount spent on planning and preparing by reducing the $ value of potential losses averted, as follows:

Optimal annual amount to be spent on planning and preparing = ($ value of potential losses actually averted − non-prepaid costs of actual deployment) × annual probability of that loss actually occurring.

3.2. A Special Case: Optimal Amount for Pre-event Protective Interventions

When considering the optimal amount to be spent on protective interventions designed to reduce (or even eliminate) the actual probability of a bioterrorist event occurring, the formula above should be adjusted to read as follows:

Equation 3:
Optimal annual amount to be spent on reducing probability of an attack = $ value of potential losses actually averted × reduction in probability of that loss actually occurring.

Table 10.1
Calculating the Optimal Amount to Spend on Planning and Preparing for a Bioterrorist Attack: Basic Framework of Chapter

Relevant section	Element	Details of elements to calculate optimal amount to spend on planning and preparing for a bioterrorist attack
2, 3	Basic equation	Optimal annual $ amount = $ value of potential losses actually averted × annual probability of event occurring
3	$ value of potential losses averted	$ value of potential losses averted = reduction in losses due to intervention × $ value of losses
3.1	Costs of deploying an intervention	Reducing estimate of $ value of potential losses averted if there will be intervention deployment costs that can not be prepaid (or a reserve fund established)
3.2	Special case: pre-event protective interventions	Calculating the optimal amount when there is an intervention(s) that could reduce the actual probability of a bioterrorist event occurring
4, 5	Reduction in losses: selecting type of intervention	First step in calculating reduction in losses is to specify type of intervention(s) to be evaluated. There are two basic categories of interventions: those designed to primarily (a) reduce the impact of an attack (reaction interventions), and (b) to prevent, or reduce the probability of, the attack occurring (protective interventions).
6	Estimating the potential losses averted: types of mathematical models	Once an intervention is selected for evaluation, estimates of potential losses with and without that intervention usually have to be produced through mathematical modeling. Selecting the appropriate type of model depends on many factors, including quantity and quality of available data (Section 6.1 discusses some of the models that could be used).
6.2	Assumed size of the attack and use of "worst case"	Mathematical models of potential casualties (with and without interventions) typically have to make some initial assumptions regarding size of possible attack. There are a number of implicit assumptions associated with each assumed attack size (e.g., motivation, technical expertise, weather conditions upon release). Each assumed size of potential attack has associated a probability of occurrence (i.e., a smaller attack probably has a higher probability of occurring than a larger attack).
7	Valuing casualties and other losses	There is often a great deal of uncertainty associated with valuing potential casualties and other losses. Even estimating direct medical costs associated with treating casualties requires assumptions regarding the type of treatment given. Indirect costs, such as time lost from work, relies even more upon who falls ill (or dies), and for how long. Loss of economic activity is perhaps the hardest to estimate.
8	Probability of event actually occurring	The probability of an event has to be carefully considered. Often, this is not known with any certainty and thus a great deal of sensitivity analyses should be done with the value of this input. Furthermore, the connection between the probability of an event occurring and assumed size of initial attack (see Section 6.2.1) needs to be carefully evaluated.
9	Selecting between the options	The calculated optimal amount to spend on each intervention is useful because it combines, in a single estimate, several variables describing both potential impact of attack and benefit of intervention. Available budget, criteria used, and other factors will also play a part in the process of selecting an intervention.

The reduction in probability of an event occurring will range from 0 (failure to have any impact on reducing probability of event occurring) to 1 (eliminate probability of event occurring).

3.3. Example 1: Annual "Premiums" for Pandemic Influenza Preparations

The formula given above can be used to calculate the amount that could be spent on preparing for any type of catastrophic infectious disease event, including the next influenza pandemic. Influenza pandemics have occurred every 10–60 years, with three occurring in the 20th century (1918, 1958/1959, and 1968/1969). Influenza pandemics occur when there is a notable genetic change (termed genetic shift) in the circulating strain of influenza, leaving large portions of the human population vulnerable to infection and illness. Meltzer *et al.* (1999) estimated that, in the United States alone, the next influenza pandemic could cause 89,000–207,000 deaths, 314,000–734,000 hospitalizations, 18–42 million outpatient visits, and 20–42 million self-care illnesses (Figure 10.1). Many of the preparations needed for pandemic influenza (e.g., planning for mass vaccinations, ensuring surge capacity in hospitals) are also needed for bioterrorist attacks potentially using other agents.

To estimate the annual "premium" for planning and preparing for the next influenza pandemic (i.e., apply the equation given earlier in Section 3), the $ value of potential losses potentially averted by a pandemic influenza vaccination program is first calculated by placing a dollar value on all the influenza-related outcomes shown in Figure 10.1, and then multiplying such losses by estimates of the possible efficacy of vaccination. The resulting estimate is then multiplied by the annual probability of such a pandemic occurring. From Figure 10.1, there are wide array of possible losses (i.e., influenza-related health outcomes). Furthermore, there are ranges of probabilities for vaccine effectiveness, costs of vaccination,* and for an influenza pandemic occurring. Table 10.2 presents a range of calculated optimal amounts, with differences due to differences in assumed gross attack rate (e.g, percentage of population that become clinically ill), the cost of vaccination per vaccinee, assumed vaccine effectiveness, percentage of the population vaccinated, and probability of a pandemic occurring (range examined: 1 in 30 years to 1 in 100 years).

Note that altering the anticipated cost of vaccination per vaccinee alters the mean annual premium for planning and preparing (Table 10.2). For any given scenario (e.g., assumed vaccine effectiveness, compliance, and probability of the pandemic occurring), an increase in the per person cost of vaccination results in a lower annual premium. This is because, with the current technology, influenza pandemic vaccines cannot be stockpiled, and thus influenza

*The cost of vaccination is more that just the cost of the vaccine. Vaccination costs, from a societal perspective, include the cost of administrating the vaccine (e.g., nurses time, supervising physician time), cost of completing and processing any paperwork related to vaccination, value of patient time waiting for vaccination, cost of patient transport to and from the place of vaccination, and the cost of treating any vaccine-related side-effects. Furthermore, because many people may never have been exposed to any strain similar to that causing the next influenza pandemic (i.e., be "immune naïve"), it is possible that a large proportion of the population will require two doses of the vaccine – a priming dose and a booster dose. Requiring patients to return to a vaccination clinic for a booster dose will significantly increase the costs of vaccination. See Meltzer *et al.* (1999) for further details regarding costs of vaccination.

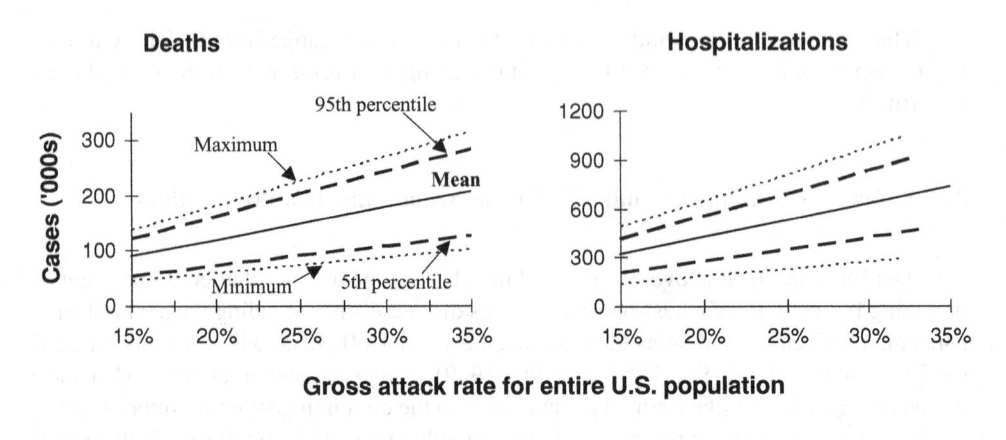

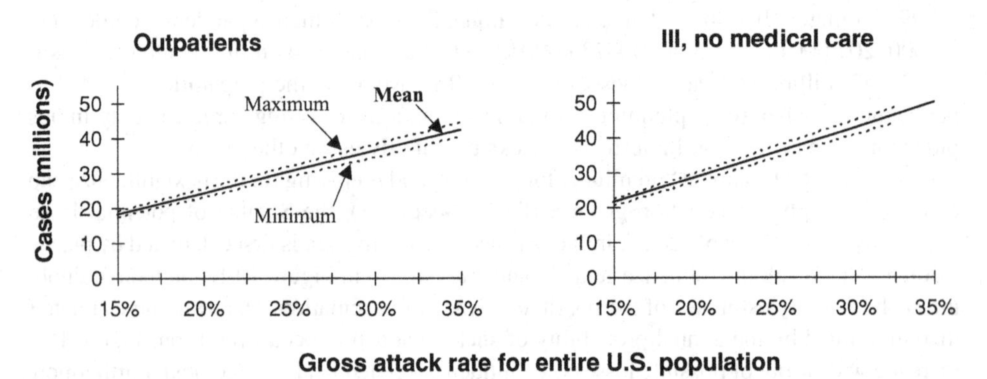

Figure 10.1. Potential impact in the United States of the next influenza pandemic. Mean, minimum, maximum, 5th, and 95th percentiles of total deaths, hospitalizations, outpatient visits, and those ill (but not seeking medical care) for different gross (clinical) attack rates. Gross attack rates refer to percentages of the total U.S. population (all age and risk groups) that may experience a clinical illness during the next influenza pandemic. From: Meltzer *et al.* (1999).

pandemic vaccine purchases will be incurred at the time of deployment of the response. The cost of deployment may be further increased if the pandemic influenza vaccine has to be given in two separate doses (priming dose plus booster – see asterisked footnote). Unless the pre-event preparations include the establishment of a reserve fund to pay for deployment cost, the larger the deployment cost, the greater the reduction in the optimal annual premium (see Section 3.1).

3.4. Example 2: Annual "Premiums" to Reduce Probability of Losses Due to Anthrax Attack

The calculated optimal amount to be spent on planning and preparing to respond to a bioterrorist release of aerosolized anthrax will depend on how quickly postexposure prophylaxis can be delivered to those effectively exposed (Figure 10.2). Unlike the

Table 10.2
Mean Annual Premium[a] for Planning, Preparing, and Practicing to Respond to the Next Influenza Pandemic (in 1995 US$)

Gross attack rate[b] (%)	Cost of vaccination $ per vaccinee[d]	Mean (std. deviation) premium ($ millions)					
		"Low" vaccine effectiveness × 40% compliance:[c] probability of pandemic			"High" vaccine effectiveness × 60% compliance:[c] probability of pandemic		
		1 in 30 years	1 in 60 years	1 in 100 years	1 in 30 years	1 in 60 years	1 in 100 years
15	21	306 (122)	153 (61)	92 (37)	872 (341)	435 (170)	262 (103)
	62	162 (122)	81 (61)	48 (37)	654 (341)	326 (170)	196 (103)
25	21	561 (204)	280 (102)	168 (61)	1,528 (569)	762 (584)	459 (171)
	62	416 (122)	207 (102)	125 (61)	1,311 (569)	653 (584)	394 (171)
35	21	815 (286)	406 (142)	245 (86)	2,184 (796)	1,089 (397)	656 (239)
	62	670 (286)	334 (142)	201 (86)	1,967 (796)	980 (397)	591 (239)

[a] Defined here as the amount of money that could be spent each year to plan, prepare, and practice to ensure that mass influenza vaccinations can take place. The mathematically optimal allocation of these funds (i.e., how these funds would be spent) requires a separate set of calculations.
[b] Gross attack rate refers to the percentage of the total U.S. population who experience a clinical illness due to pandemic influenza. Each clinical illness, as defined here, will cause a measurable economic impact (e.g., at least ½ day missed from work).
[c] For each of these two scenarios, vaccine effectiveness differs by age and by outcome. For example, "high" vaccine effectiveness is assumed to prevent 75%, 70%, and 60% of deaths among 0–19 year-olds, 20–64 year-olds and +65 year-olds, respectively. Similarly, "low" vaccine effectiveness is assumed to prevent 40%, 40%, and 30% of deaths among the same age groups. Percentage compliance refers to the total population vaccinated such that the assumed levels of effectiveness apply. See online appendix in source for further details.
[d] The cost of influenza pandemic vaccine deployment may be increased if the pandemic influenza vaccine has to be given in two separate doses (priming dose plus booster). Furthermore, with the current technology, influenza vaccination cannot be stockpiled, and thus influenza pandemic vaccine purchases will be incurred at the time of deployment of the response. These vaccination-related deployment costs will reduce the total value of potential losses that could be averted due to planning and preparing. Unless the pre-event preparations include the establishment of a reserve fund to pay for deployment costs, the larger the deployment costs, the greater the reduction in the optimal annual premium (see Sections 3.1 and 3.3).

Source: From Meltzer *et al.* (1999).

influenza pandemic example, it is possible that some counterterrorist preattack interventions could reduce the probability of an attack. However, the success of such interventions may be hard to measure, and thus the actual reduction in probability of an attack of a given size may be hard to estimate (i.e., the inputs for equation in Section 3.2 may be hard to actually quantify). Thus, the policy makers and researchers have to consider that a dual approach may be needed – responses that reduce the probability of event, plus pre-event planning to ensure as rapid a postevent response as possible.

Figure 10.2 demonstrates that time after an attack is a critical element, and that there is a notable difference in returns to response 0–2 days postattack and 5–6 days postattack. A response that starts only on day 6 postattack is almost equivalent, in terms of deaths, as having no organized program. By combining the impact-over-time data from Figure 10.2 with estimates of costs of outcomes and cost of response, it is possible to calculate the optimal total amount to be spent on combined threat reduction and response planning and preparation (Table 10.3). Table 10.3 provides estimates for a variety of scenarios, relating two levels of efficacy of postexposure prophylaxis (the minimum and maximum ranges of

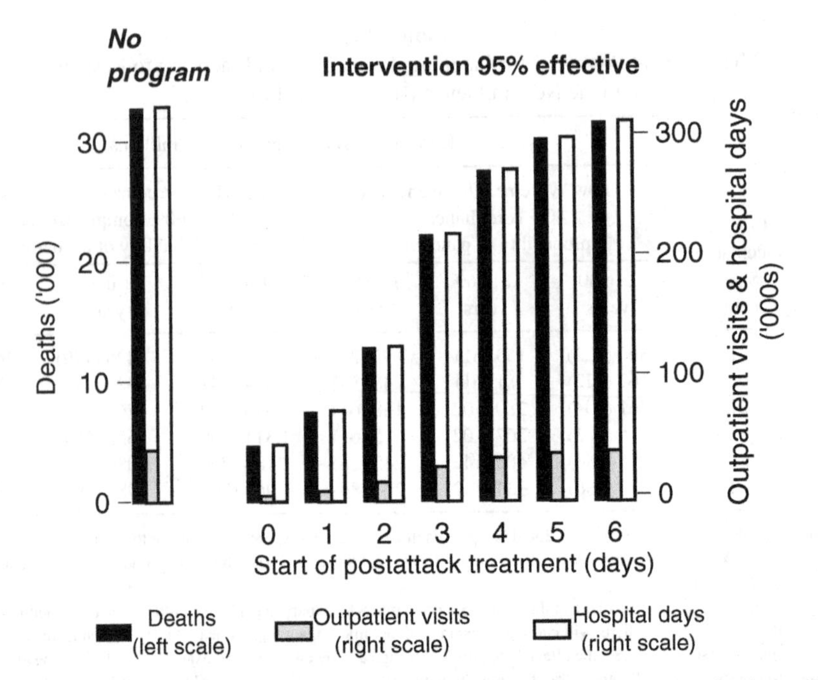

Figure 10.2. Total deaths, hospitalizations, and outpatient visits associated with a release of anthrax effectively exposing 100,000 persons. Reduction in impact by day of postattack prophylaxis. Intervention was assumed to be 95% effective in preventing disease if therapy is initiated before clinical symptoms appear. Source: Kaufmann *et al.* (1997).

possible preventable losses), days postattack when postexposure prophylaxis begins (from 0 to 6 days), and reductions in probability of event occurring (a 10-fold and 100-fold reductions). For example, assume that there is a set of interventions assumed to both reduce the probability of an attack effectively exposing 100,000 individuals by 10-fold (from, say, 0.01–0.001) and ensure that postexposure prophylaxis could be delivered by day 2 postattack. Then, the optimal amount to spend each year on these interventions would be between $90–$143 million (Table 10.3). Note that the calculated amounts do not detail how the "premiums" should be split between counterterrorist interventions designed to lower the probability of actual release and the pre-event planning to allow for rapid deployment of postexposure prophylaxis. The optimal division of the totals shown in Table 10.3 would require a separate set of calculations and probably a great deal of qualitative input.

For the remainder of the chapter, I will discuss aspects of calculating the inputs used in the equations presented in this section, and illustrated in the two examples.

4. CATEGORIES OF INTERVENTIONS

Money spent on interventions designed to reduce the impact of a potential bioterrorist attack or a catastrophic infectious disease event (e.g., influenza pandemic) can essentially be divided into two basic categories: actions that are planned to reduce the impact of an

Table 10.3
Maximum Annual Premium (in 1993 US$) for Reducing the Probability of Losses Associated with an Anthrax Attack: 100,000 Effectively Exposed[a]

Days postattack: start of response[b]	Range of preventable losses: minimum to maximum[c] ($ billions)	Range of annual premiums to prevent losses due to anthrax attack with 100,000 exposed: ($ millions)[d]	
		Reduction in annual probability of attack[d]	
		0.01 reduced to 0.001	0.01 reduced to 0.00001
0	14.4–22.4	129–201	143–224
1	12.8–20.1	115–181	128–201
2	10.0–15.9	90–143	100–159
3	5.2–8.4	47–76	52–84
4	2.4–4.2	22–38	24–42
5	1.0–2.1	9–19	10–21
6	0.4–1.0	3–9	4–10

[a] Effectively exposed was defined for these calculations as being a 2-hour duration such that 100,000 people would inhale the equivalent of an Infectious Dose$_{50}$ (ID_{50}), which means that those inhaling the anthrax have, on average, a 50% chance of becoming clinically ill. The ID_{50} was set at 20,000 spores of *B. anthracis*.
[b] Number of days from attack (release) to start of effective mass postexposure prophylaxis.
[c] Minimum preventable losses (potential net savings) calculated assuming ciprofloxacin used for postexposure prophylaxis, and 15 unexposed persons treated per person actually exposed, for an assumed effectiveness of 90% in preventing illness. Maximum preventable losses (potential net savings) calculated assuming a combination of doxycycline and anthrax vaccine used for postexposure prophylaxis, and 5 unexposed persons treated per person actually exposed, for an assumed effectiveness of 95% in preventing illness. Losses are termed "preventable" because not all illnesses, hospitalizations, and deaths can be prevented. Even if all the exposed were given postexposure prophylaxis on day 0 (day of the attack), such prophylaxis/treatment cannot be assumed to be 100% effective. See source for additional details.
[d] Range of money that could be spent annually to reduce threat of preventable losses actually occurring. If, for example, an intervention can be shown to reduce the threat of losses occurring to 1/10 of what it was (i.e., from 0.01 to 0.001), and if the intervention would be effective on day 0, then the annual premium would range from $129 to $201 million. However, for the same reduction in risk, if an intervention would only be able to be effective on, say, day 4 postrelease, then the annual premium would range from $22 to $38 million. See source for additional details.

Source: Kaufmann *et al.* (1997).

attack (reaction interventions) and those that are designed to prevent or reduce the probability of the attack occurring (protective interventions). Within these two categories, interventions can be sub-divided by when they are carried out. Pre-event protective interventions could include law enforcement or counterterrorist-type activities. Money spent on interventions before the event, to ensure the most efficient response should a bioterorrist attack occur (pre-event reaction interventions), can include mass vaccinations of the populace against a potential pathogen (e.g., smallpox), stockpiling vaccines, drugs, and other medical supplies (e.g., mechanical ventilators), and training of "first responders" and medical personnel on how to diagnose and treat a large number of persons in a short time. Postevent reaction interventions can include quarantining those that are suspected to be infected, and thus potentially could become infectious, and isolating those with clinical symptoms. A total plan to respond to the threat of a bioterrorist attack could actually be a combination of interventions. Distributing prophylactic drugs and postrelease vaccination campaigns, for example, require both pre-event stockpiling and training of those who will dispense the drugs and administer the vaccines.

I will now further discuss some aspects of postevent medical interventions, pre-event medical interventions (both reaction-type interventions), preventive preattack interventions, and how to calculate the potential savings in postevent interventions due to pre-event interventions. The later subject, in essence, considers the interaction between pre-event protective interventions and postevent reaction interventions.

4.1. Postevent Medical Interventions (Reaction Interventions)

Pre-event stockpiling of medical supplies will not automatically reduce the number of casualties. Perhaps the most difficult task following a bioterrorist attack will be the rapid distribution of stockpiled medical supplies (Figure 10.2) and the allocation of trained medical staff. One of the most difficult items to assess is the potential of hospitals to successfully treat the sudden surge in cases that will follow a large attack or a large catastrophic infectious disease event, such as an influenza pandemic (Figure 10.1). The sudden increase in numbers of persons requiring acute care will force hospital staff to triage patients by severity of illness and likelihood of surviving if treatment is supplied. This will require hospital staff to alter the accepted standards of care. Patients who, in normal times, would receive all manner of treatments may have to be denied access to certain medical treatments. This will be especially true for services and supplies relating to critical care, such as intensive care units. Rationing will undoubtedly be difficult for both patients and medical staff. In developed countries, neither group is accustomed to large-scale denial of access to medical technologies.

Another postevent distributional problem will be the hundreds, if not thousands, of people who will need or request to be medically screened because they are concerned that they were exposed to the released pathogens. Some of these people may even have some of the symptoms of the disease(s) caused by the released pathogen. The demand placed on the medical system by the "worried well" and the "walking wounded" could be enormous, and further disrupt the postevent distribution of medical supplies. Thus, in the stress of a postattack situation, it is reasonable to assume that available supplies and capacities will probably not be allocated in the manner that will maximize lives saved.

4.2. Pre-event Medical Interventions (Reaction Interventions)

The potential difficulties with postattack distribution have lead some to advocate various forms of preattack distribution. Suggestions have ranged from vaccinating large segments of the population against smallpox, to handing out boxes of antibiotics that would be taken as postexposure prophylactic treatments following an attack with a pathogen such as anthrax or tularemia. But vaccinations are rarely without side effects, and the current vaccinia-based smallpox vaccine is probably the least safe of all modern vaccines. Some vaccines have a limited time during which they provide maximum protection. Thus, if the population is vaccinated as a pre-event strategy, there has to be consideration of when booster shots will be given. The resources needed to give booster shots must be accounted for, as well as the costs associated with the side effects due to those booster shots. Pre-event distribution of antibiotics faces similar problems. Such medicines usually have a limited shelf life, after which they must be replaced. Furthermore, the longer such boxes of medicines are kept unused, the more likely it is that households will lose them. Over time, these factors will decrease the effectiveness of any pre-event medical-based intervention.

Plans for pre-event options also have to consider the impact of future population cohorts. These could be children who were too young to be vaccinated during the first round of vaccinations. Or, they could be new immigrants or persons who avoided vaccination when initially offered. Plans to deal with these issues will obviously increase the cost of such interventions.

4.3. Pre-event Protective Interventions: Reducing the Probability of Attack

It is possible to argue that some pre-event medical interventions, such as mass pre-event smallpox immunizations of large segments of the population, could potentially reduce the probability of an attack actually occurring. The assumption would be that such medical pre-event interventions would be protective, reducing the possibility of an attack by reducing the size of the vulnerable population (potential target). Other, non-medical, interventions could also reduce the probability of an attack. For example, security operations could eliminate a terrorist cell that was planning the release of a biological agent. Active counterterrorism activities might reduce the scale of the threat by disrupting the terrorist operations such that a small-scale attack occurs rather than the originally planned large-scale attack (the same might occur due to certain pre-event medical interventions).

The essential difference between pre-event protective activities and pre- and postevent medical intervention activities is that the former are designed to reduce the probability of an event actually occurring, while the latter are designed to reduce the impact after an event has occurred. To calculate the optimal amount to spend on pre-event protective interventions, the researcher would apply the special case version of the basic formula, presented earlier in Section 3.3. The problem in using that formula is determining the numerical reduction in the threat brought about by active counterterrorism or pre-event protective medical interventions. Does active counterterrorism activities reduce the threat of a large-scale bioterrorist attack from, say, 1 in 100 to 1 in 1,000 (a 10-fold reduction in threat, as in Table 10.3)? It is conceivable that even the agencies running the counterterrorism operations could not actually provide such estimates.

The lack of accurate estimates of altered risk forces the researcher to consider a range of possibilities. For example, when considering that part of the equation in Section 3.2, "reduction in probability of that loss actually occurring," a modeler might use a range of estimates of reductions in probability of event occurring.

4.4. Calculating the Savings in Post-event Interventions Due to Pre-event Interventions

Logically, if money is spent on protective interventions, and the interventions are a success (i.e., the probability of an event occurring is markedly reduced or even eliminated), then the optimal amount to be spent on postevent interventions to limit impact can be reduced. The potential savings in postevent medical interventions due to pre-event protective interventions (e.g., counterterrorism operations) can be calculated by using the formula in Section 3.2, as follows:

Equation A: Optimal annual amount to be spent on planning and preparing postevent medical interventions before pre-event protective interventions = $ value of potential losses actually averted by post-event medical interventions × annual probability of that loss actually occurring before pre-event protective interventions

Equation B: Optimal annual amount to be spent on planning and preparing by postevent medical interventions after pre-event protective interventions have been enacted = $ value of potential losses actually averted by postevent medical interventions × annual probability of that loss actually occurring after pre-event protective interventions have been enacted

Savings in reductions on expenditures on postevent medical interventions due to effective pre-event protective interventions = Result from Equation A − result from Equation B.

Note that the same set of equations apply for estimating the potential savings in any other interventions (pre- or postevent) that can be reduced in size and scope due to successful pre-event protective interventions.

5. SELECTING INTERVENTIONS FOR EVALUATION FOR FUNDING

It is possible to imagine a near endless of list of potential interventions designed to prevent or reduce the impact of a bioterrorist attack, with each having its own advocates and detractors. Selecting the initial list of potential interventions to be considered for evaluation is as much an exercise in politics as a formal exercise in mathematical optimization. There is little point, it can be argued, in spending much time considering a possible intervention for which there is little popular support.

To use the formulas given in Section 3, an intervention has to be carefully specified. It is insufficient to merely state something as general as "vaccinate everybody now against smallpox" (a pre-event intervention that could be considered either a reaction or protective intervention) or "stockpile drugs." Does the phrase "vaccinate everybody" include, for example, those less than 2 years of age? If not, will those who are currently younger than 2 years of age be vaccinated when they reach the target age? And, how long will vaccine-mediated immunity be assumed to last, and will booster shots be given (and if so, when)? To calculate resources that could be spent annually on a drug stockpile, there first has to be assumptions made regarding who will receive such drugs and under what circumstances. Will stockpiled drugs, for example, be given out principally as postexposure prophylaxis, or only to those who have symptoms?

Thus, one general intervention concept, such as pre-event vaccination, can produce several carefully defined/described interventions. Some of these will, again, be deemed infeasible for a variety of reasons, with perceptions of popular support often defining feasibility. However, there is a potential problem of using only political and popular sentiment (however those are measured) to select interventions for evaluation. The problem is that popular sentiment may be driven by a less-than-complete set of facts. A particular intervention may have a large measure of popular support because of insufficient understanding of the costs of the popularly supported intervention and the costs of alternative interventions. Thus, further evaluating at least some options that may appear as infeasible can provide decision makers and the public with a set of references that can be used to compare the cost of the currently popular intervention. Such comparisons could either cause popular opinion to change, or cause the decisionmakers and the public to be more accepting of the currently popular intervention.

6. CALCULATING THE NUMBER OF CASUALTIES AND CASUALTIES AVERTED

The next task in evaluating the amount to spend annually on planning and preparing for a bioterrorist attack is to determine the number of causalties and size of other losses that could be averted by each intervention that has been selected for evaluation. This is highly problematic because history is rarely a direct help. There are no instances in the 20th century of large-scale attacks (i.e., thousands exposed at one instance) using biological agents on civilian populations, and thus there are no data to guide us as to the impact of a pathogen on an unsuspecting populace.† To study the potential impact of a release of a pathogen on a susceptible population, epidemiologists have used mathematical models to create potential attack scenarios and their aftermaths.

6.1. Types of Mathematical Models

The potential mathematical tools available to analyze the possible casualties from a bioterrorist attack can be divided into two broad categories: statistical and mathematical models. In this categorization, statistical models encompass all mathematical techniques used to analyze existing data (only) and formally test well-defined hypotheses. Mathematical models are used when the amount of existing data is inadequate to answer the questions posed, and assumptions have to be made. As described previously, usually there are insufficient data to calculate possible bioterrorist attack-related causalities using only

†It could be argued that the letter-based anthrax attacks that occurred in the United States in October 2001 exposed thousands of people to anthrax, particularly postal employees. However, since the postal employees were given postexposure prophylactic antibiotics, there is no way to be certain (thankfully) of how many actually were exposed to an infectious dose, sufficient to cause a clinical case of anthrax. Up to 60% of these workers reported that they did not complete the recommended drug regime. Yet, this group of people did not experience any cases of anthrax. This raises the possibility that very few of these employees were actually exposed to an infectious dose of anthrax.

statistical methods. The types of mathematical models used to calculate can be broadly divided into three groups: deterministic, stochastic, and a mixture of the two.‡

Regardless of the category of mathematical model used, most models typically describe how disease-susceptible persons became infected, and then move into the incubating, infectious, and recovered (or dead) states. In the infectious stage, of course, the disease can be spread to other susceptible persons. Additional substates, such as the prodrome state in smallpox (when a patient has a high fever, but typically does not have pustules), can be added as deemed suitable. The other feature usually common to all relevant models is that the models are constructed to calculate the daily numbers of victims in each disease state (i.e., the "time step" of the output is one day).

6.1.1. *Increasing Complexity*

One method used to make both deterministic and stochastic models more complex is to divide the model population into subgroups. Perhaps the ultimate degree of complexity occurs in models that track each individual in a population. Such individual agent models (as they are sometimes called) potentially have a great deal of intuitive appeal among nonmathematical users of the results. The overwhelming problem with such models, however, is that they require an enormous amount of information. It is not sufficient to use, for example, the average number of persons infected per infectious person. Individual agent models require data describing who infects whom, and (very often) where the transmission occurs. To illustrate, on a given day, an infectious school-aged child may have one probability of infecting a fellow classmate at school, and another probability of infecting a sibling at home. The parents of that infectious school-aged child face different probabilities of being infected at home and at work (and perhaps an additional set of probabilities of becoming infected during the commute to and from work, and while shopping in the community). The problem, then, is not building such models, but finding a suitable set of data that can provide reliable estimates of such probabilities.

6.1.2. *Deterministic Mathematical Models*

Deterministic models typically use a series of equations to describe how an infectious disease spreads through a society. Such models are called deterministic because both the equations and the rates, describing how members of a population move between the disease stages, are fixed (the models are often a series of partial-differential equations, with time as the differential). For example, a modeler working on estimates of casualties due to a release of smallpox might decide to fix the incubation period at 10 days, which has the equivalent incubation rate of 1/10 per day.

‡The categorization and description of all types of mathematical models that could conceivably be used to calculate potential losses due to a bioterrorist attack is a subject for a separate text book. For this chapter, I will concentrate on the two types most commonly used in the published literature: deterministic and stochastic (described in the main text). Detailed descriptions of subsets of models within these two subcategories are also beyond the scope of this chapter. For detailed descriptions of various types of epidemiological models, particularly deterministic models, readers are referred to Anderson and May (1991). See Koopman *et al.* (2001) for a discussion of the differences between discrete and stochastic mathematical models, how they may be combined, and the problems of such combinations.

6.1.3. Stochastic Models

Instead of using equations to describe movement of victims between the disease states, stochastic models use probabilities. For example, a person who has been incubating smallpox for 7 days may have a 90% probability of incubating for the next day, and a 10% probability of moving into the next disease state (i.e., becoming infectious). Furthermore, if the same person is still incubating on day 9, then the probability of moving into the infectious state on day 10 may have increased to, say, 20%. An advantage of such models is that there is no need to consider why a person incubates for 10 days and another incubates for 12 days. You do not need to know how nature "works," but just know the consequences in terms of probabilities of movement between disease states.

6.2. Model Limitations

By definition, mathematical models are simplifications of systems whose complexity we may not fully appreciate. It would be foolish to believe that any mathematical model can always accurately predict the size of a possible bioterrorist attack. There are just too many variables that could influence the size, location, and timing of an attack. It is quite possible that a prediction produced by a given model will accurately match a given attack. The problem is that it is practically impossible to know, before the event, which prediction will be accurate and which will not.

This lack of known, pre-event accuracy does not mean that it is a waste of time building mathematical models as part of the process to determine how much should be spent to plan and prepare for a bioterrorist or catastrophic infectious disease event(s). The equations presented in Section 3 are still worthwhile calculating. The problem with being uncertain about the accuracy of predicted impact of an attack means that there must be due caution when interpreting the results produced from using the equation. The main point of using the equation is to provide policy makers with some estimates of resources that could be allocated to planning and preparing, as well as a description of the limitations of such estimates. Modelers and economists presenting estimates of size of attack and options for planning and preparing should give policy makers a clear idea of the limits of accuracy of their models.

The variables most likely to influence any mathematical model of a bioterrorist attack are those describing the size of the initial attack and the impact of possible interventions. The latter, of course, is dependent on the amount of resources invested in planning and preparing for an attack.

6.2.1. Size of Attack

Many modelers start their estimations of total number of potential casualties by using a set number of persons who have been exposed to a pathogen and subsequently become infectious (e.g., Figure 10.2 and Table 10.3). For example, they may assume that due to a release of a pathogen such as smallpox, 1,000 persons become infectious. The modelers may then consider the implications if that number were altered (but many do not publish the results of such an exercise). For noncontagious pathogens – such as anthrax, or toxins such as ricin – the modelers may assume a release of a given amount (e.g., in an aerosol or in the water supply), and then estimate how many people may become clinically ill due to exposure to the pathogen or toxin.

6.2.2. *Numbers Initially Infected and Implicit Assumptions*

Assumed size of a biological attack contains several implicit assumptions, which are infrequently discussed. The assumptions implied in any estimate of attack size are as follows: (a) terrorist organizations can produce sufficient quantities of a pathogen to be used in an attack; (b) the pathogens produced are of a predictable quality (i.e., the infectiousness of the agents are actually known); (c) known intention of the terrorist organizations to use the pathogens that they produce; and (d) the terrorist-produced agents are dispersed in a manner that will cause the assumed number of persons to actually develop clinical illnesses. Full consideration of these seemingly simple assumptions enormously complicates any realistic estimate regarding the potential number of persons initially infected during a biological attack.

Even the first assumption, regarding ease of producing biological weapons can complicate estimates of impact. There are some examples (discussed elsewhere in this book) in which terrorist organizations and the like have both developed biological weapons and used them. What is uncertain is the number of terrorist organizations that have either not seriously tried to acquire and use such weapons, or have abandoned such developments before the weapons can be used. Even among those groups that produce biological weapons, it would seem unreasonable to assume that each group would produce weapons of identical quality, and thereby pose identical threats. There is a world of difference between slurry containing a highly infectious pathogen or toxin and a biological weapon that has been "engineered" so as to maximize the number of persons exposed.

Although it can obviously affect the numbers initially infected, motivation for releasing a biological weapon is perhaps the hardest thing to predict. Although the media and the scientific literature carry extensive debates on this subject, it is difficult to distill such discussions into a single probability, or even range of probabilities. Complicating such analyses is the fact that terrorist organizations are smaller, less democratic (if at all democratic), and presumably more homogeneous than the targeted societies. Hence, the terrorist organizations can rapidly change their proposed target, potentially changing the number of persons both at risk, and numbers actually made clinically ill.

Finally, even if terrorists produce a biological weapon and do wish to release it on a target population, there can be a large difference between the number of persons that theoretically would be infected and the number that would actually be infected. There are a number of possible reasons why this may happen, including unsuitable (for the terrorists) weather conditions to ensure maximum dispersal and incompetence by the terrorists (e.g., not knowing how to properly use spray equipment).

6.2.3. *Why Not Use "Worst Case?"*

One obvious solution to the problems outlined previously is to use the "worst case," or maximum possible, estimate of the number of people who could potentially be infected upon release of a given amount of a given agent. However, there is a problem in defining a "worst case" scenario. If one assumes, for example, that 1 kg of pathogen could infect 1 million persons (worst case), why not assume five simultaneous or even ten simultaneous attacks each using 1 kg of material? Defining a worst case scenario is actually not that easy, and the supposed size of a worst case scenario is limited only by "realistic appraisal." The latter term, of course, can be entirely subjective.

The other problem with using only the "worst case scenario" is that it can cause a serious mis-allocation of resources. Applying the formula for optimal allocation of resources under an assumption of (only) worst case conditions can quickly result in all available resources (e.g., monetary budget) being used to prepare against only one or two pathogens. This would leave society vulnerable to attacks using other pathogens or any other catastrophic infectious disease events, such as the next influenza pandemic.

In conclusion, when considering the number of persons that may initially be infected during a bioterrorist attack, analysts must allow for a wide range in possible scenarios. Each scenario, in turn, has a set of probabilities of occurring.

6.3. Realistic Expectations and Keeping It Simple

As discussed in the previous sections, calculating the possible casualties with and without an intervention(s) can be a difficult task. Often, the researcher/modeler and policy maker have to accept that there is a lack of readily available credible evidence. For example, what is the probability of an attack with aerosolized anthrax that would effectively expose 100,000 persons? And, how would the factors discussed in Section 6.2 impact that probability? To attempt to incorporate a large amount of the unknowns, complex mathematical models can be built (see Section 6.1), and a large number of scenarios can be constructed. The problem with such an approach is that answers become almost as difficult to interpret as the original problem.

In such situations, modelers and policy makers should consider the option of using deliberately simple models, in which one model variable could be considered as a proxy for a number of variables. Simple models – in which one variable may act as a proxy for a number of possible inputs – have several advantages, including reducing the "black box" syndrome in which only a handful of people (usually the modelers) really understand the interactions of the variables. This "black box" syndrome can reduce a policy maker's confidence in the results. For example, Meltzer *et al.* (2001) built a simple model of a smallpox bioterrorist attack. The model did not explicitly include different impacts by different age groups, the potential for protection due to pre-existing immunity, or separately consider the impact of vaccination or even who would be vaccinated. Instead, the model allowed for different rates of transmission at different stages of the outbreak. That is, changing the rate of transmission at different stages could act as a proxy for all of these variables. The simple model was built to answer a single question regarding the minimum size of a stockpile of smallpox vaccine. A number of other researchers have subsequently published models that individually consider these and other variables.

6.3.1. Sensitivity Analyses and Policy Levers

Given the number of unknowns associated with probability of the size and occurrence of a bioterrorist attack (see Sections 6.1 and 6.2), it is extremely important that a researcher consider several "what if" scenarios. That is, there is a need to conduct sensitivity analyses to determine how robust the conclusions and implications from the original set of results. What if, for example, a 10% change in input A produces a very different set of conclusions? Simple models have the "virtue" of readily allowing the researcher to conduct a series of sensitivity analyses to demonstrate the impact of changes in any of the inputs. Obviously, sensitivity analyses can be conducted on more complex models. But, as the number of inputs increase, the number of alternative combinations of different-valued inputs rises exponentially.

With sufficient forethought, the researcher can design such sensitivity analyses to identify which input, relative to all others in the model, is the most important. That is, the researcher should strive to identify which 2–3 inputs are "driving" the results. In my experience, no matter how complex the model, within a given range of values for each input it is usually 2–3 inputs (and no more than 5) that have most impact on determining the final result of a model. Identifying such inputs can be very valuable for a policy maker. Such critical inputs could be "policy levers," representing variables that policy makers may be able to do something to change. For example, in the simple model used to draw Figure 10.2, it is obvious that a policy maker must focus pre-and postevent medical interventions on plans to ensure a rapid distribution of postexposure prophylaxis following an anthrax attack (see Section 3.4).

7. CALCULATING THE VALUE OF CASUALTIES AND OTHER LOSSES AVERTED

In the equation presented in Section 3, a crucial input is estimates of "$ value of potential losses actually averted." Measuring total lost economic activity after a catastrophic infectious disease event may not be an easy task. For example, in 2003, sudden acute respiratory syndrome (SARS) caused a great deal of disruptions in a number of countries. Some researchers estimated, based on surveys of business persons in the affected regions, that the economic impact of SARS in some countries was equivalent to 0.5–1.0% of Gross Domestic Product (GDP). However, by the end of 2003, many economists came to the conclusion that it was very difficult to estimate/find the impact of SARS in the annual estimates of GDP for the affected countries. One of the reasons was that there was a lot of "background" noise in many economies, such as the end of an economic recession and the war in Iraq. Furthermore, SARS may have delayed, but not prevented, a great deal of economic activity. At the height of the SARS epidemic, business travelers may have canceled their trips to areas affected by the disease. But, they could either have placed purchase orders (or made sales) via mail and electronic communications, or traveled to those areas at a later date. So, certain business may have experienced a delay, but not a loss, in business transactions. Such delays can, of course, be critical to certain businesses – some might close or have to reduce staff (even if temporarily) because of shortages in current revenues (cash flow).

Unlike natural disasters, such as an earthquake or a hurricane, or an attack with an explosive device, large-scale catastrophic infectious disease events (regardless if natural or human initiated) typically incur little or no damage to physical infrastructure. The damage done by biological agents is measured in terms of illnesses and deaths. And, seemingly paradoxically, a catastrophic infectious disease event, whatever the cause, can actually increase the GDP for certain sectors of the economy. GDP measures overall activity of goods and services within an economy, and an increase in medical services due to a bioterrorist attack or an influenza pandemic will result (for the health care sector) in an increase in value of services delivered. Thus, researchers and policy makers wishing to apply the equations presented in Section 3 may find that using estimates of GDP is an unsatisfactory measure of potential losses averted due to any possible investment in an intervention.

Faced with the problem of estimating total losses to an economy due to a catastrophic infectious disease event, a researcher can realistically estimate the direct medical and indirect costs (e.g., lost productivity) associated with each clinical case. This approach may be simpler than trying to estimate impact on total economy, but typically it still requires a researcher to make many assumptions. For example, when calculating the potential economic impact in the United States of the next influenza pandemic (used to produce the estimates in Table 10.2), we had to make assumptions time spent in hospital and days lost posthospitalization while convalescing (Meltzer *et al.*, 1999). Similarly, for the estimates in Table 10.3, we had to make assumptions regarding the care that would be given to those exposed to aerosolized anthrax (Kaufmann *et al.*, 1997). Haddix *et al.* (2003) provide detailed definitions and discussions of costing direct and indirect medical and nonmedical costs.

Perhaps the most difficult aspect is the valuation of human life lost. As can be readily appreciated, this is a topic with a lot of qualitative valuation (i.e., emotions). Not surprisingly, economists rarely reach consensus regarding the optimal method of measuring the value of life lost. Some economists, for example, use the value of life implied by differences in wages due to differences in work-related risk-of-death. Using such reasoning, some have estimated that the value of a life in the United States is between $3–$7 million. However, other economists point out that such estimates only cover those working, ignoring the retired, disabled, and young. Valuing lives lost in terms of wages forgone produced the estimates shown in Tables 10.2 and 10.3. The estimates of average wages foregone were weighted by the age and gender composition of the U.S. workforce (see Grosse, 2003).

In summary, estimating the "$ value of potential losses actually averted" is not a trival task. Researchers and policy makers using the equations given in Section 3 will have to make a number of assumptions while valuing losses that might actually be averted due to a given intervention. These assumptions should be clearly stated. Furthermore, similar to the discussion regarding sensitivity analyses for the estimates of casualties (scc Section 6.3.1), the researcher should conduct sensitivity analyses around the value of outcomes prevented.

8. PROBABILITY OF AN EVENT OCCURRING

In the equation presented in Section 3, the variable "annual probability of that loss actually occurring" is critical to calculating the optimal annual amount to be spent on planning and preparing for a catastrophic infectious disease event. Some of the factors affecting this probability were discussed in Sections 6.2.2 and 6.2.3. It cannot be stressed enough, however, that both the researcher and policy maker must be fully aware that estimates of a given attack occurring are fraught with a great deal of uncertainty. Thus, the policy maker should request, and the researcher readily provide, estimates of the optimal amount considering different probabilities of occurrence (e.g., as in Tables 10.2 and 10.3).

9. SELECTING BETWEEN THE OPTIONS

A researcher can, using the equations given in Section 3, evaluate the optimal amount to be spent for a number of different options (e.g, stockpiling drugs, vaccinating population now, counterterrorism activities, some combinations of these). Such calculations do not automatically mean that the "best" option(s) will be identified. Calculations using Equation 3 may, for example, identify that a particular option would be too "costly" (i.e., exceed available budget). But, the set of final estimates of optimal amounts for a variety of interventions do not typically easily identify which intervention(s) should be actually funded (i.e., chosen).

However, some of the intermediate calculations can help the selection process. For example, estimates of casualties avoided may identify one or more interventions as being "unattractive" (e.g., prevent too few deaths to be easily a popular choice). Similarly, when defining the interventions to be analyzed (see Sections 4 and 5), the need to be very specific in defining an intervention can highlight potential problems and benefits not necessarily directly included in the calculations. For example, as mentioned previously, distributing postexposure prophylaxis following an attack with aerosolized anthrax could be highly problematic, resulting in a significant and costly delay (Figure 10.2). To make such an option work, it will probably require additional funds to ensure the setting up, and smooth operation, of distribution centers. Such a cost should be included in the estimates of effectiveness and cost of the entire intervention.

10. SUMMARY

In this chapter, I have demonstrated that it is possible to calculate the optimal annual amount that should be spent on a defined intervention to prevent or reduce the impact of a bioterrorist attack or other catastrophic infectious disease event. Such calculations, however, are not trivial and require a great deal of information. The information includes the epidemiology of the disease (e.g., who gets ill, what happens to them), the cost of such impacts, the potential effectiveness of proposed interventions, and an understanding of the annual probability of the event actually occurring. There are a great number of uncertainties associated with each of these inputs, and the impact of such uncertainties must be explored through systematic sensitivity analyses. One of the goals of such sensitivity analyses should be to identify the 2–3 most influential inputs. These 2–3 influential inputs then represent potential "policy levers" that policy makers can focus interventions. Finally, it must be appreciated that calculating the optimal annual amount to be spent on an intervention is not the complete set of decision-making points. There are many other factors influencing the choice of intervention. However, because calculating the optimal amount to be spent on a given intervention combines many different variables into a single estimate (the optimal amount), the calculated amount provides valuable information by which to start the decision-making process.

References

Anderson, R.M., and May, R.M. (1991). *Infectious Diseases of Humans.* Oxford University Press, New York.

Grosse, S.D. (2003). Productivity loss tables. Appendix I. In: Haddix, A.C., Teutsch, S.M., and Corso, P.S. (eds.), *Prevention Effectiveness, 2nd ed.* Oxford University Press, New York, pp. 245–257.

Haddix, A.C., Corso, P.S., and Gorsky, R.D. (2003). Costs. In: Haddix, A.C., Teutsch, S.M., and Corso, P.S. (eds.), *Prevention Effectiveness, 2nd ed.* Oxford University Press, New York, pp. 53–76.

Kaufman, A.F., Meltzer, M.I., and Schmid, G.P. (1997). The economic impact of a bioterrorist attack: are prevention and postattack intervention programs justifiable? *Emerg. Infect. Dis.* 3:83–94.

Koopman, J.S., Jacquez, G., and Chick, S.E. (2001). New data and tools for integrating discrete and continuous population modeling strategies. *Ann. N.Y. Acad. Sci.* 954:268–294.

Meltzer, M.I., Cox, N.J., and Fukuda, K. (1999). The economics impact of pandemic influenza in the United States: priorities for intervention. *Emerg. Infect. Dis.* 5:659–671.

Meltzer, M.I., Damon, I., LeDuc, J.W., and Millar, J.D. (2001). Modeling potential responses to smallpox as a bioterrorist weapon. *Emerg. Infect. Dis.* 7:959–969.

Index